T0300521

Biotechnology – the Making of a Global Controversy

Biotechnology is one of the fastest growing areas of scientific, technical and industrial innovation; it is also one of the most widely publicised new technologies. Innovations as varied as the development of genetic testing and therapies, genetically modified food crops, animal and stem-cell cloning have given rise to increasingly prominent public debates. Genetically modified soya and Dolly the sheep – to mention the two most prominent events – led to world-wide controversies. While the biotechnology industry initially assumed that regulatory processes were the sole hurdle, it is now apparent that a second hurdle, national and international public opinion, must be reckoned with. *Biotechnology – the Making of a Global Controversy* brings together key findings from a comparative study of public perceptions, media coverage and regulatory frameworks of a newly emergent technology. With chapters from leading international experts in the field, the book contributes important empirical and conceptual analyses to this crucial public debate.

Martin W. Bauer is Senior Lecturer in Social Psychology at the London School of Economics and Research Fellow at the Science Museum London. He co-coordinated the research presented in this volume. His research interests are the functions of resistance in social processes; longitudinal indicators of public understanding of science through media and survey research; the changing representations of science and technology in texts and images. His publications include *Resistance to New Technology: Nuclear Power, Information Technology, Biotechnology* (Cambridge, 1995).

George Gaskell is Professor of Social Psychology at the London School of Economics and Director of the School's Methodology Institute. His research interests include risk as a representation; expert and lay understanding of risk and uncertainty; the structure of public perception of biotechnology; how public opinion is shaped by and shapes policy and regulation; public opinion and regulation; and the democratisation of science and technology policy. He is currently coordinator of the research project 'Life Sciences in European Society', funded under the European Commission's 5th Framework Programme.

The coordinators' co-publications include *Qualitative Researching with Text, Image and Sound* (2000), *Biotechnology 1996–2000: the Years of Controversy* (2001) and *Biotechnology in the Public Sphere: a European Sourcebook* (1998).

Biotechnology – the Making of a Global Controversy

edited by

M. W. Bauer and G. Gaskell

Published in association with the Science Museum, London

CAMBRIDGE UNIVERSITY PRESS
Cambridge, New York, Melbourne, Madrid, Cape Town,
Singapore, São Paulo, Delhi, Tokyo, Mexico City

Cambridge University Press
The Edinburgh Building, Cambridge CB2 8RU, UK

Published in the United States of America by Cambridge University Press, New York

www.cambridge.org
Information on this title: www.cambridge.org/9780521773171

First published 2002

A catalogue record for this publication is available from the British Library

Library of Congress Cataloguing in Publication Data
Biotechnology : the making of a global controversy / M. W. Bauer &
G. Gaskell, eds.
p. cm.
Includes bibliographical references and index.
ISBN 0-521-77317-2 – ISBN 0-521-77439-X (pb.)
1. Biotechnology – Europe – Public opinion. 2. Mass media – Europe.
3. Biotechnology – Social aspects. I. Bauer, Martin W. II. Gaskell, George.
TP248.23.B563 2002
660.6 – dc21 2001052636

ISBN 978-0-521-77317-1 Hardback
ISBN 978-0-521-77439-0 Paperback

Contents

Plates

Figures

Tables

List of contributors

AGNES ALLANSDOTTIR – Department of Communication Science, University of Siena, Italy

NICK ALLUM – Methodology Institute, London School of Economics and Political Science, UK

*MARTIN W. BAUER – Department of Social Psychology, London School of Economics and Political Science, UK

MARIE-LOUISE VON BERGMANN-WINBERG – Faculty of Interdisciplinary Research, Mid Sweden University, Oestersund, Sweden

ANNE BERTHOMIER – Laboratoire Communication et Politique, Centre National de la Recherche Scientifique (CNRS), Paris, France

HEINZ BONFADELLI – Institute of Mass Communication and Media Research, University of Zürich, Switzerland

DANIEL BOY – Maison des Sciences de l'Homme, Centre d'Études de la Vie Politique Française (Cevipof), Paris, France

ELEANOR BRIDGMAN – Public Understanding of Science Group, Science Museum, London

AIGLI CHATJOULI – Department of Biological Research and Biotechnology, Office of Science Communication and Bioethics, National Hellenic Research Foundation, Athens, Greece

SUZANNE DE CHEVEIGNÉ – Laboratoire Communication et Politique, Centre National de la Recherche Scientifique (CNRS), Paris, France

CARMEN DIEGO – Department of Sociology, ISCTE, University of Lisbon, Portugal

ROBIN DOWNEY – Faculty of General Studies, University of Calgary, Canada

* Co-ordinators of the project

JOHN DURANT – Director, At-Bristol, Bristol, UK

EDNA F. EINSIEDEL – Faculty of General Studies, University of Calgary, Canada

BJÖRN FJÆSTAD – Mid Sweden University, Stockholm, Sweden

HELLE FREDERIKSEN – Department of Communication, Education and Computer Science, Roskilde University, Denmark

JEAN-CHRISTOPHE GALLOUX – University of Versailles Saint-Quentin and Director of the Centre de Recherche en Droit de Technologies de l'Information et du Vivant (CRDTIV)

*GEORGE GASKELL – Methodology Institute, London School of Economics and Political Science, UK

ALEXANDER GOERKE – Faculty of Social and Behaviour Sciences, Friedrich Schiller University, Jena, Germany

PETRA GRABNER – Department of Political Science, University of Salzburg, Austria

JAN M. GUTTELING – Department of Communication Studies, University of Twente, Enschede, The Netherlands

JÜRGEN HAMPEL – Centre of Technology Assessment in Baden-Württemberg, Stuttgart, Germany

MARCUS HEINßEN – Centre of Technology Assessment in Baden-Württemberg, Stuttgart, Germany

PETRA HIEBER – Department of Social and Economic Psychology, University of Linz, Austria

ERLING JELSØE – Department of Environment, Technology and Social Studies, Roskilde University, Denmark

MERCI WAMBUI KAMARA – Department of Environment, Technology and Social Studies, Roskilde University, Denmark

MATTHIAS KOHRING – Department of Media Science, Friedrich Schiller University, Jena, Germany

NICOLE KRONBERGER – Department of Social and Economic Psychology, University of Linz, Austria

JESPER LASSEN – Department of Development and Planning, Aalborg University, Denmark

MARTINA LEONARZ – Institute of Mass Communication and Media Research, University of Zürich, Switzerland

MILTOS LIAKOPOULOS – Europäische Akademie zur Erforschung von Folgen wissenschaftlich-technischer Entwicklungen, Bad Neuwahr-Ahrwiler, Germany

FREDERICA MANZOLI – Department of Communication Science, University of Siena, Italy

ATHENA MAROUDA-CHATJOULIS – Institute of Applied Psychology, Department of Communication and Mass Media, University of Athens, Greece

CEES J. H. MIDDEN – Department of Technology Management, Eindhoven University of Technology, The Netherlands

JON D. MILLER – Centre for Biomedical Communications, Northwestern University, Chicago, USA

ARNE THING MORTENSEN – Department of Communications, Roskilde University, Denmark

TORBEN HVIID NIELSEN – Centre for Technology and Culture, University of Oslo, Norway

MARIANNE ODEGAARD – Centre for Technology and Culture, University of Oslo, Norway

SUSANNA ÖHMAN – Mid Sweden University, Oestersund, Sweden

ANNA OLOFSSON – Mid Sweden University, Oestersund, Sweden

ANDRZEJ PRZESTALSKI – Institute of Sociology, Adam Mickiewicz University, Poznan, Poland

BIANCA RIZZO – Department of Communication, University of Siena, Italy

GEORG RUHRMANN – Department of Media Science, Friedrich Schiller University, Jena, Germany

MARIA RUSANEN – Department of Social Sciences, University of Kuopio, Finland

TIMO RUSANEN – Department of Social Sciences, University of Kuopio, Finland

GEORGE SAKELLARIS – Institute of Biological Research, National Hellenic Research Foundation, Athens, Greece

MICHAEL SCHANNE – AGK Communication Consulting, Zürich, Switzerland

FRANZ SEIFERT – Institute of Advanced Studies, University of Vienna, Austria

G. CARLA J. SMINK – SWOKA Institute for Consumer Research, The Netherlands

ANGELIKI STATHOPOULOU – Department of Qualitative Research, Metron Analysis S.A., Athens, Greece

PAUL THOMPSON – Department of Philosophy, Purdue University, Lafayette, Indiana, USA

HELGE TORGERSEN – Institut für Technikfolgen Abschätzung, Österreichische Akademie der Wissenschaften, Vienna, Austria

TOMASZ TWARDOWSKI – Institute of Bioorganic Chemistry, Polish Academy of Sciences, Poznan, Poland

WOLFGANG WAGNER – Institut für Pädagogie und Psychologie, Johannes-Kepler-University, Linz, Austria

Acknowledgements

The research for this book was funded by the Fourth (EU) RTD Framework, Concerted Actions BIO4-CT98-0488 (DG12) and BIO4-CT95-0043 (DG12-SSMA), granted to John Durant at the Science Museum, London, UK. The book was written with the support of the European Commission, Research Directorate (QLRT-1999-00286). The authors and contributors acknowledge a debt of gratitude to John Durant, who played a vital role in the research but who has moved on to new challenges. Thanks also to Jane Gregory for her English language work on the chapters.

Additional funding was received from the following institutions:

Austria: Österreichischer Fonds zur Förderung der wissenschaftlichen Forschung, P11849-SOZ, and Institute of Technology Assessment, Austrian Academy of Sciences.

Canada: Social Sciences and Humanities Research Council (Science and Technology Policy Strategic Grants Program).

Denmark: Danish Social Science Research Council for research within the project 'Biotechnology and the Danish Public'.

France: Ministry of Agriculture (DGAL), INRA and CNRS (SHS and Programme 'Risques collectifs et situations de crise').

Germany: Deutsche Forschungsgemeinschaft (DFG), contracts 467/2-1 and 467/2-2.

Poland: Polish State Committee for Research.

Sweden: Mid Sweden University, the Freja Foundation, the Erinaceidæ Foundation, the Megn Bergvall Foundation and the Swedish Natural Science Research Council.

United Kingdom: Leverhulme Trust, contract F/4/BG, 'Biotechnology and the British Public'.

1 Researching the public sphere of biotechnology

Martin W. Bauer and George Gaskell

Biotechnology is one of the fastest-growing areas of scientific, technical and industrial innovation of recent times, and it is also one of the most prominent in public discussion. Following the development of recombinant DNA techniques in the early 1970s, modern biotechnology has burgeoned in diverse areas including pharmaceuticals, diagnostics and testing, cloning and xenotransplantation, genetically modified seeds and foods and environmental remediation. Such is the breadth of impacts across previously unrelated sectors that a new collective category, 'the life sciences', has been adopted within the industrial and scientific communities. Accompanying this research, development and commercial exploitation has been a widening range and growing intensity of public debates. These have featured issues such as the use of genetic information, the labelling of genetically modified foods, intellectual property rights, the privatisation of research activities and biodiversity, but these have also been paralleled by more fundamental considerations of the rights and wrongs of modern biotechnology as a whole.

While debates on these issues have appeared, and indeed disappeared, in different countries and at different times in Europe and the United States, modern biotechnology has become increasingly sensitive, socially and politically. In contemporary times, public opinion is not merely a perspective 'after the fact'; it is a crucial constraint, in the dual sense of the limitations and opportunities for governments and industries to exploit the new technology. Whereas the biotechnology industry assumed that regulatory processes were the sole hurdle prior to commercialisation, it is now apparent that a second hurdle, national and international public opinion, must be taken into account.

This book takes up themes explored at a conference at the Science Museum, London, in 1993, which was convened to explore the structures and functions of resistance in the development of new technologies. At that meeting, three base technologies of the post-war years were contrasted: nuclear power, information technology and genetic engineering. The main thesis of the conference was that resistance is not a problem residing in the public, rather it is a signal that something is going wrong with the technology; and that resistance acts as a catalyst for organisational and institutional learning (Bauer, 1995). With an exclusive focus on biotechnology, this idea is further developed and expanded in this volume.

Here we present results of a four-year international research project conducted between 1996 and 1999. The project, 'Biotechnology and the European Public', brought together social scientists from a variety of different disciplines, including science and technology studies, sociology, social philosophy and psychology, consumer behaviour, communication science and political science. All the members of the project, based in fourteen countries – Austria, Denmark, Finland, France, Germany, Greece, Italy, the Netherlands, Poland, Sweden, Switzerland and the UK, with associates from Canada and the USA – had at least one thing in common. They shared a keen interest in monitoring and understanding the reception of modern biotechnology in the public spheres of Europe and North America. In the research, the public sphere is defined as the intersections between public opinion as evidenced in public perceptions and media coverage, and regulation and policy-making. The objective was to chart the dynamics of public opinion and regulatory activity that accompanied the development of biotechnology, from its beginnings in 1973 until 1996, in a multinational and comparative framework.

In our previous book (Durant, Bauer and Gaskell, 1998), we published the basic empirical data together with descriptive commentaries on the national developments in regulation and public opinion. In the present book, we step back from the data and reflect on biotechnology in Europe and North America in the years up to 1996/97. With the benefit of hindsight, this proved to be a watershed in the development of this strategic technology. Towards the end of 1996 the annual cargo of American soya was shipped into Europe. For the first time this was a crop of soya that included genetically modified soya (GM soya) grown from engineered seeds made resistant to Roundup herbicide. The seeds for this new GM soya were developed by the American multinational company Monsanto, whose name became synonymous with GM products. In the heady days of new developments of genetically modified seeds, with their promise to introduce a second green revolution, Monsanto may have been pleased to see their name as the brand leader. But this unusual cargo, intended or unintended as it may have been, had consequences that changed the image of biotechnology among the European public, and spilled over into the other parts of the globe. A few months later, in February 1997, the Scottish veterinary research station at Edinburgh claimed to have achieved what hitherto had been thought impossible: it had transferred the genetic material from an adult sheep to a uterus cell, and raised a cloned, genetically identical, offspring. 'Dolly the sheep' turned science fiction into a reality. Both these events, though local in character, became global markers and symbols of the genetic society cultivating contrary visions of progress and awe against doom and anxiety.

At other times, these two events might have quickly evaporated into thin air following the knee-jerk reactions of sensationalist mass media. But it is significant that they followed a slow build-up of public debate and concern about biotechnology that had been rumbling since the early 1970s. This book demonstrates how the build-up of public awareness and information, and the contrasting euphoria and gloom that accompanied the early developments of biotechnology, set the context for the reception of these two events. By understanding this earlier period, we can appreciate why the two events achieved their significance, and how that significance influenced the development of a global biotechnology controversy in the latter part of the 1990s. In sum, we explore the preconditions of what historians may come to call the 'great European biotechnology debate' which unfolded in the last few years of the twentieth century concerning a novel technology with commercial applications in the fields of crops and food production and pharmaceuticals and medicines.

Throughout this study, the term biotechnology is generally used to mean 'modern biotechnology', i.e. those processes, products and services that have been developed on the basis of interventions at the level of the gene. In the literature, modern biotechnology thus defined is generally contrasted with 'traditional biotechnology', i.e. those processes, products and services that have been developed on the basis of interventions at the level of the cell, tissue or whole organism. Although these are justifiable distinctions from the point of view of the biotechnologist, it is important to note that, in any particular social situation, they may or may not accord with the representations of biotechnology in the public sphere. Since the public sphere is our principal object of interest, it is vital that we acknowledge what the public understands by biotechnology, irrespective of whether this 'lay definition' complies with scientific definitions.

The fact that the project dealt with eight different languages led to some semantic challenges. The denotation of 'biotechnology and new genetics' has to be recovered from a changing lexicon of words and phrases both across languages and over time. It became clear, for example, that as the technology has developed so too has the vocabulary that denotes it. In English, for example, the term 'recombinant DNA' (rDNA) was current in the 1970s but disappeared later on; the term 'biotechnology' was not commonly used until around 1980; and terms such as 'genetic engineering', 'genetic manipulation' and 'genetic modification' all appear and disappear later on, in what seems to be a complex game played with semantics in the public sphere. Indicators for this semantic uncertainty are the Ernst & Young bi-annual reports on the state of the European biotechnology business which over recent years used different terms from 'biotechnology' (1996) to 'life sciences' (1998) to 'Evolution' (2000).

The public sphere: conceptual framework and empirical foundations

We consider biotechnology as an emerging scientific-industrial complex – a growing activity complex of research, development, production and service provision. By this, we do not mean to imply that biotechnology is a unified field, complete with a single, hierarchical mechanism of command and control; rather, we regard it as a heterogeneous coalition of many different actors, institutions and interests engaged in a competitive game over the control of this complex for purposes of commercial advantage. The biotechnology complex evolves alongside and within established societal spheres – economic, legal, mass media, political, religious, and so on – that collectively constitute its environment. Developmental change occurs in part through 'challenges' of one societal system upon another, and responses to these challenges. For research purposes, any particular societal system may be foregrounded as the focus of attention. For our purposes, it is not biotechnology itself – its locations, business logic, manpower, capitalisation, and so on – but certain aspects of the political systems as part of its environment that are foregrounded in this way. In particular, we are interested in the way in which old and new structures of a modern public sphere (Habermas, 1989) shape the contents and trajectory of a new technology.

The economic, legal, political and media environments each give more or less attention to biotechnology at different times, frame the technology according to a particular logic, and all have 'eyes' on other issues. As each turns its gaze to biotechnology, it may construct a different representation of the 'object'. Thus, for the financial system the representation is likely to emphasise investment opportunities, risk and stock market performance; whereas for the mass media it may consist largely in the 'news value' of particular developments tied to novelty, human interest or scandal. In this sense, the symbolic environment of biotechnology is made by a set of observers with different levels of attention and different ways of seeing. But these are more than merely passive observers. Their gaze is an active process of selection and framing that may facilitate and/or constrain the development of biotechnology in particular ways. The presumptions underlying this research are that, in the course of its twenty-five-year development: first, biotechnology regularly presented challenges to observers within the public sphere; and, second, these observers at times responded with counter-challenges or resistance that contributed to shape the continued development of biotechnology itself.

By systematically observing the observers of biotechnology in the public sphere, we aim to document the presence and the potential influence of the public counter-challenges upon this emerging technology. Our research observes the public sphere as a tripartite structure of policy, media and perceptions, and through what we term 'representations' of biotechnology (see chapter 13). For present purposes, a public representation is simply conversation and writing within the public sphere referring to biotechnology, which is 'objectified' for the conduct of research (Bauer and Gaskell, 1999).

The research was organised in the following way. First, each participating country conducted a longitudinal (historical) analysis of the development of public policy for biotechnology over the period 1973 to 1996 (a similar longitudinal study of policy developments at the European level has also been undertaken). This period was chosen to embrace the entire history of modern biotechnology from the discovery of rDNA technology up to 1996, the year in which our research project commenced. Second, each participating country conducted a longitudinal analysis of media coverage of biotechnology in the opinion-leading press, also from 1973 to 1996. Third, all participating countries contributed to the development and analysis of a representative sample survey of public perceptions of biotechnology, Eurobarometer 46.1. This survey was carried out in October/November of 1996 in each member state of the European Union (EU). Similar surveys were also developed and carried out by affiliated teams in Norway, Switzerland, Canada and the USA.

Public policy: chronology and domains of regulation

Public policy is an important expression of the aspirations, attitudes and values of a country. Public policies may have various explicit or implicit aims: they may seek to promote public goods (for example, through the encouragement of innovation), or to prevent public harms (for example, through the imposition of health and safety regulations); they may seek to protect the interests of producers (for example, through the patent laws), or those of consumers (for example, through product labelling requirements); and they may seek to reconcile conflicting ideals or interests (for example, in the provision of guidelines for the acceptable conduct of research on human embryos). Public policy is the outcome of activities in political forums. In pluralistic democracies, these activities are necessarily multiple and multi-valent. In other words, at any particular time no single actor or interest group is likely to dominate the policy-making process to the exclusion of all others. Instead, different actors and interest groups vie for influence in a political process (part private, part public) involving competition, cooperation, lobbying, public relations campaigns, coalition-making and breaking, and compromise. In the European Union, policy-making for biotechnology takes place at both the national and the European levels. To the complex of actors and interests operating at the level of the individual nation-state must therefore be added a second complex of actors and interests (including the European Commission, the European Council and the European Parliament) operating at the level of the fifteen member states. In the end, what transpires as official policy may be something that no single actor or interest group originally intended.

In reviewing the history of biotechnology policy-making, we have concentrated in the main on formal policy-making processes; that is, on the institutionalised activities by which official public policy has been established. However, wherever possible we have also paid attention to informal influences (such as lobbying by business organisations, or the opposition activities of non-governmental organisations) on the formal sector. We have been interested in questions of three main types: those that concern the characterisation of 'frames' of biotechnology within the policy field; those that concern the mechanisms by which policy has been framed; and those that concern the relationships between individual nation-states and the European Union.

In the first category, we considered questions such as: which issues have been debated? how have these issues been framed by the selection of themes? which have been the principal sponsors/constituencies of particular themes? how has the policy process dealt with opportunities and risks in relation to biotechnology? have policy-makers concentrated on the control of processes or products? In the second category, we considered questions such as: what have been the distinctive 'policy cultures' for biotechnology in Europe? what have been the principal mechanisms for generating biotechnology policy in Europe? what has been the influence of public opinion upon policy-making? have policy-making processes tended towards the 'technocratic' or the 'participative' mode? and is there evidence of 'institutional learning' as policy-makers develop new instruments and forums in light of previous experience? In the third category, we considered the timing of policy processes: how early or late do particular countries become engaged with biotechnology policy-making?

how far do particular countries 'lead' or 'follow' in particular areas of policy-making? and what are the relations between national and European initiatives on biotechnology?

We developed a chronology of key policy developments in each country for the period 1973–96. This chronology was based partly on published sources and partly on interviews with key actors from different arenas, including government, industry and non-governmental organisations. The aims of the chronologies were: to document concisely the most important policy initiatives in each country; to provide a base-line of data for comparison with the chronology emerging from the mass media study; and to provide a base-line of data for purposes of international comparison. Once the chronologies had been completed, these were converted into a 'policy template', providing in a standardised form a concise summary of policy developments in each country over almost a quarter of a century. Policy events were classified into ten areas: reproductive technologies; gene therapy; genetic screening; transgenic animals; genetically modified food; releases of genetically modified organisms (GMOs); GMO contained use; health and safety; research and development policy; and intellectual property rights.

These chronologies and templates were complemented by a review of the 'policy culture' in each country. By policy culture in this context we mean the prevailing styles of policy-making at the national level. These were judged to be vitally important for the interpretation of national similarities and differences (see, for details, Durant, Bauer and Gaskell, 1998).

Media coverage: intensity and contents of coverage

The mass media constitute a major arena of the modern public sphere. There is general agreement in the literature that the mass media are influential, but much less agreement about the exact nature of this influence. It is variously argued that the mass media serve to 'frame' issues in the public domain, that they serve an 'agenda-setting' role, and that they pander to and therefore, by way of appeal, express public perceptions. For our purposes, the mass media are viewed as one of several modes of representing biotechnology in the public sphere. They function both to explain and legitimate formal policies ('top–down'), and to signal issues and themes to policy-making that arise from informal political forums ('bottom–up'). Throughout, it is the complex interrelationships between media discourses, policy discourses and public perceptions with respect to biotechnology that are the focus of our attention. In order to study this interrelationship empirically, we constructed a media database following the paradigm of 'parallel content analysis' (Neuman, 1989).

The media analysis comprised two elements: first, an indicator of intensity of coverage over time; and, second, the characterisation of this coverage in a longitudinal content analysis from 1973 to 1996. We established an indicator of the intensity of media coverage by estimating the number of all relevant articles on a year-by-year basis. The second key element was the creation of a corpus of media material in each country for purposes of comparative analysis. As we were dealing with the emergence of a new technology in the public sphere, we selected the opinion-leading press for study. By 'opinion-leading' we refer to outlets that are read by decision-makers for information and by other journalists for inspiration. We assumed that for each country it is possible to identify outlets that stand as proxies for the nature and intensity of media coverage more generally. In most participating countries national newspapers still act as opinion leaders. If this is doubtful, the criterion 'opinion leader' provides a functional reference for selection independent of other characteristics such as circulation or quality. In some contexts, this criterion may lead to the use of more than one newspaper within the longitudinal study. One newspaper alone may not cover the opinion-leading function, or the newspapers may change their function over time. Furthermore, newspapers make convenient and reliable sources for purposes of data collection.

Establishing a comparable sampling frame for media analysis across twelve countries was a challenge in itself. Our strategy was to establish functional equivalence across the countries. Some newspapers offer a historical index of articles. This constitutes a self-classification by journalists. For the early years, several of us relied on this entry point, although we were aware that this classification was not necessarily exhaustive. Such indices were checked by manual scanning, under a protocol according to which the number of issues scanned was inversely proportional to the amount of relevant material they were expected to contain (the smaller the number of expected articles, the greater the number of issues that need to be checked in order to establish a reliable intensity index). With on-line resources such as FT-Profile or CD-ROMs from certain newspapers, the complexity of sampling is reduced to the question: what are relevant keywords or search strings? To answer this question, the project defined a core set of key words translatable into all of the eight languages involved. These were: *biotech**, *genetic**, *genome*, and *DNA*.

The coding method indicated for our purpose was classical content analysis (e.g. Bauer, 2000). We chose this approach from a multitude of textual analysis techniques because: first, it allows for systematic (i.e. publicly accountable and replicable) comparisons on the basis of a common coding frame; second, it can cope with large amounts of material; and, third, it is sensitive to symbolic material, albeit through a process of coordinated local interpretation. The aim of the analysis was to deliver a systematic and comparable interpretation of the mass media traces of biotechnology since 1973.

The coding frame provided a grid of comparison of coverage in terms of framing, thematic structure and evaluation of biotechnology. Frames, themes and evaluations were further differentiated. The unit of analysis was the 'single press article', which was read by the coders and interpreted in the light of the questions posed by the coding frame. As most coders were highly educated members of the national research teams, their readings are likely to reflect the subcultural features of those who produced the articles and thus to constitute in this sense a valid, albeit not a universal, reading.

For each article, the coding frame assessed journalistic features such as the section of the newspaper in which it appeared; the size of the article, as an indicator of news importance; the format of the article; and whether the article appeared to be controversial. The news event was characterised by authorship, the actors identified with biotechnology, the themes, their location, the attributed consequences in terms of risks and benefits, and the implicit evaluation of biotechnology. A key feature of the coding was the identification of 'frames' of coverage (see Gamson and Modigliani, 1989).

Quality management of the process included careful negotiation of sampling and coding procedures, familiarisation with the procedures in the context of local constraints, revision of the coding frame to take account of local pilot work, and formal reliability checks for both within-country and cross-country consistency (see Durant, Bauer and Gaskell, 1998: 297–8 for reliability assessments).

Public perceptions: knowledge, attitudes and images

The third module of the research was concerned with measurement of public perceptions by means of random sample social survey. By 'public perceptions' in this context we mean all of the considerations, expressed in interviews, that people may have concerning biotechnology. As such, the term embraces interest and involvement in, understanding of and attitudes towards biotechnology; but it also includes the images, hopes, fears, expectations and even forebodings that people may experience when they think about biotechnology. The term 'perception' includes the processes of imagination at any moment in time. The importance of imagination lies in its capacity to go beyond the present reality by re-presenting 'things' independently of space and time: to locate events that happen at other places, to recollect or link past events, and to anticipate a negative or positive future that inspires present-day actions, even to play on fantasies without any constraints such as science fiction. In the case of biotechnology – literally, the technology of life itself – the importance of individual and collective imagination can scarcely be exaggerated. The cultural resonance of key phrases – 'test-tube baby', 'genetic engineering', 'cloning', 'the blueprint of life', 'frankenfood', 'Boys from Brazil' – has as much to do with their metaphorical and mythopoeic powers as with their scientific and technological significance.

From the outset, our research was organised around the opportunity to conduct a social survey through the Eurobarometer Office of the European Commission. After extensive qualitative research using individual and focus group interviewing in the spring of 1996, a survey instrument was designed and pilot tested. Following necessary modifications, the Eurobarometer on Biotechnology (46.1) was conducted during October and November 1996. The survey was carried out in each member state of the European Union, using a multi-stage random sampling procedure providing a statistically representative sample of national residents aged 15 and over. The total sample within the European Union was 16,246 respondents (i.e. about 1,000 per EU country). In addition, similar samples were achieved in Norway and Canada (1996) and in Switzerland and the USA (1997).

The survey drew on the questionnaire employed in 1991 and 1993 in previous Eurobarometers on Biotechnology (35.1 and 39.1). Where possible, questions were repeated for purposes of trend analysis; but changes both in biotechnology itself and in the public debate about biotechnology dictated the need for a number of new question sets in the survey instrument. The revised questionnaire included items on the following topics: optimism/pessimism about the impact of specified technologies (including biotechnology and genetic engineering); elementary scientific knowledge relating to biotechnology; beliefs about the role of nature and nurture in the development of human attributes; specific attitudes on six applications of biotechnology measured on four dimensions of usefulness, risk, moral acceptability and support; general attitudes towards the regulation of biotechnology and its agencies; confidence in different institutions to tell the truth about biotechnology; future expectations about the contributions of biotechnology to society; the importance of the issue, sources of information and attentiveness to the issue; political orientation; and, finally, socio-demographic characteristics such as age, religious orientation, sex, income and level of education.

By integrating the results from the 1996 Eurobarometer (46.1) with the systematic media and policy analyses, the project maps the main contours of the different 'national public landscapes' within which biotechnology is developing in Europe and in North America. The project was designed to study public opinion in the public sphere. This necessitates going beyond social survey data, thus avoiding the trap of equating public opinion with public opinion polls and social survey data.

The structure of the present volume

The book presents a series of empirical studies exploring the interrelationships between mass media coverage, public perceptions and public policy in the development of biotechnology to 1996/97. Together these provide an analysis of the preconditions of the recent public controversies about biotechnology across Europe. The book is divided into five parts.

The first part provides a longitudinal perspective dealing with the long-term cultivation of symbolism around biotechnology in Europe since 1973.

Chapter 2 reviews the development of biotechnology regulations across Europe and in the European Community from the 1970s onwards. In their extended *tour d'horizon*, Torgersen et al. depart from the traditional notion that societies adapt to a technology after the fact, with a cultural lag. The struggle for national and international regulation has the double character of preparing for, as well as reacting to, developments. They offer a carefully crafted periodicity of European regulatory events since the Asilomar Conference in 1975. They explain the puzzle of why the controversies that flared briefly during the 1980s, and were widely thought to be closed, re-emerged in Europe, with due national variation, during the first half of the 1990s and fully reopened during the watershed of 1996/97. They explore several explanations and weigh the evidence: the failure of risk regulation in the face of diverging expert observations and public perceptions; the specificity of biotechnology in dealing with the enigma of life; the ethical implications of the technology, which make it unsuitable for a simple technological fix; and, finally, the possibility that in many contexts biotechnology is only the pretext, while people's real sources of grievance lie elsewhere, for example in the realities of European integration. Public opinion projects onto this new technology, which acted almost as a sounding board for a range of cultural discontents at the end of the twentieth century.

In polemic mode, the media are often blamed for causing the 'difficulties' in which some sectors of the biotechnology movement find themselves: the media misinform the public and thereby stir up irrational anxieties. That this is short-sighted as well as just factually wrong is the topic of the third chapter. At an early stage a technology has mainly a symbolic existence in the form of visions, images and ideas. Most citizens depend on the media to raise their awareness and to provide information and images of things to come. The chapter analyses the logic of selection and framing that is characteristic of media reportage, using national elite press as a comparative proxy. The dynamics of the coverage of biotechnology in elite newspapers in different European countries are compared for the period between 1973 and 1996. The description of the evolution of the intensity of press reportage, and the featured actors, themes and location of actions, provide a detailed picture of the shifting frames within which biotechnology is represented in public. Gutteling et al. finally explore some correlations between the contents of press coverage and people's awareness and evaluation of biotechnology as at 1996.

Chapter 4 by Galloux et al. explores the emergence of the ethical concern in the biotechnology discussion. In a comparison of Denmark, France, Italy and Greece, they reconstruct the institutionalisation of ethics as applied to issues of biotechnology over the past twenty-five years. They identify three different frames or ethical arguments that have come to bear on issues in biotechnology: arguments of utility, democracy and veneration. They explore the trends towards, and the dangers of, intentionally or unintentionally reintroducing a technocratic logic under a different guise.

The last chapter with an explicit diachronic outlook explores the interrelations between controversy, media coverage and public knowledge about biotechnology. Bauer and Bonfadelli take up the long-standing discussions on mass media effects, in particular the knowledge gap hypothesis, and put several propositions to the test. This hypothesis argues that media coverage on issues far from everyday experience, such as emerging technologies, will increase the inequalities of knowledge and awareness in society owing to differential public attention to the mass media. However, public controversies will increase the penetration of the mass media and thereby reduce the disparities in the representation of biotechnology in public. Mass media coverage of new issues increases the educational disparities that already exist in a country, while large-scale controversies will reduce them by speeding up and deepening the circulation of information. Controversies thus raise awareness and educate the public. The authors show that knowledge and other gaps are associated with the intensity of media coverage, and that the level of controversies is associated with a narrower knowledge gap. However, the longitudinal test shows that intensified coverage over time is associated with widening and narrowing knowledge gaps, not directly related to the level of controversy; other factors come to play a role. This leads to the conclusion that the knowledge gap hypothesis is underspecified as a dynamic model.

The second part of the book explores the structure and functions of public perceptions across Europe at the end of 1996, as evidenced mainly in survey data. The chapters do this from several theoretical and analytic orientations.

Chapter 6 takes up the idea of a social movement that is forming against biotechnology. Nielsen et al. explore, with the help of survey data, the constituencies of such resistance. By analogy to trends in environmentalism, they hypothesise that two ideological positions can be typified: a 'blue' or conservative and backward-looking resistance, and a 'green' or modernist and forward-looking resistance. They are able to demonstrate a cluster of ideas consistent with this distinction in most, but not all, European countries. In profiling the socio-demographic make-up of the two positions, they identify a north–south diversity. A blue in the north is not the same social constituency as a blue in the south; the same is true for the greens. The chapter closes with the construction of the prototypical arguments for a blue and a green position, drawing on different key metaphors and coming to different conclusions over the regulation of biotechnology.

Chapter 7 considers people's attitudes towards biotechnology. Classical social psychological attitude theory considers the strength, based on knowledge, and the structure or dimensionality of attitudes. It explores schematic information-processing in making judgements. Midden et al. draw a map of favourability and expectations towards biotechnology in Europe, and show that Europeans' attitudes are structured (based on a Lisrel model) along two independent dimensions: a utility–morality judgement, and a risk assessment. The low level of knowledge involved in these judgements suggests that, as is to be expected in the early stages of a public debate, these attitudes are not-yet-attitudes – not yet crystallised into a stable mental disposition. Their character as non-attitudes makes them inherently unstable and prone to change in the future. The relationship between knowledge and attitudes shows a low overall correlation. However, knowledgeable people are likely to hold extreme positions on the attitude scales, either positive or negative. This observation contradicts the popular notion of a knowledge deficit underlying public concerns over biotechnology. Or, as common sense has it: to know is to love it; on the other hand, familiarity breeds contempt.

Chapter 8 takes a different look at the same survey data. Allum et al. engage with the discussions in public understanding of science over the attitude–knowledge complex that struggle to prove or disprove the simplistic deficit model. They explore the 'post-industrial model of public understanding of science', according to which the knowledge deficit model (to know is to love it) is neither true nor false in general, but may have or may not have validity in particular socio-economic contexts.

They consider work on the economic regionalisation of Europe and test the variation of knowledge and attitudes, and the relationship between attitudes and knowledge, in a multilevel analysis, with the individuals, the regions and the countries as sources of variation. All three levels contribute to variations in knowledge of biotechnology, and therefore the authors conclude that regions matter. They map perceptions of biotechnology into a four-fold typology of European regions and show characteristic patterns of perception along these lines; however, these differences are not in line with the expectations of the post-industrial model of public understanding of science. Here the authors meet the limitation of the hypothesis and of the attempts to test a diachronic hypothesis on synchronic evidence.

It is widely known and accepted that attitudinal judgements depend on knowledge and images of the object to be judged. The conceptualisation of this phenomenon is theoretically controversial. Chapter 9 demonstrates the representations that underpin attitudinal judgements of biotechnology. Wagner et al. investigate this with data from the same survey as the previous three chapters, but use in addition a particular type of data, free associations. Respondents were asked to answer the open question 'What comes to your mind when you think about modern biotechnology in a broad sense, that is including genetic engineering?' The authors systematically analysed several thousand answers to this question as an index of common sense in Austria, the UK, France, Germany, Norway and Sweden. The idea is that discursive frames evidence the process of symbolic coping by which the public takes up the challenge of biotechnology in different contexts. The ambiguous evaluation evident in the data is related to the images of biotechnology: curing diseases, unnatural food, diagnostic testing and monsters. The identification of general images of biotechnology is complemented by particular national images with iconic quality.

The third part of the book presents two case studies from the watershed years of 1996/97 that triggered an international and synchronous public controversy over biotechnology.

Chapter 10 systematically analyses the events of late 1996 and early 1997 when Monsanto's shipments of soya that contained admixtures of a GM variety entered European ports. Lassen et al. show that official approval of Monsanto by the European Commission was undermined by the issue of labelling, which was left unresolved and which provided an entry for public controversy. Different patterns of immediate events are analysed for Denmark, Sweden, Italy and the UK in terms of socio-economic standing, political tradition, the strength of the actors involved and the discursive environment. The soya controversy had three effects:

it reopened the debate simultaneously in many countries; it reframed the discourses; and it left Monsanto as a corporate pariah on the European stage.

The first ten days of mass media coverage of Dolly the sheep across eleven European countries plus Canada are the focus of Chapter 11. In February 1997, eighteen months after its conception, a sheep was presented to the globally alerted mass media in Edinburgh, highlighting an achievement that had been deemed impossible. Whereas Round-up Ready soya was the trigger for the first European-wide biotechnology controversy, Dolly the sheep put the whole world into synchrony over biotechnology for the first time. Einsiedel et al. document the elaboration of a global trigger event in a cascade of mass media stories, evidencing a confluence of the operating logic of journalist practices and the logic of social representation – the metaphorical anchoring and pictorial objectification of strange, and potentially threatening, happenings. The proliferation of stories fluctuated wildly between the frames of progress and doom, and documented the ongoing struggle to represent technology in the late-modern age. The global moral panic over potentials for human cloning masked not only the national peculiarities in the imagining of cloning, but also what the Roslin research was intended to achieve: the efficient production of specific pharmaceuticals through live factories.

The fourth part of the book comprises the penultimate chapter, which takes a look across the Atlantic. A constant puzzle in our project was the apparent absence in the United States of concerns and voices analogous to those in Europe. Gaskell et al. explore this issue with a comparison of public perceptions and media coverage in Europe and the USA to 1996, supplemented by an excursion into the history of biotechnology regulations in the USA. The authors typify three logics of public reasoning: opposed, risk-consciously in favour, and in favour of biotechnology applications. Comparison of different applications reveals generally lower concerns in the USA than in Europe, although sensitivities are different across the Atlantic. The authors explore three putative explanations of the observed transatlantic differences: differences in media coverage, in public knowledge and in trust in the regulatory system. The characterisation of the history of debates and policy-making in the United States argues that, whereas the 1990s were a period of relatively low controversy in the USA, the 1970s and 1980s were marked by more lively debates. Through a mixture of normal and revolutionary policy-making the USA is found to have secured high levels of trust in the checks and balances of the regulatory system as a whole. However, there seem to be signs that Dolly and GM crops and food might reopen the public debate.

In the final part, chapter 13 offers a reflection on the ideas that developed in the course of this study, and proposes the integrative concept of a 'biotechnology movement' that carries this new technology through history. Centred on the scientific-industrial complex, the biotechnology movement is an actor network characterised by conflicting visions of the future. The biotechnology movement presents itself in material form in products and services and is symbolically re-presented in the tripartite public sphere of mass media coverage, policy and regulations, and public perceptions. Bauer and Gaskell identify a lacuna in the social studies of science and technology, namely a non-specified notion of the dynamics of the public sphere. They elaborate a 'three-arena model of the public sphere', comprising regulation, mass media and everyday conversations. Biotechnology constitutes a challenge for the public sphere, which in turn poses the counter-challenge for the biotechnology movement. The public sphere modelled as a tripartite arena allows us to characterise the changing symbolic environment of the biotechnology movement and, at the same time, the potentials for organisational and institutional learning that may result from public resistance. Two forms of collective learning – assimilation and accommodation – are introduced for this purpose. The traditional technological strategy has been to get society to adopt technological innovations by way of assimilating society to the corporate image of technology. However, resistance from the public sphere may lead to a change in the trajectory of technology, by way of accommodating corporate activities to societies.

In toto, this volume draws on empirical observations from policy-making, media reporting and public perceptions. It addresses conceptual and theoretical issues to document and explain the ways in which different European public spheres converged on controversies over biotechnology triggered by the watershed events in 1996/97. Beyond the particular focus on biotechnology, the book offers a conceptual framework and an empirical paradigm for observing new technologies as socio-technical movements within the constraints of the modern public sphere in the twenty-first century. In so doing, it is intended to contribute both to the continuous public debates over biotechnologies as well as to a growing corpus of scholarly interdisciplinary literature on science, technology and society.

REFERENCES

Bauer, M. (1995) 'Towards a Functional Analysis of Resistance', in M. Bauer (ed.) *Resistance to New Technology – Nuclear Power, Information Technology, Biotechnology*, Cambridge: Cambridge University Press; 2nd edn, 1997.

(2000) 'Classical Content Analysis – a Review', in M.W. Bauer and G. Gaskell (eds.), *Qualitative Researching with Text, Image and Sound – a Practical Handbook*, London: Sage, pp. 131–51.

Bauer, M.W. and G. Gaskell (1999) 'Towards a Paradigm for Research on Social Representations', *Journal for the Theory of Social Behaviour* 29: 163–86.

Durant, J., M.W. Bauer and G. Gaskell (eds.) (1998) *Biotechnology in the Public Sphere – a European Sourcebook*, London: Science Museum.

Gamson, W.A. and A. Modigliani (1989) 'Media Discourse and Public Opinion on Nuclear Power: a Constructivist Approach', *American Journal of Sociology* 95: 1–37.

Habermas, J. (1989) *The Transformation of the Public Sphere*, Cambridge: Polity Press.

Neuman, W.R. (1989) 'Parallel Content Analysis: Old Paradigms and New Proposals', *Public Communication and Behavior* 2: 205–89.

Part I

The framing of a new technology: 1973–1996

2 Promise, problems and proxies: twenty-five years of debate and regulation in Europe

Helge Torgersen, Jürgen Hampel, Marie-Louise von Bergmann-Winberg, Eleanor Bridgman, John Durant, Edna Einsiedel, Björn Fjæstad, George Gaskell, Petra Grabner, Petra Hieber, Erling Jelsøe, Jesper Lassen, Athena Marouda-Chatjoulis, Torben Hviid Nielsen, Timo Rusanen, George Sakellaris, Franz Seifert, Carla Smink, Tomasz Twardowski and Merci Wambui Kamara

Introduction: biotechnology – a chequered field

By the end of the twentieth century, disputes about biotechnology and its agricultural products had reached a peak in Europe. Industry had failed to win the hearts of European consumers and, de facto, there was still no real harmonisation of product assessment within the European Union (EU). This is noteworthy since, after a quarter of a century of rapid scientific and technological development on the one hand, and contentious debates about real or possible implications on the other, biotechnology has 'come of age'. Industry has finally engaged in biotechnology on a broad base, forming large conglomerates that deal with seeds as well as with pharmaceuticals under the heading of 'life science'. Ten (or even five) years ago, most spectators expected disputes over biotechnology to be settled imminently; nevertheless, issues are again subject to conflicts to such a degree that the future of some applications of biotechnology in Europe is uncertain. Obviously, traditional ways of introducing this new technology have failed. But can we establish why?

Classically, authors understood the relationship between technology and society to involve society adapting to technological developments (see Ogburn, 1973; Ellul, 1964). They saw technology as relatively independent of society, following its own rationality, or even forcing society to adapt to it. According to this view, it is inevitable that society 'culturally lags' behind technology, because technological development is faster than social and cultural adaptation (Ogburn, 1973). Society seeks to catch up, while new technological developments constantly open new gaps.

Indeed, the introduction of a new technology often seemed to follow this schema. For example, communication and information technologies raised debates in the 1980s (stimulated by Orwell's famous novel *1984*); but now, after the 'cultural lag' has been narrowed, they constitute welcome elements of everyday life. However, in the case of biotechnology, this classical model of explaining technological innovation obviously falls short. Here, even an apparent calming of the debate in the early 1990s proved to be only temporary and conflicts have reappeared.

From early on, uncertainty over possible hazards nourished struggles over biotechnology. It is a truism that every technological innovation is associated with risks or even hazards – remember explosions involving steam engines, or airplane crashes in the pioneer days of aviation. The difference here was that there was uncertainty as to whether there were any hazards at all. Opponents of biotechnology strenuously advanced claims of risks, while supporters denied them with equal assiduity. It is not our objective to determine whether any of these claims have been substantiated, or whether the risks identified are only hypothetical. Rather, in this chapter we attempt to trace *opinions* about whether there are risks, and, if there are, which kind, as well as the political consequences thereof.

At the heart of many controversies about risk lies a difference between public and expert experience of what biotechnology 'really' is. For the general public, this is not a technology like any other, since specific installations and artefacts are not easily discernible.[1] An atomic power plant serves as a symbol for nuclear technology, and computers are typical products of microelectronics. Without these practical applications, the technology would not have existed. Biotechnology, in contrast, is usually invisible to laypeople. Although applied in many fields, its techniques are employed behind closed doors in laboratories indistinguishable from outside. As such, at first glance, its products are not at all specific. Biotechnology consequently remains enigmatic, at the same time as dealing with 'the secret of life'. For experts, on the other hand, its character is no different from that of other technologies. Biotechnology laboratories are, for them, definitely recognisable, containing, as they do, specific apparatuses for research and production. To the expert, their products are clearly distinguishable by their properties, and no other technology could have produced them. Moreover, for the skilled, biotechnology did not pose intrinsic problems that could not be understood, and controlling the mechanisms of heredity, in principle, does not present insurmountable problems.

In order to deal with uncertainty, and to make risks in many fields calculable and manageable, engineers and insurers have developed (technical) risk assessments. Although twenty-five years have elapsed without a single major accident caused by biotechnology,[2] expectations among scientists that they would be able comprehensively and credibly to identify, control, compare and evaluate the technical risks of biotechnology have, nonetheless, failed. One major reason for this is that, owing to the subject of biotechnology being 'life' itself, ethical considerations played a significant role from the very beginning. Hence, debates went beyond the discussion of technical aspects, and especially those involving risks that could be technically determined. Indeed, in its status as a contested issue, biotechnology served as a sounding board, or a projection screen, for deep differences in interests and world-views. This resulted in a wide opening of the debate and the forging of links with other issues considered to be at stake.

A similar broadening of the debate could be observed in the case of nuclear energy and, to a much lesser degree, with microelectronics. Both technologies had acquired the position of strategic technologies long before biotechnology, although the latter was assigned this status soon after its birth (Bauer, 1995). Differences arose in the perceived need to regulate these technologies: with nuclear energy, the demand for regulation by the government was obvious; all but a few, however, agreed that the regulation of microelectronics was hardly feasible. In contrast, the struggle over biotechnology led to an ambiguous result, since different actors failed to agree on the issue of regulation, and different groups sought to delay or to support the use of biotechnological applications. Debates went beyond the question of whether or not a law should be passed, to the issue of the appropriate contents of regulatory policies. The latter became an especially contested topic not only within Europe, but also between the USA – as a promoter of a regulatory approach based on scientific evidence alone – and the EU, which was opting for a broader basis of argument.

Responding to competing interests, governments adopted different permutations of a two-sided policy: on the one hand, they provided regulation as a means of containing risks; on the other hand, they backed research and application in order to stay, or become, economically competitive (Jasanoff, 1995). This strategy seemed appropriate in order to 'make biotechnology happen'.[3] Obviously, governments thought that biotechnology was something worth developing and they supported it with alacrity. Yet they also styled themselves as impartial regulators of what many perceived to be a risky endeavour. This ambiguity later proved to be one of the sources of public distrust.

Despite this general pattern, the style and background of the debates and their political outcomes differed markedly from country to country, and subjects, actors and forums changed considerably over time. Even among EU member states,[4] we encounter an astonishing multitude of reactions to the challenges that biotechnology presented, in terms both of public debate as well as of regulation. Even differences in regulatory style influenced how biotechnology was perceived and which of the problems raised by different actors were deemed legitimate areas of state involvement. In order to gain an understanding of what may appear, at first glance, an odd and almost impenetrable jungle of rhetoric and facts, policies and interests, opportunities and risks, wishes and fears, attitudes and world-views, we have to examine this varied picture and explore its determining factors.

Much has been written about this issue over time.[5] As such, this chapter draws upon a wealth of data, resulting from four years of concerted research by an international group of scientists from various disciplines. This research covers policy initiatives, media coverage and public attitudes towards biotechnology in thirteen European countries, as well as in the USA and Canada (compiled in Durant et al., 1998).[6]

Concepts and methodology

In the following, we will trace the tangled history of biotechnology debate and policy-making during the last quarter of a century. We will keep in mind the four general issues summarised in table 2.1: first, the transgression of the boundaries of technical risk issues in the debates, which indicate that biotechnology acquired the role of a sounding board for the articulation of deeper concerns; secondly, the challenging of the technocratic paradigm of an autonomous chain of innovation; thirdly, the prevalence of regulatory responses intended to make biotechnology happen in the context of a two-sided policy of regulation and support; and, finally, the difficulties for the EU common market presented by the existence of profound differences among the nations of the EU in their attitudes towards biotechnology. These themes are, of course, interrelated; for example, the transgression of risk issues in the debates contributed significantly to those delays and detours (even dead-ends) encountered in attempts to implement biotechnology that served to challenge the

Table 2.1 *General issues in the biotechnology debate*

General issues in the biotechnology debate
Sounding board: the transgression of risk issues
Challenging the traditional innovation chain paradigm
Dual strategy in order to make biotechnology happen
National variations across the EU

Table 2.2 *Definitions of terms*

Terms	Definition
Biotechnology	Recombinant DNA techniques as well as methods for their practical application in various fields
Recombinant DNA techniques, genetic engineering	Combination of methods for splitting, sequencing, splicing and constructing genes
Genetically modified organisms	Organisms that carry genes modified by these techniques
Products of biotechnology	Results of production processes applying such methods
(General) public	Societal entity that carries public perceptions
Regulation	Means to put an aspect of biotechnology under rules

innovation chain paradigm. And many governments' dual strategies for encouraging biotechnology were inevitably embedded in national political styles. Different emphases in the various countries on the perceived problems concerning biotechnology made the sounding-board play different tunes, and facilitated or hindered governments falling back into their perception of the innovation chain.

Over time, we observe developments not only in biotechnology per se, but also in the meaning of the word. So, before looking at the history of this technology we should provide some provisional definitions of terms, since at different times stakeholder groups have used different terminology (see table 2.2).[7] For the sake of simplicity, we will stick to the term 'biotechnology' even if we refer to developments that scientists would consider to fall outside this categorisation. In general, however, we will focus on applications made possible by research on nucleic acids, which emerged during the 1970s as 'genetic engineering'. However, because biotechnology, even in this sense, is a very broad field, we have to restrict the scope of this investigation somewhat. We will therefore concentrate on those issues that are commonly covered by national biotechnology regulation, and give particular emphasis to the problems surrounding the release and marketing of genetically modified organisms (GMOs), since it is such products that have triggered the most intense public debate in recent years (BEP, 1997; Durant et al., 1998).

Debates arose not only among regulators and those involved in research and development (R&D), but also among the general public. It is not easy to define this 'public': there are many conceptualisations of the 'public' among social scientists, and there are probably also different 'layers' of the 'public'. Thus, our pragmatic approach is to adopt an understanding of the public as an entity that carries 'public perceptions' about biotechnology (see Gaskell et al., 1998: 9–10).[8]

The political problems associated with biotechnology changed over time as the technology matured from a set of methods for basic research to techniques for applications and products in various fields. Consequently, 'biotechnology policy' meant something different in each of these fields. Debates arose within many political arenas, such as parliaments, political parties and professional societies, but also in newspapers and other media. Actors, such as scientists, politicians, industry representatives, activists from non-governmental organisations (NGOs), and well-respected intellectuals, demanded different degrees of regulation.

Since the political reasons for regulation varied among countries, a multitude of ways of regulating biotechnology emerged, including special laws, adaptations of existing regulation, statutory guidelines and self-regulation. Clearly, regulation could take on very different forms, according to the type of actor each intervention was aimed at, and the functions it was intended to fulfil. As with most new technologies, from the very beginning the overall aim of attempts to control aspects of biotechnology was to provide a reliable frame within which it could be pursued safely while it was being implemented and promoted. Thus, keeping in mind the varieties of approaches involved, when we speak of regulation we refer to any means of applying rules to an aspect of biotechnology; if this is by law, we call it legal regulation.

Phases

In this chapter we are dealing with the social and political reactions to, and conflicts over, technological, social and political developments with respect to biotechnology. This task is complicated not only because of the complexity and variability of the technology itself, but also because the conflicts, debates and political reactions occurred in a bewildering array of temporal and spatial contexts. If we compare the history of debates and policy-making across Europe (Durant et al., 1998), we see numerous periods in which public attention and policy initiatives were negligible, alternating with more active periods. Likewise, we see activity in different countries at different points in time.

There are many reasons for this variegated picture. In Europe in this period, a loose economic community was changing its character to become an increasingly integrated system of member states. Yet, in spite of this integration, the political cultures and legal systems in European countries remain very different. Some legal systems are oriented towards Germanic law, whereas others follow Roman law. Countries with a first-past-the-post electoral system sit alongside countries with proportional representation, which lowers the hurdles for the formation of new political parties. Some countries saw the development of green parties in this period; others did not. The way in which the political systems reacted to public unease varied considerably; and the responsiveness of governments may have changed over time, since, during the twenty-five years covered, each European country experienced at least one change of government.

Initially, the asynchrony across Europe was most noticeable between the north and the south. Countries in the vanguard of technological development, political debates and regulation from the north and west of Europe had been dealing with the issue since the 1970s. Some of them witnessed widespread debate and others tried out various styles of regulation. These nations set the agenda for other countries, but some of the latter followed only after a delay of up to twenty years. Regarding the form of regulation, almost every country adopted a different position. The variety of approaches and responses in the European countries, and the fact that the EU institutions added another layer of regulation in an attempt to impose some harmonisation, make generalising extremely difficult. Thus, at first glance, producing a comparative analysis of the political debates and regulation of biotechnology in Europe might seem to be a hopeless enterprise.

However, regardless of all the intra- and international differences, there are some common characteristics. Some similarities and general patterns emerge with the help of comparative media analysis (Durant et al., 1998), which reveals common themes, or 'frames', representing the main interpretations of the opportunities and challenges for biotechnology. These frames were characterised by a shared set of problems, reference to similar stages in the development of the technology, and the expression of specific views that predominated among important actors. From our analysis of the main issues in public debates and policy-making over the past twenty-five years, we have discerned four phases, exemplified by distinct and overarching frames (table 2.3). Thus, it seems that, after a short period of time, the media emphasis shifted from laboratory science to industrial and economic aspects, subsequently moving on to the question

Table 2.3 *Phases in the debate about biotechnology*

Phase	Main issues
Phase 1 (from1973)	Scientific research
Phase 2 (from 1978)	Competitiveness, resistance and regulatory responses
Phase 3 (from 1990)	European integration
Phase 4 (from 1996)	Renewed opposition: consumers

of the EU's harmonisation of regulation, and, finally to the issue of consumer market products.

Biotechnology as a *scientific endeavour* was the first frame. Early 1970s coverage was, almost universally, characterised by a conceptualisation of biotechnology as an area of promising scientific research. Although the issue of risk was stressed, it was discussed by scientists themselves, who soon concluded that the risks were manageable. When, in some countries, public opposition emerged, scientists made attempts to reassure the public.

Later, this view of biotechnology gave way to a more practically oriented perception. It came to be recognised that biotechnology might gain enormous *economic* importance in an increasingly competitive world. Hence, it had to be defended against *resistance* arising out of safety fears. Scientists from other fields had developed counter-arguments, and these were taken up by critics. Opponents also asserted links with other contested issues, and the moral status of biotechnology – rendering life itself malleable – gave rise to concerns. It therefore became clear to many governments that self-regulation by scientists could not provide a satisfactory degree of protection against the perceived risks. In an attempt to confine the issues at stake, some governments began to *regulate* the field on a national basis.

The divergence of views about the appropriate way of regulating biotechnology eventually led the EU's institutions to take responsibility. It was hoped that this would engender some degree of *harmonisation* in the common market. This proved to be problematic, since regulatory styles and the history of the debate in the member countries varied so profoundly. Eventually, however, resistance faded and it appeared as if biotechnology had become a well-regulated and *accepted technology*.

Yet, when *consumer* products materialised, old conflicts reopened over new issues and the *opposition* was renewed. As consumers, citizens had a powerful voice, and it became obvious that existing modes of regulation appeared inadequate to many, and that national governments and EU institutions would have to find new solutions. Conflicts reached an unprecedented level even in countries where people had previously been fairly positive towards biotechnology.

The phases of the debate outlined above not only reflect the progression of the technology, as conventional innovation theory would predict, but also represent changes in the political, social, cultural and economic environment of biotechnology. This is illustrated by certain prevailing frames in the debates and by the character of particular policy initiatives at different stages. These frames were certainly not synchronous everywhere; rather, they overlapped differently for each country and field of application. Indeed, challenges did not occur at the same time in different parts of Europe and problems had not necessarily been solved when the new phase began. The subjects of earlier phases retained relevance in later stages. As such, the phases we stipulate indicate the addition of further layers of complexity rather than fundamental changes in the issues at stake (see table 2.3).

The debates expanded and contracted during each period. When biotechnology was seen as a scientific endeavour, debates could mostly be settled by scientists voicing safety reassurances. When biotechnology entered society at large, and commercial interests gained momentum, debates expanded once again. Controversy then began to cover applications and to emphasise world views, which prompted regulatory responses that often appeased resistance and led to a contraction of debates. The EU took up the issue at this juncture and attempted to harmonise regulatory frameworks. This became a new subject for debate and inflamed existing controversies; although, when biotechnology was formally endorsed, debate contracted once more. However, the commercial release of GMOs again triggered widespread opposition, and today we are in another phase of widespread debate.

This 'phase' structure, underpinned by concrete events, allows us to develop a framework for analysing the development of biotechnology debates and regulations in Europe. It also permits us to trace the four general issues mentioned above as characterising the four phases. In the following we will examine the four phases in turn, elaborating those aspects that are important for understanding this rich and complex field. In the conclusion we will then revisit the general issues and, finally, speculate on what the future might hold.

Phase 1 (1973–1978): scientific research

When biotechnology first appeared as a public issue, debate was clearly shaped by existing notions of scientific progress and by demands for scientific accountability. From the beginning, the issue of risk was emphasised and diffused from scientific discussions into public debate. The objective was to promote scientific research, while minimising technical risks and coping effectively with an emerging public debate. We will see how some countries, with a strong tradition in biotechnology, major industry interests and/or an alerted public, met the new challenges.

A call for scientific accountability

Methods of splitting, sequencing and splicing genes were developed in the USA. The first recombinant DNA (rDNA) result, in 1973, involved the successful *in vitro* transfer of a gene between species. With the advent of these new techniques, concerns arose in the scientific community about possible safety hazards. This resulted in a moratorium being agreed in order to allow time for scientists to think about the consequences of further research. Hence, biotechnology was accompanied, from birth, by the attribution of risk. This perception continues to this day.

Details published in a letter, entitled 'Potential Biohazards of Recombinant DNA Molecules', to the magazine *Science* (185, 1974: 303) outlined the scientists' concerns. The authors, Paul Berg and colleagues, called for a cautious approach until the hazards were better understood and an international conference could agree on how those risks should be contained. The scientific community felt that controls were necessary in order for work to be continued safely, and they accepted a voluntary moratorium. After eight months, at the Asilomar Conference, it was agreed that the moratorium could be lifted, and it was left to national governments to act upon the recommendations discussed there. In 1976, the US National Institutes of Health (NIH) issued technical guidelines for recombinant DNA research in line with the proposals, which were generally accepted world-wide as a standard for ensuring laboratory safety. Covering all work that was funded by the federal government, these famous NIH guidelines specified norms for risk evaluation in genetic engineering, for the safe handling of GMOs and for the adequate design of laboratories. With the help of these guidelines, funding agencies in the USA and other countries, as well as national governments, could easily adopt an internationally compatible frame of what to consider risky or safe. However voluntary it was, this was the first instance of a regulation being introduced in order to ensure that biotechnology could be developed.

From 1974 on, researchers and regulators responded promptly to the concerns raised in the Berg letter in an international, collaborative approach to regulating certain aspects of biotechnology. They perceived the need for rules long before the politicians. Scientists appreciated that research in biotechnology had huge potential but that this potential would be realised only if controls were put in place. Thus, the new NIH standard was welcomed. Yet it raised the political question of how to implement regulation on a national level, since countries differ hugely in their scientific landscapes. Some countries, such as Denmark, France, Germany, the Netherlands, Sweden, Switzerland and the UK, had a strong tradition in molecular and cellular biological research and major industrial interests, and some of these 'forerunners' also had an alerted public that would react to the new challenge. Other countries lacked such traditions, interests or a similar level of public alertness.

The solution was to leave it to the scientists to decide. The US debate that led to the moratorium had taken place mostly among the scientific community, and this group constituted the major actor in Europe as well. Since many researchers had for long been committed to international collaboration, they reacted in a fairly uniform manner in defining the problem in terms of science-based risk assessments. Most of the 'forerunner' countries quickly opted for a voluntary approach based on the NIH guidelines, and advisory committees were set up in these countries in similar ways. Since the emerging public policy issues focused on the perceived health and safety risks, it was no wonder that the different regulatory solutions were strikingly uniform from the practical point of view. The European countries had a common interest in safeguarding the future scientific development of biotechnology, so (voluntary) regulation aimed to provide the conditions in which the new and exciting field could flourish. Scientists themselves, it was felt, should narrow the gap between technology and society by virtue of their technical expertise.

Despite a general impression of uniformity, however, there were already some differences in these early days. Some of the 'forerunner' countries tried to keep the problem strictly within the scientific frame of technology and left the solution to technical experts alone, while others adopted a more inclusive approach, soliciting views from other sections of society. Indeed, it was soon clear that biotechnology implicated issues beyond the purely technical aspects of risk and safety.

From science to politics

Reactions among European 'forerunners' Among the first countries to react to scientific concerns was the UK, where the Department of Education and Science had implemented its own moratorium. Research Councils were keen to ensure that appropriate guidelines were introduced in order to allow researchers to return to work quickly. These guidelines were based on the existing regulatory framework for hazards from such sources as chemicals, and they even preceded the NIH guidelines. Although the UK did not see any significant public debate in this period, and discussions remained firmly within the scientific arena, the members of the Genetic Manipulation Advisory Group founded in 1976 (the term 'manipulation' was changed later to 'modification' owing to its perceived contentiousness) were recruited in a very British mode. Although they only advised on technical matters, such as risk assessment and physical containment levels, not all of them were scientific or medical experts – some were drawn from trade unions, management, industry and public interest groups. This composition provided for some information flow into other sectors of society (Cantley, 1995: 516), but the scientific frame was not questioned. Much as in the USA, the media focused on scientific (particularly medical) progress (see Gutteling et al. in chapter 3 in this book). Nevertheless, the UK was one of the first countries to introduce a regulation (in 1978) seeking to ensure a safe approach to genetic modification. Under existing legislation, laboratories had to notify the Health and Safety Executive of experiments. Scientists were advised on a 'voluntary' case-by-case basis, but they were not required by law to follow the advice; they only had to submit the information.

The Swedish Scientific Committee, too, set up in 1975 by the Academy of Sciences, was composed not only of scientists but also of representatives from government agencies, industry and NGOs. Their tasks were slightly different from those of the British: besides the review of research proposals, they had a mandate to keep the general public informed and to advise the Cabinet. This indicates that the issue was held to be of general concern and not confined to the scientific community. This perception proved to be a realistic assessment, although, according to another interpretation, it helped to trigger the first public debate on biotechnology in Europe when the announcement to build a special laboratory attracted media attention in 1978. For some time, as in the USA and other European countries, biotechnology had been a story of a promising new technology that focused on the USA. Now surveys showed strong public opposition to the plans, as biotechnology appeared dangerous and morally questionable. The Swedish government's answer to public concerns was regulation. In 1980, biotechnology was brought under existing laws, still following the NIH guidelines. The director of the powerful Committee for Industrial Safety and Occupational Health chaired the new governmental Recombinant DNA Advisory Agency, indicating that the new issue had found its place in the existing framework of risk management. A general feeling that trusted public institutions had taken care of the matter allowed public debate to wane; however, it never ceased altogether.

Comparing Britain and Sweden, we may identify similarities in that both were 'forerunners' in science; both saw the need for public and government involvement in the issue; both acknowledged, implicitly or explicitly, legitimate interests beyond the purely technical; and they both set up advisory committees. In the UK this resulted in rapid regulation, while in Sweden it resulted in the first public debate on biotechnology in Europe (regulation came later). The difference obviously lay in the fact that in Sweden a special incident (the new laboratory) elicited public attention. In the UK there was no such event in phase 1 or, at least, similar events were not covered in the same way, and public debate arose only much later. Such 'key events' of various kinds played a significant role in many countries because they raised public interest and eventually kicked off public debates, in turn leading to regulatory initiatives. However, this happened at different points in time, which contributed to the striking European diversity in the response to biotechnology.

Elsewhere, politicians (as well as the public) saw any problem related to biotechnology as an issue to be resolved exclusively by the scientific expert community. This also resulted in the idea that matters with which scientific expertise was unable to deal were to be considered irrelevant. France's regulatory approach came closest to a system of scientific self-regulation. Institutions performing rDNA research were to sign a contract with a research agency promising that they would follow French and international guidelines (Gottweis, 1995: 150). Although ethical considerations were emphasised early on (Cantley, 1995: 587 ff.), they were to be dealt with only in special committees. This approach was complemented by government's state-centred modernisation efforts. The institutionalisation of molecular biology was guided by a deliberate science-push policy, in line with a traditional understanding of the path of technical innovation. The French regulatory system, with its tradition of long-term planning and strong top–down orientation, succeeded for a long time in keeping the issue within the realms of the scientific and regulatory expert community. In order not to jeopardise this official policy, the French government was reluctant to respond to any concerns on issues other than physical risk.

Switzerland and Germany, which were among the technological 'forerunners' with strong pharmaceutical industries, also followed the path of scientific self-regulation, but initially with less emphasis placed on coordinating state action than in France. As early as 1975, the Swiss scientific community adopted the NIH guidelines and the Academy of Sciences established an Advisory Committee. Also in 1975, the German Research Association founded a Commission for Recombinant DNA. Later, the chemical engineers' association extended their concept of biotechnology to include the 'new genetics'. In 1978, at the same time as in the UK, the German government enacted safety guidelines for biotechnological research. Attempts by the German government to be the first in formulating a draft law for 'the protection from risks associated with modern biotechnology' in the same year triggered resistance by scientific organisations and industry (Gill, 1991). Scientists perceived it as an insult to their freedom of research when government tried to impose regulations on a scientific field. Otherwise, during phase 1, there was little indication of the serious conflicts that would later emerge.

'Latecomer' countries did not exhibit particular strategies with respect to biotechnology, and consequently adopted regulation only later (this group included the Mediterranean countries, Ireland, Finland, Austria,

Belgium, Luxembourg and Norway[9]). They had either a less important research base or an industry that had no genuine interest in lobbying for support. Alternatively, they simply followed, after a short delay, the path of the 'forerunners'. As long as there was no public debate or reaction to perceived risks, 'latecomer' countries avoided actively addressing this issue, adopting an attitude of 'wait and see'. Eventually, some of the 'latecomers' adopted strategies resembling those of phase 1 in the 'forerunner' countries, at a point in time when the latter had already reached subsequent phases. As we shall see, others went their own way.

Europe and the USA parting The debate on genetic manipulation had culminated in the USA in 1977 when an increasing number of scientists concluded that the risks had been exaggerated and the moratorium was a mistake. What had begun as a small group of scientists attempting to act responsibly towards the public was seen to have spiralled out of control because it resulted in the public perception of biotechnology as a field replete with risks. Consequently, and pointing to the fact that no major accidents had yet been reported, the US scientists started publicly to deny that there were any novel 'biohazards', i.e. any risks that would justify special legislation, let alone a moratorium. Having opened the debate on biotechnology, over the next couple of years the scientific community also managed to close it. At least for the time being, genetic technology ceased to be a hot topic for public debate in the USA (if it ever had been one).

At this juncture, the European and the American scenes diverged. At the end of phase 1, when public debate was beginning to wane in the USA, events attracting media attention led to new debates in Sweden, and at a later point in time in Denmark, Germany, the Netherlands and other countries. European countries also took up various issues beyond those of technical risk, thus widening the scope of the debate. It is important to note that until the mid-1990s these debates were diverse and strictly national, and they were often confined to local contexts. At the EU level, however, the Directorate General XII responsible for R&D (DG XII), accepted the dominant framing of biotechnology as an entirely scientific endeavour, much in line with the US approach (Cantley, 1995: 518 ff.).[10]

Summary

An initial wave of concern originating from the scientific community in the USA led to a self-imposed moratorium that subsequently gave rise to the NIH guidelines. Doubts arose about these concerns – and about whether the price to be paid (the self-imposed restrictions) was too high. In the USA, public debate waned after scientists had reassured the public that risks were both marginal and manageable if voluntary guidelines were followed. In certain European countries, public opposition was mobilised when a particular event or project attracted enough media attention. The conflicts often resulted in demands for the regulation of biotechnology under special laws. However, the US example prompted many regulators to abandon any such plans. By 1980, self-regulation by the scientific community in line with the US guidelines, and made statutory because of links to the existing regulation of other fields (for example, chemicals), seemed to provide a flexible and sufficient means of ensuring an acceptable safety level for the handling of rDNA. Concerns other than those of technical risk were dismissed as unscientific or were left to be dealt with separately. This arrangement served effectively to assuage public debates in most countries. For the regulators, the problem appeared to have been settled. However, contrary to the situation in the USA, it was only a matter of time before new conflicts arose.

Phase 2 (1978–1990): competitiveness, resistance and regulatory responses

Phase 2 saw the general entry of biotechnology into the public sphere. New frames arose and new debates emerged as it became clear that biotechnology had great economic potential. But, at the same time, the issue began to attract the interest of new actors. Starting from controversies over risk, the debates opened up, borrowing arguments from other contested fields in society. In particular, when GMOs were grown in large quantities or were to be released into the environment, biotechnology was discussed from a variety of perspectives and no longer within the scientific frame alone. Political responses themselves were different according to the regulatory styles and policy traditions of each nation. Over time, regulatory events and arguments raised in the various debates influenced each other across borders and boundaries, and biotechnology became a European issue.

Biotechnology meets society at large: the new economic frame

In phase 1, it was mainly scientists who were concerned about the risks of their daily work in the laboratory. In phase 2, biotechnology began to branch out from basic research into applied science. In pharmaceutical, chemical and, later, also plant-breeding companies, biotechnology improved both R&D as well as production processes. Its promise was to create products that had never before been available, or at least not in such quantities or at such high quality.[11] This produced an enlargement in the range of legitimate points for consideration in discussions of regulation. Economic expectations became more important, and in order to speed up the handling of applications, and thus to enhance national competitiveness, there was pressure for regulation to be simplified.

Another factor supporting this development was the general political climate. The 1980s witnessed a 'paradigm shift' according to Peter Hall (Hall, 1993), manifested in the rise of conservative governments and the expansion of supply-side economic thinking. International competitiveness became important, and the early commercialisation of biotechnology fits neatly into this paradigm. Unemployment increased in many countries and started to become a serious political problem. Competing in the high-technology innovation race appeared to be the only means of maintaining established levels of welfare in the face of rising unemployment and strained social security systems. In this context, biotechnology gained the reputation of being a 'key technology' and an 'industrially strategic area' (Sharp, 1985).

Starting in the USA in the early 1980s, public attention also shifted to the industrial potential of the technology (Pline, 1991). The press emphasised the importance of biotechnology in a new era of the scientific-technological race. Biotechnology was framed as a lead technology, along with information technology, which at that time had already started to have an immense impact on industrial production processes.[12] Accordingly, it was believed that those who missed the chance to be at the forefront of technological innovation would put the wealth of their nation at peril: biotechnology had to be made to happen. Thus, supporters of biotechnology warned national governments not to delay its application by 'unnecessary' regulation. Especially in the USA, where commercial biotechnology was most advanced, regulation was seen as a threat to economic competitiveness. Research had been going on without incident for several years, and the NIH had continuously modified the guidelines with a view to facilitating R&D and, implicitly, encouraging the successful application of its results. Small start-up firms, founded as spin-offs from academic research, struggled to develop their techniques while keeping up the promise for new products in order to raise money on the stock markets (Oakey et al., 1990).

Following this imperative, by 1978 the UK government had recategorised much rDNA work as low risk and advised that it should take place under 'good microbiological practice'.[13] In general, whereas in the 1970s regulation in the form of guidelines for laboratory safety had been brought to the fore, now regulatory efforts were reduced. However, an encouraging regulatory environment was only one element in securing a successful future for biotechnology. Promoters demanded more: namely an explicit and dedicated support strategy for translating scientific insights into commercial products, in other words, to accelerate the speed of the innovation chain.[14]

So, support for R&D gained momentum with the complementary strategy of governments' two-sided approach to fostering biotechnology. In most European 'forerunner' countries, governments funded huge programmes to promote research.[15] However, paramount in realising the promised potential of biotechnology was support for new firms. Some European governments invested in new biotechnology start-up firms or forged relations between academic researchers and industry. Companies such as Celltech in Britain and Kabi in Sweden, but also Novo in Denmark, Hoechst in Germany and Gist Brocades in the Netherlands, embraced biotechnology as a field of basic innovation with enormous economic opportunities.[16]

Emerging debates: commercial applications of genetically modified organisms

New safety problems arose with the possibility of commercial application. 'Large scale' and 'deliberate release' were terms that came to symbolise concerns that went beyond the conditions of basic scientific research (Cantley, 1995: 528).[17] In particular, the regulatory challenges presented by large-scale production had to be reconciled with the desire to support industrial biotechnology.[18] The growth of modified micro-organisms in large fermenters, as opposed to tiny articles of laboratory apparatus, posed a considerable risk. If an accident occurred, a huge number of GMOs, usually bacteria, would be released into the environment, with unknown consequences. However, there were tools at hand to deal with such risks. The NIH guidelines had already defined containment measures for various classes of organisms according to their potential hazard. The first products, available from the mid-1980s, were biologically active substances such as human growth hormone, insulin, interferon and vaccines against an increasing variety of diseases. Other types of products entering the market stage were enzymes for washing purposes and chymosin for cheese production.

The second area of commercial application was agriculture, which implied the release of GMOs into the environment without any containment. There were no blueprints for guidelines at hand, in contrast to the situation with large-scale production. Consequently, uncertainty was much greater, and scientists too had some doubts about their ability to predict risks. The opposition of anti-biotechnology NGOs was more far-reaching (see Beck, 1986). They considered the risks posed by the genetic modification of living organisms to be 'new', because such organisms could not be encountered in nature. The release of such an organism was seen as an irreversible experiment with unforeseeable consequences. In the light of the precautionary principle,[19] opponents claimed that such uncertainty demanded a ban. Proof of the absence of risk was argued to be a proper precondition for the use of GMOs. And, since large-scale production posed the risk of unintentional release, the same argument was advanced for other applications.

Proponents of biotechnology considered the NGOs' demands to be illogical nonsense, since the absence of something cannot be proven. They pointed to the fact that every innovation to date had entailed some uncertainty, otherwise it would not have been innovative. According to proponents, in the case of biotechnology in general, and even releases of GMOs, a risk/benefit analysis would reveal huge benefits combined with minimal risks. This opened up a second path of argumentation. Yet opponents also claimed that the new biotechnological products had disadvantages compared with traditional ones, which meant that the risk/benefit relation would always be negative in their view.

The new discussions about the deliberate releases of GMOs emerged across the Atlantic, when in 1982 researchers in the USA for the first time sought permission for a release, which was granted after years of litigation and conflict (Krimsky, 1991: 113–32).[20] The photograph of a worker wearing a protective astronaut-like suit while spraying bacteria on a field became one of the governing images of what the release of GMOs could mean. While the debate in the USA waned,[21] or became confined to scientific circles, interest was intensified in Europe in a political climate shaped by several other conflicts over technologies. In the early 1980s, some European countries, such as Sweden, Denmark, Germany and the Netherlands, saw frequent public debates about technology, and the struggle over nuclear power was at its height. In a climate in which NGOs and sections of the public were geared to engage in debates, new developments in the commercial application of biotechnology swiftly evoked opposition. For example, when the Hoechst Company announced its plan to construct a plant for recombinant insulin production, this became the German 'key event' and triggered widespread public debate, NGO action and ligitation.

However, we have to keep in mind that reactions varied from country to country. For example, the question of whether or not GMOs should be released elicited virtually no reaction in France but led to a heated controversy in Germany.

Biotechnology as a sounding board

It was not only the risk issue involved in the move from basic research to applied science that had huge consequences for debates on biotechnology. Once elicited, debates rarely stuck simply to the technology itself: they came to inhere in much broader societal discourses. Together with interests in commercial application, those interests focused on biotechnology that ran counter to the kind of innovations the new technology would bring about.

An example was the conflict over recombinant Bovine Somatotropin (rBST).[22] This case sheds light on the fact that, from early on, there was acute sensitivity about applications of biotechnology in agriculture, and revealed that more was at stake than the health of people or animals. Here, socio-economic considerations about the structure of agricultural production were clearly at issue. Opponents considered biotechnology to be spearheading an agricultural rationalisation and industrialisation oriented solely towards economic criteria.[23] Although such arguments were dismissed as irrelevant when it came to the approval of single veterinary drugs, these concerns were nevertheless considered implicitly and played a role in the decision-making process (Tichy, 1995).[24]

Even more fundamental was the debate on medical diagnostics, because this clearly touched on issues of medical ethics. As the German sociologist Ulrich Beck (1988) speculated, such applications would not only permit, but reinforce, eugenic sentiments.[25] In some (Romanic) countries, ethical considerations had played a role for some time in the discourse among experts and churches. But Beck's concerns were derived from both ethical and societal considerations, since improved medical diagnostics might have major consequences for health insurance systems and for social solidarity.

Characteristic of this phase was also the fact that arguments were not focused on the different forms of application, but often transgressed the boundaries and conceived of biotechnology as a general entity. Critics tried to construe, in all applications, the underlying rationale of biotechnology as a technological and reductionist perception of life itself, one that seeks to instrumentalise life for the sake of profit. For their part, promoters tried to demonstrate the contributions to the common good that would emerge from the new methods. They stressed enhanced precision, widening possibilities for medicine, environmental remedies and industrial production. Ethical problems were sometimes acknowledged, but were usually left to expert advice.

Opponents often sought advice from scientists opposed to the analytical/synthetical approach of molecular biology, and often found them in the ecological sciences. Hence, in the struggles over biotechnology, scientific expertise of two different kinds was deployed, either denying the possibility of hazards or emphasising uncertainty and risks.[26] This contributed to the bewildering array of laypeople and politicians who demanded 'impartial' expertise from the sciences, since for most of them it was inconceivable that there was more than one version of 'objective' truth. Ultimately, this would contribute to the erosion of trust in scientific advice.

As the debate grew more intense, the argumentation patterns became even more concentrated on social and environmental aspects and the implications of the technology than on the technological methods themselves. In dealing with social, economic, ethical and environmental subjects, the frame became enlarged. As well as debating new technological opportunities and risks, the basic relation between man and nature was questioned. And, by looking at the performance of the actors involved, the relation between scientific and political elites and the general public came into focus (Kliment et al., 1994).

The media reported two different stories about biotechnology. Industry emphasised commercial applications including large-scale fermenting and deliberate release. Exaggerations of future promises were commonplace as companies tried to attract venture capital, despite the fact that very few of them had an actual product. On the other hand, environmentalists put forward not only risk arguments – whether or not they were substantiated – and abstract ethical considerations, but also concrete and serious doubts about the track record of the chemical and agro-industry and the financial entanglement of senior researchers with start-up companies (Cantley, 1995: 547).

Thus, the image of biotechnology as a potentially controversial technique became established in part of the elite's mind. In phase 1, the critical arguments originated from within the scientific community; in phase 2, all sorts of organisations, including churches and NGOs, contributed to the debate. Biotechnology became a widely discussed political topic, where underlying interests and world-views played a significant role. Also the meaning of 'risk' changed: in phase 1, a technical risk definition was still adequate, whereas in phase 2 critics linked the issue to other contested areas, such as reproductive technology, feminism, animal rights and even the desired path of future development. The argumentation patterns reflected the technology in its social and environmental context. Along with new technological opportunities came new questions: who should have the profit? And what should the relationship be between the scientific and political elites and ordinary citizens? By going beyond the narrow range of issues that most scientific experts deemed legitimate for discussion, the seemingly 'natural' way of implementing the new technology and entering the next stage in the innovation chain was seriously challenged.

In order to understand the evolving controversies, we have to look at how the political systems in the European countries reacted to this challenge. Not surprisingly, we meet a stunning variety of responses. Clearly, there is no such thing as a standard political response; rather, each country developed its own way of dealing with the emerging challenge posed by different stakeholders' demands.

Political responses

Proponents wanted to continue the expert-oriented self-regulatory practice established in phase 1. This was perceived to be sufficient for all legitimate concerns. After all, according to scientists, genetic engineering methods were inherently safe; and they discounted the potential for intentional misuse or remotely possible unintentional hazards. Critics tried to abolish industry's privilege of self-regulation, which, they argued, inevitably entailed a conflict between commercial interest and safety. They wanted to broaden the scope of regulation to cover all applications, or even to ban the technology altogether.

There was major variation in these lines of thought across Europe, owing to differences in the development of the biotechnology industry and research bases, but also because of differing political situations. Some countries had green parties in their parliaments, or NGOs integrated into the national political system, and others had not. The role of churches had been different, and governments were confronted with public awareness of the potential risks of genetic engineering to differing degrees. Some countries experienced severe economic crises during the 1980s, and the period witnessed the rise of conservative governments and an increased emphasis on economic issues.

Accordingly, variations in national regulation at the beginning of phase 2 continued or even increased over time.[27] The task for each government was to realise the promises of biotechnology as a 'key technology', but the strategies employed differed. Governments did not heed public demands for a ban on biotechnology; rather, they looked for possibilities of integrating concerns in a way that did not threaten the economic status of national industries or the competitiveness of the research base. They also sought to maintain efforts to support R&D activities. These political strategies, designed to avoid widespread conflicts, were more or less successful in most countries. This does not mean that, in reality, a national 'consensus' was achieved. Rather, the opposing forces – with some notable exceptions – did not intend, or were unable, to provoke an open conflict fought out in a public arena.[28]

Countries and policy styles

Within the range of European diversity, some countries provide exemplars of the different types of political response. We can distinguish four (idealised) ways of dealing with the task of regulation and with public concerns:

1. exclusive or elite decision-making (as in France and the UK);
2. co-option (as in the Netherlands and Sweden);
3. public participation (as in Denmark);
4. delegation to the European level (as in most southern countries, among others).

Model I: exclusive or elite decision-making France had experienced a relatively intensive and controversial debate on nuclear energy, which had even been 'significantly more aggressive' than that in Germany (Rucht, 1995: 282). In comparison, public awareness of biotechnology remained negligible. France was the only 'forerunner' country that succeeded in maintaining an entirely expert-based regulatory style without the incorporation of public concerns. The partnership between public authorities, industry and research institutes was supported by the recruitment system dominated by the 'Grands Corps', which permitted common orientations and personal ties to develop among the different societal elites. Until 1996, other institutions such as churches, consumer or environmental organisations never succeeded in making (or never intended to make) biotechnology a controversial issue. In phase 2, this closed regulatory process was not confronted by any public mobilisation as far as biotechnology is concerned.

Like France, the UK had a closed political culture, supported by the electoral system. But British science and technology policy-making, although oriented towards expert knowledge, favoured a pragmatic case-by-case approach, in which the main actors cooperated in the introduction of a series of more-or-less ad hoc regulatory mechanisms through a complex network. 'The British style of policy-making . . . tends to be informal, co-operative, and closed to all but a selected inner circle of participants. Disputes are resolved as far as possible through negotiation within this socially bounded space' (Jasanoff, 1995: 325f.). The major difference from France was that this closure was handled pragmatically, with channels provided for the integration of NGOs. Advisory committees to government (a key part of UK technology policy-making) could take account of wider public concerns because of their membership (they often included public interest representatives), as well as through various forms of public consultation.[29] Thus, government could reconcile the critics' demands comparatively well without jeopardising its elitist decision-making style and without becoming truly participatory. When experiencing early difficulties with the public acceptance of biotechnology, the British authorities included environmentalists on the advisory committees. This pattern secured the co-option of moderate groups to include 'a voice of reasoned dissent that could be internalised without seriously jeopardising the evolution of technology' (Jasanoff, 1995: 319). Such committees were also flexible enough to adapt to new policies: following recognition of the industrial use of biotechnology, in 1984 the advisory body was replaced, marking the shift from a regulatory phase to that of a concrete promotion strategy.

Early in phase 2, the German public was mobilised. This was the peak of a development, which began with the student movement of the late 1960s, questioning the traditional path of technological modernisation. Different concerns 'merged together over the years and formed a highly active and politicised movement sector with considerable overlaps by the end of the 1970s without losing its political and organisational heterogeneity' (Rucht, 1991: 185). The government tried to channel the emerging public debate back to the scientific experts (Gill, 1991) by concentrating on technical details.[30] But these attempts to close the debate failed. With the Green Party, debate moved to parliament, and there a Commission of Enquiry discussed the (economic) opportunities versus the (technical) risks of biotechnology on a rather broad base. When attempts to ban the technology were dismissed, the Green Party started to mobilise the public (Gill, 1991: 172). German industry, at first strictly opposed to a special law, came to appreciate the need for a clear legal framework as a protection against litigation. The German gene law of 1990 offered, as a partial concession, participation on the government's key advisory committee and created a new public hearing process for deliberate release applications. These innovations seemed responsive to the enactment. In practice, however, the first hearings deteriorated into administrative wrangles and rhetorical stand-offs that led the government to revise the law in 1993 (Jasanoff, 1995: 323).

Model II: co-option The situation in the Netherlands was quite different from that in the UK and in France. As in Germany, radical groups had been involved in the debate on biotechnology, and in the late 1980s fields with GM crops were destroyed (van Praag, 1991). But social support in general for radical political action was low. 'All movements contain important reformist factions, primarily concerned with influencing decision-making, which therefore attach great importance to a pragmatic bond with political parties, trade-unions and various ministries' (van Praag, 1991: 312). Dutch biotechnology policy had always been explicitly two-sided: technology stimulation to harness the potential benefits was combined with attention to potential risk and public accountability.[31]

> [T]he relationship between [new] social movements and the government... differs strongly from countries such as the Federal Republic of Germany. In the Netherlands, this relationship must be seen against the background of the traditional bargaining model which has always governed labour relations and the political practice... of subsidising various social groups and, if necessary, involving them in policy preparation. After the initial shock of the various protest manifestations in the sixties, the government soon decided to modify its politics with regard to the... [different] movements. Cooperation and consultation, rather than repression, were the government's first priorities.... Up till now, this policy has been rather successful; the Netherlands is one of the least [politically] violent countries in Europe.... One could say that the Dutch political system is relatively open to new demands and social needs. (van Praag, 1991: 311)

In social democratic Sweden, the relatively tough regulatory regime was seen as a barrier to competitiveness and international collaboration despite certain relaxations of the regulations (Sharp, 1985). Concerns still focused on the need to ensure the management of health and safety risks, but in an environment that would not restrict R&D, since technological development was seen as being part, or even the basis, of societal progress. As it became clear that defining parameters for risk and support required a close relationship between the academic and industrial communities, the regulators' answer was to bring stakeholders from all backgrounds together in several scientific advisory committees, thus enlarging the scope of the problem definition while keeping it under strict control.

Model III: public participation Denmark, according to the results of the Eurobarometer survey (Durant et al., 1998), had one of the most 'sceptical' publics in Europe. Biotechnology issues were hotly discussed and environmental movements had been important in creating a public debate (Gundelach, 1991: 278). Yet public debate seemed to be more differentiated, and Denmark appeared to be the showpiece for public participation. This participatory political approach also reflected the regulators' inclusive political tradition. 'In Denmark... politicians have been very efficient in taking the wind out of the movements by entering into discussion with the movements and accepting several of their demands' (Gundelach, 1991: 274). Using the consensus conference model,[32] Denmark engaged in a public debate that was deemed legitimate and was able to ensure at least public tolerance of biotechnology. In 1986, it became the first country in the world to pass a special 'horizontal' law on genetic engineering, covering various applications in research and industrial production under a single piece of legislation. This largely followed the NIH guidelines for work with recombinant organisms in the laboratory, but it also provided a means to apply biotechnology on a large scale. Concerning releases, there were certain regulatory idiosyncrasies, which met the critics' demands in a rhetorical way.[33]

Model IV: delegation to European regulators In 'latecomer' countries – mostly southern ones such as Italy, Spain and Greece, but also countries at the European periphery such as Ireland – biotechnology was not as developed as in the 'forerunners', and public awareness was more or less absent. Hence they could choose simply to proceed with the application of the NIH guidelines until such time as the EU developed regulations aimed at harmonising the common European market.[34]

During the 1980s, as at other times, most governments aimed to promote the research base, and there was no need to develop their own regulatory activities, which could have led to the public's becoming interested or suspicious. Debates about biotechnology, if they arose, remained national, and even in countries sharing a common language, such as Germany, Switzerland and Austria, the situations were entirely different. Whether or not the public 'caught fire' remained largely a matter of chance, or depended on a mobilising event such as the first release of a GMO or the construction of a particular production facility. In most 'latecomer' countries, there was no such crucial event during the 1980s.

Cross-country repercussions and EU regulation

Although debates remained confined to national settings in phase 2, any national regulation – such as the German gene law – was, of course, not totally detached from what was going on elsewhere in Europe. This piece of legislation, for example, took into account the precedent of the Danish law as well as those of the USA, the UK and the Netherlands in attempting to legislate for the entire range of biotechnological applications. The German law also built on the 'Recombinant DNA Safety Considerations' of the Organisation for Economic Cooperation and Development (OECD, 1986; see below),[35] and mirrored the discussions at the EU level of proposals for several special directives on biotechnology.[36] In fact, the EU proposals on the idea of separate legislation may have weakened those forces in Germany that still resisted a special national gene law.

Conversely, regulatory developments in Germany also had repercussions in other European countries. In France, in the late 1980s, policymakers attempted to counteract what they perceived as rising public distrust (Gottweis, 1995: 394–402).[37] What happened in Germany was considered anathema to the promoters of biotechnology. But, unlike the British and German governments, the French did not pass any legally binding regulations, in particular because a regulatory decision was soon to be expected at the EU level. Rather, they followed a 'business as usual' strategy, keeping matters within the realm of expert decision-making.[38] From 1986, the Commission Génie Biomoléculaire at the Ministry of Agriculture promoted the objective of France becoming the leading European country in the field of GMO releases. Indeed, the number of French releases did increase, and by the end of 1998 they were almost double those in Italy, the country with the second-largest number of releases in Europe. Only in Canada and the USA were there more releases.

Regulation and the economic frame

The European Commission had long been watching the emergence of regulations in European countries. It feared that this trend would jeopardise the common market in a future field of high technology, and over the 1980s it made serious efforts to define a common European legal standard (for reviews see Cantley, 1995; Galloux et al., 1998). The first such attempt, a proposal in 1979 by DG XII, had been abandoned because at that time non-regulation was the favoured option (Cantley, 1995: 526; Galloux et al., 1998: 180). Since the methods of biotechnology had not resulted in any apparent negative incidents, the intention was not to endanger the 'key technology's' promise of industrial growth and new products.

In response to changing perceptions of the need for regulation, in 1982 the Council recommended the development of oversight structures. The general feeling over the next three years was that new applications would not pose insurmountable problems as far as safety was concerned. To enhance European competitiveness,[39] the Commission put more emphasis on a coherent strategy to promote R&D through DG XII and its Biotechnology Steering Committee.[40] However, with the small budget available, this proved to be difficult for the Commission, and industry hesitated to cooperate. 'At the end of the 1980s, the Commission saw itself confronted with the fact that its aim, namely to promote cooperation between public and private research and to construct a European research network, could not be reached' (Bongert, 1997: 127, our translation).

Extension of the frame of regulation

The political efforts for the promotion of R&D changed in the mid-1980s. Paramount was 'the entry of environmental interests into the policy debates on biotechnology . . . , as the public authorities at national and Community levels interpreted their general responsibilities for the protection of the environment in relation to the challenges of the new processes and products resulting from biotechnology' (Cantley, 1995: 546). This marked a major widening of the scope of what were perceived to be legitimate problems involved with biotechnology regulation. Although the 'precautionary principle' was not explicitly adopted, the environment became a focus of attention, along with economic and agricultural interests, while the question of workers' protection was still unresolved. The views of the 'sounding boards' in the individual countries were heard in the European Commission.

This was even more the case with the European Parliament, which in 1985 came up with a view more extreme than that of the Commission: it demanded special regulation.[41] In the same year, the Commission set up a committee with members from several Directorates General (Galloux et al., 1998: 181), with the aim of reviewing existing rules and determining whether they adequately dealt with commercial applications. Its composition (representatives of the rival DGs for industry, social affairs, agriculture, environment and science, internally competing for influence) suggested a broader understanding of the issues at stake in terms both of the direction of commercial application and of environmental concerns. At the same time, an amendment to the Chemicals Directive provided a blueprint for much of the possible biotechnology regulation.[42]

The way to the EU directives

The 'hinge year' proved to be 1986 (Cantley, 1995: 549). The first genetic engineering law, passed in Denmark in order to provide an overarching legal frame for the activities of biotechnology research and industry, put the Commission under pressure to act. The German Commission of Enquiry, influential across the borders of the country, also had the implicit aim of devising a national biotechnology law. Both the Danish and German initiatives, as well as the European Parliament's report, stressed the need for special legislation.

On the other hand, both Britain and the Netherlands felt comfortable with their pragmatic approach under existing sectoral legislation. They received backing from the USA, which defined competencies among government agencies under existing legislation in its Coordinated Framework for the Regulation of Biotechnology. In the same year, the OECD issued its 'Recombinant DNA Safety Considerations' (OECD, 1986), introducing recommendations for the risk assessment of GMOs, e.g. the 'step-by-step' principle and the idea of Good Industrial Large Scale Practice (GILSP), while postponing guidelines for deliberate releases. The OECD report was in line with the US Framework and held strongly the position that 'there is no scientific base for specific legislation to regulate the use of recombinant organisms'; this was also the way sectoral DGs of the European Commission, as well as industry, preferred to see the issue. Inadvertently, however, the report helped to pave the way for viewing recombinant organisms as constituting a distinct category.

The question of whether a transgenic organism should fall under special regulation solely on the grounds of its genetic modifications, and irrespective of its application, was hotly debated. It is still a moot point whether such 'horizontal' regulation is adequate, as opposed to 'vertical' regulation for each product according to its properties. Although industry and scientists repeatedly denied the need for – and the feasibility of – a special regulation for GMOs, the Commission finally proposed 'horizontal' regulation in a 'Community Framework for the Regulation of Biotechnology'. The reasons for this step (see Cantley, 1995: 550 ff.) were threefold and may be conceptualised in terms of harmonisation, risk reduction and dealing with uncertainty. First, the Common Market was to be protected from idiosyncratic national regulation. Secondly, the intention was to apply appropriate guidelines for containing levels of production universally. Thirdly, there was still uncertainty about the criteria for the risk assessment of GMOs to be released to the environment, so a case-by-case approach was designed to systematise the accumulation of experiences.

Clearly, one additional, and openly admitted, aspect of the pressure for regulation was 'public concern, in part a spillover effect from other areas of technology where accidents of an unexpected nature or scale had occurred'.[43] In line with the European Parliament's position, and rhetorically fortified by public concerns, the Commission decided to issue a regulation on GMOs. The proposals for the Directives on Contained Use, as well as on the Deliberate Release and the Placing on the Market of Products, exclusively dealt with GMOs.[44] So, although other directives or drafts had been issued that were more sectorally oriented,[45] this approach was abandoned in favour of a more horizontal one.

Industry strongly opposed such horizontal regulation, which it feared would lead to the 'stigmatisation' of the technology of genetic modification. Consequently, the proposals prompted harsh criticism from industry and scientists. These critics were supported by the USA, which claimed that there was no scientific basis for regulating GMOs as a special category; that the directives would hinder R&D; and that they would introduce non-tariff trade barriers on future products.[46] The opposition prompted the EC to construct the Directive on Deliberate Release and the Placing on the Market of Products (which became known as Directive 90/220/EEC) in such a way as to facilitate future 'sectoral' legislation (in spite of the 'horizontal' scope of the proposal). The Commission immediately started work on formulating such legislation for 'novel food'.

Two opposing demands had to be balanced. On the one hand, the European Parliament as a representative of a fictitious 'European public' conveyed the plea for a very cautious approach following the public unease expressed in several member countries, where fears of long-term and indirect effects clearly played a role. On the other hand, science and industry demanded a 'purely science-based' approach focusing only on organismic and product properties, in order not to jeopardise innovation by regulatory demands that remained ambiguous to many. Industry wanted a narrowly defined legal framework so as to enter the race for the application of a 'key technology'. It was sympathetic to the explicitly deregulatory US approach of officially relating the degree of regulation to the amount of 'objective risk' posed by the organism.

The political solution seemed to be to stress uncertainty, and the Commission defended the need for a special regulation with the argument that the technique was novel. So the purpose of the regulation was announced to be proactive (i.e. put in place before any accidents had occurred), in accordance with (at least a certain understanding of) the precautionary principle. The directives were definitely not enacted on the basis of proven risks from the methods of genetic engineering; rather, the rationale was to use discretion until more experience was gained and regulation could be relaxed.[47] By emphasising uncertainty, this regulatory solution tried to force together two apparently incongruent approaches to risk assessment, namely one that built on scientific evidence ex post, and one that built on scenarios of hypothetical risks ex ante. It also acknowledged the dual nature of the 'biotechnology problem' – technical as well as a matter of public perception – without openly addressing issues beyond risk that could not be dealt with by scientific experts. This artistic and delicate balancing act attempted to bridge the gap between the different regulatory styles and public attitudes in various European countries, in order to provide a unifying framework for future technological innovations. Since the stakes were set so high, it was no wonder that difficulties later arose.

Summary

In phase 2 biotechnology and the debate surrounding it diversified and extended, as it became clear that the new methods could be applied in various fields and contexts. The argumentation reflected two lines of thought prominent in the 1980s. On the one hand, the economic exploitation of research results (as of anything else) was prevalent in a climate shaped by the increasing dominance of supply-side and market economic thinking. On the other hand, technology critics placed biotechnology on the list of contentious technologies, together with nuclear energy, that were considered to play a crucial role in the 'risk society'. Uncertainty about both possible or perceived risks and promises of economic opportunities led some European governments to 'balance chances and risks' via national regulation. The European Commission recognised several challenges, all of which highlighted the need for unified conditions in the application of biotechnology. First, industry demanded a defined legal framework for its activities. Secondly, national regulations threatened to jeopardise the common market for new products. Thirdly, since public opposition in mostly Scandinavian or German-speaking countries supported the demand for a special law, there was a need for compromises in order to contain criticism and to prevent its spread.

Phase 3 (1990–1996): European integration

The European Community's biotechnology directives were an attempt to arrive at consistent and homogeneous regulation across the EU. The aims were to prevent the passage of idiosyncratic national laws that would jeopardise the implementation of this key technology in a common market, and to provide universal safety standards. Aside from this, the directives had a variety of impacts in a period in which the EU was considerably enlarged, and when biotechnology regulation became a hotly debated issue across the Atlantic. Although the homogenisation of regulatory approaches proved to be difficult, public debates waned as biotechnology proceeded and became entrenched in science and the economy.

Directives

In 1990, after long deliberations, the European Commission passed a set of directives on biotechnology.[48] The Contained Use Directive imposed a categorisation scheme for micro-organisms and the split between the application of such organisms in 'small-scale' (e.g. for research purposes) and 'large-scale' devices (e.g. for production). The Deliberate Release

Directive rendered mandatory the principles of 'step-by-step' (from laboratory experiments to small research releases to releases of a large number of organisms and, finally, to commercialisation) and 'case-by-case' (which implied that each release application had to be considered separately and, although some provisions for a later exemption were made, there was no categorisation). Permissions for the commercialisation of products were granted Community-wide, whereas permissions for releases remained within national competencies. The directives did not permit socio-economic criteria to be considered during the assessments of planned releases or products to be commercialised; only risk issues were intended to be dealt with. This should have served to restrict the range of permissible arguments and to keep the regulatory process firmly within the scope of scientific risk assessment. However, it turned out that this expectation could not be entirely fulfilled.[49]

This programme was, in fact, a compromise. Although influenced by public debate, it fell short of a comprehensive regulation that took into account all the concerns that had been voiced in the various countries.[50] The directives provided a legal framework, but industry was not very enthusiastic: it perceived the scientific base of the technology-centred regulation to be flawed, and feared difficulties in implementation as well as an 'unnecessary burden' for biotechnology. Despite their perceived shortcomings, and the Commission's reluctance to see the directives as a final piece of legislation, their importance was enormous.

Now biotechnology had become a truly European matter. Until then, problem awareness among scientists and regulators had frequently been triggered by developments in the USA. Public debates, on the other hand, had arisen almost exclusively in response to certain national events triggering a reaction in the media and arousing the interest of NGOs.[51] The relative failure of the European Commission to create a Community-wide research network by the end of the 1980s had mirrored the national preoccupation with the research base as well. Hence, before 1990, matters of biotechnology had largely been either transatlantic or national issues in the 'forerunner' countries and non-issues everywhere else. The directives marked a significant shift, since the intervention of the EU in the regulatory process had consequences for all EU members as well as for other countries.

- The 'forerunners' had to adapt their national regulations to the new European standard.
- Other member states of the EU had to adopt the regulation of a hitherto unregulated field.
- New member states had to implement the European regulation.

- After the collapse of the Soviet system, East and Central European states became liberal democracies and developed market economies. As part of their drive towards membership, they adopted relevant EU regulations.
- In order to prevent future trade wars due to 'non-tariff trade barriers', when GM products were to be placed on the market ways had to be found to reconcile the directives with the US and Canadian way of (de-)regulating biotechnology.

Impact of the directives on European countries

'Forerunners' The 'forerunner' countries had, over the years, set up rules in a system of national interest intermediation. The delicate balance of power between the various actors arose from written and unwritten rules and structural conditions varying from country to country. These rules and conditions changed when regulations started to be negotiated in Brussels instead. The strategic balance in biotechnology conflicts changed dramatically thereafter. Interest groups lost their influence unless they joined European organisations, where there were different rules and different actors had their say.

In general, a multi-layered regulatory system like the EU tends to favour more potent actors such as industry, since they have better resources (Bandelow, 1997). Consequently, the shift of the decision-making power to Brussels had a negative impact on the influence of national NGOs, whose resource base was relatively weak. Industry radically reorganised their lobby apparatus in Brussels in the early 1990s, creating a new common organisation called Europa-Bio that served to increase their visibility for the Commission and other actors. Consumer and environmental organisations were much slower in securing representation at the European level. This might have been an additional reason why regulation in phase 3 primarily followed industry's interests, and why regulatory changes were oriented to deregulation rather than to public concerns as communicated by NGOs.

The Deliberate Release Directive mentioned the possibility of public participation in national decision-making on some issues, but this was not mandatory and was implemented in only a few countries. Jülich (1998: 56) analysed the formal possibilities for participation in such processes and distinguished two groups of countries. The Netherlands, Luxembourg and, somewhat differently, Denmark and Norway (not an EU member, see below), but also new members of the EU such as Sweden and Austria, offered guaranteed channels for public participation. In other member states such as Belgium, France, the UK, Ireland, Italy,

Portugal and Spain, opportunities for public involvement in decision-making processes were minimal. Germany was in the first group until 1993, when the revised gene law restricted participation.

The directives elicited different reactions in each member state. In some countries, implementation entailed stricter rules or even initiated regulation; in others the existing regulation was relaxed. Some countries felt the necessity of emphasising areas that were not formally regulated by the directives, for example, ethical considerations, but did so by means of committees and not by law. Thus, though EU regulation was confined to scientific assessments of risks to health and the environment, this did not prevent the broadening of risk issues in public debate and even as a subject to be dealt with by official institutions.

The UK's answer to the directives, beyond some legal changes, was to establish an Advisory Committee on Releases to the Environment (ACRE). The UK's flexible political mode also allowed for adaptation to new questions that were not dealt with by the directives: in 1991 the launch of the Nuffield Council for Bioethics signalled the increasing importance of ethical issues. Since then, a plethora of new Advisory Committees to Government has been created in order to deal with ethical and social issues arising out of particular medical applications.[52]

In Denmark, the most important change in the Genetic Engineering Act brought about by the implementation of the directives was the endorsement of the EU approval system for marketing GMOs, which replaced the general prohibition on releases. This demanded free marketing of all products approved in the EU, balanced (rhetorically) by environmental guarantees. After the revision of the Danish regulation, political and public activities around biotechnology abated. It appeared as if Denmark had found a level of regulation that was acceptable to most actors; though some frustration on the part of the opponents may have persisted. Later on, this temporary frustration manifested itself in renewed opposition.

In France, the directives forced stricter regulation than previously and ended the particular French method of scientists' self-regulation. Genetic engineering as a topic found its way into political debate, but the French system was also confronted by new questions. As in Britain, problems of bioethics officially entered the political arena, although they were not formally regulated. Ethical issues had been debated for a considerable time, and French contributions to this field had been influential both at the EU level and within the United Nations Educational, Scientific, and Cultural Organisation (UNESCO).

In the Netherlands, some adaptations to existing laws were necessary, but did not arouse much public interest compared with the advent of the first 'transgenic' (i.e. genetically modified) animal, 'Herman' the bull.[53] In 1992, shortly after the directives had been implemented, this event revealed a shortcoming in the Deliberate Release Directive: its inadequate treatment of transgenic animals. This case elicited a heated public debate: there were things at stake other than risk–benefit calculations, such as animals' rights and their intrinsic dignity, which many considered were being violated by genetic manipulation.[54] The political reactions within the pragmatic and open tradition in the Netherlands consisted of parliamentary interventions, technology assessments,[55] and the establishment of a new ethics committee.

In Germany, the European directives had a considerable impact on national regulation. Soon after it had been passed, the German gene law was criticised as being too strict and a threat to competitiveness. Promoters of biotechnology saw the directives as a chance quickly to get rid of some of the most unwelcome paragraphs. The 1993 revision of the gene law, to comply with the directives, led to considerable deregulation.[56] In the early 1990s, debates saw a shift in focus from environmental problems to economic perspectives (Gill et al., 1998: 22), and the government eventually succeeded in linking biotechnology and economic perspectives within the frame of competitiveness. Consequently, over the 1990s, Germany became the most prominent promoter of deregulation at the European level.

'Latecomers' Before 1990, regulation was perceived as a problem mainly in the 'forerunner' countries. After the directives had been passed, the situation changed, especially in the Mediterranean countries, where the public had hitherto taken little notice of biotechnology. However, developments were not at all uniform. Although public debates comparable to those in 'forerunner' countries could still not be observed after the implementation of the directives, over the 1990s GMOs slowly became an issue in Greece, and involved international NGOs at the beginning of the debate. The situation in Italy was different: the issue was hardly on the agenda, although the Catholic Church had argued publicly against biotechnology. The impact of the different economic situation in each country was more decisive for the position of biotechnology than was public debate. In Greece, EU programmes and cooperation increased the funding and strengthened the position of biotechnology, whereas the crisis of the pharmaceutical industry in Italy during the same period slowed down its development.

'Newcomers' The directives also became binding for the new member countries Sweden, Finland and Austria, which joined the EU in

January 1995. In Finland, the gene law passed in 1995 closely reflected the EU regulation. This law, however, did not involve a simple adoption of EU policy, but resulted from extensive discussions between governmental bodies and a plethora of different institutions such as universities, trade unions and NGOs. The openness of the regulative process reflected the intentions of official bodies rather than of public pressure, since biotechnology had never been a subject of public or political debate in Finland.

This contrasted with neighbouring Sweden, where the first public debates about biotechnology had taken place in the 1970s. As in Finland, NGOs and a wide variety of interest groups were vocal in 1994. The Swedish gene law exhibited a substantial difference from other European regulations in that it also considered ethical aspects. Other attempts to establish stricter regulation failed. Nevertheless, in spite of the generally critical attitude of the Swedish public towards biotechnology, the debate receded once more after the enactment of the law in 1995.

In Austria, the European directives had already been integrated into the national regulations before Austria joined the EU, solving the conflict over whether or not a specific law was necessary. A parliamentary Commission of Enquiry was set up, based on the German example, with the task of anticipating public concerns. Eventually, however, industry interests became more influential, as reflected in the law of 1994. The Commission's work had hardly any impact beyond the integration of a paragraph on the 'social sustainability' of products. This raised doubts about its compatibility with Directive 90/220, since it also employed a socio-economic criterion (Seifert and Torgersen, 1997). Such criteria were perceived by the EU as raising non-tariff trade barriers, which were contrary to the overall aim of the Common Market. Since the Austrian paragraph on social sustainability was in practice irrelevant, it remained in the law.

Although the Norwegians decided in a referendum not to join the EU, Norway remained a member of the European Economic Area, which required that it comply with the Community's biotechnology regulations. Part of the Norwegian law was in even more striking contrast to the Release Directive's focus on scientifically assessable risk issues than the Austrian social sustainability criterion. The Norwegian law demanded that GMOs be used in an ethically and socially justifiable way, and that special emphasis must be given to the benefit to society and the contribution to sustainable development.[57] Being a 'latecomer' compared with its neighbours Sweden and Denmark, Norway thus adopted the most restrictive gene law in Europe. Obviously, industry interests were unable to counter such moral considerations and national peculiarities (Nielsen, 1999).

The new democracies Phase 3 also saw the decline of the Soviet system and a new political order in Europe. The general situation switched from post-war to post-Cold War, and circumstances in East-Central and Eastern Europe changed dramatically. As a consequence, former Soviet 'satellite' states adopted the Western system and tried to become members of the EU. Whereas most of them had followed a US style of regulation, applying guidelines inspired by the NIH, it now became necessary to adopt the EU directives. Thus, in states associated with the EU (the Czech Republic, Poland and Hungary), the directives led to a reorientation and served as a model for national regulations.

The Czech Republic had no law on biotechnology until 1998, but when joining the OECD it agreed to introduce regulation within two years based on the EU Directives 90/219 and 90/220. In Hungary, the competent authority secured EU conformity by explicitly following the OECD guidelines. In Poland, regulation started earlier with a Patent Law (1993, reflecting the influence of US views) and the Biodiversity Convention (1995). In 1997, Poland submitted the first draft of a new biotechnology law on the basis of the directives, to be implemented with the assistance of international organisations (the OECD, the United Nations Environment Programme and the United Nations Industrial Development Organisation). In all three states, there were no signs of substantial public or political debate on biotechnology. Regulation was left to experts, whose role was questioned neither by domestic NGOs nor by a concerned public.[58]

Political implications

US point of view: product versus process From the American point of view, European regulation was perceived to be process oriented, since biotechnology as such was regulated in the directives, irrespective of what kind of product would result from the process (i.e., the technology) involved. The American position was that it is only the product that can be regulated – irrespective of the process applied in its production. This, they argued, was essential to obviate discrimination against novel technologies. The complicated US system of diverse regulatory rulings and various authorities with competence in this area closely mirrored this attitude. Yet, the EU insisted that its approach was not actually process based. It claimed that, because there was not enough experience with the new technology, as a temporary measure 'horizontal' regulation had been put in place. The EU argued that it expected the horizontal regulation to be replaced, in due time, by sectoral or 'vertical' laws according to the type of product.[59]

The USA was constantly suspicious that the EU would introduce forms of regulation based not upon 'sound science' (i.e. on evidence of harm), but rather on a doubtful precautionary principle open to voluntary and unpredictable interpretations. By and large, the US position was that the products of modern biotechnology were not substantially different from traditional products. And, by invoking this claim, under the heading of deregulation the USA had started, early on, to exempt certain crop plants from the requirement of obtaining permission prior to release. This was hardly acceptable to the EU with its case-by-case principle. Marketing authorisations that proved virtually unobtainable in the EU were more readily granted in the USA. Consequently, the EU was accused of protectionism, because the products in question could not be proven, by 'sound science', to pose any significant risk. Hence the USA perceived any objections posed by the EU as attempts to protect its internal market from US imports. The manifestation of these different regulatory styles in the EU and the USA led eventually to a renewed public engagement in phase 4.

Problems integrating European diversity However, the European stance on GMO releases and marketing conditions was by no means as uniform as it appeared to the USA. There was profound variation in how the obligatory risk assessment was performed; what factors were considered; what 'familiarity' meant; how the step-by-step principle was interpreted; and how to proceed on a case-by-case basis in the face of an increasing number of release applications (Levidow et al., 1996).[60] National regulators implementing the EU directives had to establish their own normative standards and in doing so they made implicit value statements (von Schomberg, 1998). Hence, strong national differences persisted in spite of the existence of common EU regulations. Disharmony was aggravated by different regulatory cultures in the EU member countries, for example with respect to the disclosure of data from the application files.[61] The Commission, aware of the implementation problems, created a high-level inter-DG Biotechnology Coordination Committee in 1991 in order to oversee the implementation of the directives. This powerful committee proved to be open to requests for a flexible revision of the regulation,[62] 'according to technical progress, easing administrative requirements' (Galloux et al., 1998: 181).

Debates waning: towards acceptance

Despite the continuing national differences, the EU regulation contributed to a calming of the debate on biotechnology in most of the European countries. The economic side and the frame of competitiveness seemed to have suborned public concerns.

Progress in medical applications, which had become generally accepted, was one of the reasons the public opposition cooled in many countries where biotechnology had been vigorously contested. It was clear that such medical progress was intimately linked to various areas of biotechnological research. A range of highly welcome new drugs and vaccines had emerged over time. Forensic identification of suspects had made huge progress and had been accepted by courts of law; and in several spectacular cases such methods had led to the identification and sentencing of criminals. The Human Genome Project made great progress during the 1990s in developing methods of establishing the base sequence of various organisms and – ultimately – of the human genome. Although in the USA voices warned of the societal implications for medical insurance of the newly developed predictive medicine, this triggered less concern in Europe with its public health systems providing universal coverage, so that debate relating to this problem was confined to academic circles.

Aside from medicine, a handful of biotechnological consumer products had entered the market in some countries.[63] In general, applications that appeared to be of advantage for the consumer seemed to have gained general acceptance by 1995. When a legal basis had been created that officially satisfied the formal regulative needs within a scientific frame of risk assessment,[64] the first products were sold on the market. Hence, existing regulation (at least in the UK) succeeded in ensuring that innovative products could enter the market.

As early as 1995 the first 'genetically engineered' food product, a canned tomato paste made from transgenic fruit, was sold in the UK by two supermarket chains. For the first time, consumers were directly confronted with such products, albeit in a very small market segment. Even when labelling was not statutory, the companies selling the tomato paste labelled voluntarily, arguing that consumers have the right to choose. These market releases, for the first time allowing consumers to choose between GM and traditional products, did not elicit any public debate in the UK or elsewhere, and the products sold well. Although survey results indicated persistent public unease towards biotechnology (INRA, 1993; Macer, 1994), it seemed that the conflict had lost its ferocity.

However, the picture was more complicated and could not simply be explained as an adaptation process, 'narrowing the gap' between society and technology. In countries such as Germany, 'green' issues had lost some of their appeal in favour of efforts to create jobs, but this was a superficial development. The themes remained in the vocabulary and, as further events showed, also in people's minds. What was actually observed (as, for example, shown by the 1993 Eurobarometer survey; INRA, 1993) was a bifurcation in the public's attitude towards biotechnology: on the one hand, applications that led to new treatments, drugs and vaccines were now rarely disputed; on the other, agricultural applications continued to arouse suspicion.[65] Although, by 1995, the progress of biotechnology in agriculture appeared inexorable,[66] general diffusion conditions for companies engaged in agricultural biotechnology became more difficult.[67]

The biotechnology industry

Not until the early 1990s did biotechnology emerge as a 'key technology' of economic significance. New developments were now being fed into the innovation chain without reference to public debate, especially in the USA, which had taken a substantial lead.

These years were characterised in many European countries by the aftermath of the late 1980s recession; the ever-increasing dynamics of globalisation; and a strong focus on technological innovation. This also led to the emergence of a 'biotechnology industry'. Up to then, pharmaceutical companies and producers of agrochemicals had strong links with organic chemistry. Now, links were being established between the pharmaceutical industry and seed producers, since both depended on biotechnological methods and the patenting of genes (this was of increasing importance), and both were facing rapidly increasing costs for R&D. It resulted in mergers of companies globally engaged in the pharmaceutical and seed sectors. These companies' engagement in particular countries was said to show the latter's competitiveness. In official language, anything that jeopardised efforts to sustain or attain this degree of national competitiveness was to be assiduously avoided. The state's role was perceived to be restricted to providing a congenial environment for industrial performance, and it was no longer considered appropriate for the state to promote other societal goals when regulating biotechnology (Bongert, 1997).

Meanwhile, the biotechnology industry underwent massive expansion. From 1993 to 1996, the numbers employed by that industry more than doubled, although it was still small in comparison with more traditional sectors. The number of patent applications increased sharply between 1990 and 1996, and the sector expected a growth rate of 20 per cent. Nevertheless, European industry still lagged heavily behind that of the USA. This was largely because of structural weaknesses: from the industrial and economic point of view there was 'insufficient collaboration between academia and industry, lack of coordination of research between EU member states, [lack of] shared access to resources and infrastructure, and inadequate venture capital' (Galloux et al., 1998: 177). The aim of the EU's policy was to obviate these shortcomings and to narrow the gap between the EU and the USA.

This disparity was especially pronounced in agriculture. In the USA and Canada, but also in China, GM crop plants, such as maize, tobacco, oilseed rape and cotton, began to cover ever-increasing percentages of the total agricultural area.[68] Aside from the more favourable regulatory climate in such countries, the EU also suffered because the sorts of plants that had been developed were most suited to large-scale agricultural systems like that of the USA, and were less suitable for the small-scale production typified by many European countries. Only France and neighbouring Spain were expected to become significant producers of transgenic crops, with other countries lagging behind.

Summary

In phase 3, all EU member states had implemented national regulations. The political elites were able to handle biotechnology within their scientific and economic frames, and these became more and more dominant. The EU made attempts to homogenise the rules for the application of biotechnology in order to create a single common market for the products to come. This proved to be more difficult than originally expected, especially since the Release Directive did not resolve, once and for all, national differences in thinking about how to proceed with risk assessment and so on. Nevertheless, the directives created a compulsory frame that all member countries had to adopt, and that non-members also took as a reference point (especially because some of them subsequently sought to become full EU members). The shift of decision-making competence to Brussels led to a new power balance in favour of industry. Public debate waned in most countries owing to other problems that had emerged after the collapse of the former socialist empire and in the light of increasing pressure from globalisation. Public resistance against biotechnology, as such, more or less waned. However, this was not due to overwhelming acceptance; rather, the applications were now being differentiated in the public mind. Thus, medical applications were more or less welcomed, whereas agricultural biotechnology was still not so well received. This development was intensified after the marketing of the first commercial food products made from GM commodity crop plants.

Phase 4 (1996–2000): renewed opposition and consumers' distrust

The backlash

In phase 3, economic competitiveness and European harmonisation had been the dominating perspectives and the division in perceptions between medical and agricultural biotechnology, along with the shift to other concerns, had led to a calming of the debate. In phase 4, the issues at stake diversified again, and debates that had been considered closed re-emerged. The importation of the first GM food crops from the USA in 1996 marked the division between phase 3 and phase 4, as consumer products of biotechnology reached the market stage. This trigger event led to renewed public and NGO protests. Later, when new cloning techniques impinged upon ethically sensitive issues, a wave of concern about the disparity between what was technically possible and what was ethically defensible spread across the globe.

So phase 4 brought a double challenge: on the one hand, renewed conflicts about health risks combined with a struggle for consumers' rights, and, on the other hand, deep ethical concerns about the overstepping of boundaries affected the image of experimental biology and also biotechnology. Essentially, the issues upon which these conflicts were based were by no means new, and their outline was already visible during phase 2. But the vigour and rapid spread of the newly arising debate had serious implications for individual countries' policies as well as for those of the EU, and especially for the relationship with the USA.

Notwithstanding the European debates, biotechnology had become a global issue. International trade as well as the struggle to secure intellectual property rights became the most important fields of conflict between the EU and the USA on the one hand, and between the industrialised world and the Third World on the other. Biotechnology had finally, and undeniably, acquired the status of a 'key technology'. Ironically, as soon as biotechnology began to take off, the term itself began to be replaced by other less contested but more comprehensive descriptions.

Prelude: BSE and the loss of trust in experts and regulators

One reason the evolving conflicts over food issues became so ferocious was the earlier scandal in the EU over BSE.[69] This was totally unrelated to the issue of GM crop plants, but it set the scene in terms of consumers' concerns about food safety.

A few months before the first arrival of GM soya in a European harbour in 1996, the British government conceded, after years of rumour, that there might be a link between human Creutzfeld–Jakob disease and Bovine Spongiform Encephalitis. This led to a collapse in the beef market and a ban by the EU on British beef. The main impact of the ensuing 'BSE crisis' was a growing distrust in scientific experts and political regulators in matters of food safety, since they had been reassuring the public for a long time that there was no evidence for such a link. Consequently, regulators had to change their way of dealing with risks: they acknowledged that regulation could not be left to industry, contrary to the approach preferred by the UK government, and they were forced to accept the fallibility of expert committees.

The UK government and the European Commission were harshly reproved for their handling of this scientifically controversial issue. In 1997 the European Parliament criticised the secretive decision-making processes and the complex and undemocratic system of scientific committees. Threatened with a vote of no confidence, the Commission conceded a strengthening of consumer representation in DG XXIV (Consumer Affairs) and more transparency, indicating increasing parliamentary influence vis-à-vis the Commission (Baggott, 1998: 70ff.).[70]

Critics in some countries quickly asserted parallels between the BSE scandal and the way in which GM food products were entering the market. The issues were obviously scientifically unrelated, but opponents considered them both to be consequences of industrialised agriculture. At the same time, organic farming began to leave the small circles of eco-sectarianism in some Central and North European countries, and emerged as a counterpoint to industrialised agriculture and biotechnology. Products from organic farming could be bought in ordinary supermarkets and served a new consumer interest. This paved the way for later cooperation between NGOs and retailers fearing consumer boycotts.

Soya and maize: the European labelling debate

The soya case changed the frames of the regulatory debate away from an emphasis on scientific risk assessment towards finding ways to deal with consumer interests. When the importation of GM soya and maize from the USA to Europe began some months after the outbreak of the BSE crisis, trust in experts who had consistently reassured the public about food safety had already been shaken. The assumption that consumers would generally accept GM food (were it available) proved to be wrong in spite of previously successful market introductions into the UK.

After the EU had granted an approval for the US seed company Monsanto's GM soya, in the winter of 1995/6, the first actual imports from the USA arrived in November 1996. Obviously in an attempt to create a *fait accompli*, the US-grown modified soya had been mixed with conventional soya. Although legal from a US regulatory point of view, this precluded meaningful labelling. However, consumer and environmental organisations had demanded the 'right to choose' between GM and traditional products. The fact that consumers would not be given such a choice was successfully used by a combination of consumer and environmental NGOs to mobilise the public in most EU member countries. As a result, in many countries retailing chains that were concerned about consumer confidence joined consumer organisations in pushing for clear labelling, and renewed public opposition led to a heated debate about the appropriate labelling of such products, which industry was unwilling to guarantee.

This event was widely reported in the media and resulted in a change of priorities for regulatory efforts at both national and European levels. The most important effect was that the EU was now pushed to finalise the Novel Food Regulation. Plans for such a regulation had existed since the Release directive was debated, and the Commission had issued the first proposal as early as 1992 (see Lassen et al., chapter 10 in this volume). From the very beginning, the Novel Food Regulation was meant to cover the introduction of all new food products – including products of biotechnology that were in principle regulated by the Directive on Deliberate Release. Since the directive did not specify labelling, however, there were major differences between the approvals already granted for maize and soya and those applicable to future products.

The policies of the member countries, the Commission and the European Parliament differed significantly on the question of labelling (Behrens et al., 1997).[71] The Commission and the Council had consistently been more sympathetic to industry's arguments than Parliament was, and they had tried to regulate the issue within the economic and scientific frame of safety. However, as in the conflict over BSE, influence began to shift from the Commission to the Parliament in late 1995. The Commission accepted a proposal for the Novel Food Regulation that allowed for compulsory labelling under certain conditions[72] – if the new product was recognisably different from existing equivalents; if the product might give rise to ethical concerns or had health implications; or if the actual GMO was present in the product. When marketing approvals for GM soya and maize were pending, the labelling question became more and more pressing;[73] and, when the Novel Food Regulation was passed in January 1997 (shortly *after* the approval for maize), it took most of the Parliament's suggestions into account. This marks a turning point in that the Parliament's views now exerted the greatest influence. However, with the passing of the new regulation, problems lingered on.[74] Owing to delayed implementation, by 1999, on shelves throughout Europe, there were still unlabelled products that contained modified soya products, if only trace amounts.

Voluntary labelling had been devised as a means to ensure the public acceptance of biotechnology food products. The reluctance of the food industry to label, and the political pressure required in compelling it to do so, resulted in a conflict that created obstacles to market entry. Public opposition had reached such a level in some countries that any comprehensive labelling appeared to be a hazard for the producer. Consequently, industry became very reluctant to issue new GM products in the EU. Lacking consumer acceptance for its food products, agricultural biotechnology was in danger of never becoming established in Europe.

In line with a new emphasis on ethics in all fields, ethical questions also began to play a role with respect to the marketing of GMOs. Critics referred to the old problems of how to deal with consumer risk and protection. The growing uncertainty concerning the merits, dangers or moral ambivalence of biotechnology fostered and energised new actor constellations, such as organisations claiming to address consumer interests and retailer chains. These actors differed in importance, means of influence and power among the European countries but, particularly after 1996, they commanded considerable trust in the public sphere. Their success was facilitated by the provocative way in which multinational companies had tried to push their GM products in the European market. They made it easier for critics to portray GMO producers as villains. Additionally, references to agriculture stimulated particular national sensitivities.[75]

However, it was not only plant biotechnology that generated fierce dispute. As the following short excursus shows, other areas of biological research elicited heated responses; and in this case the moral dimension was writ large.

Dolly and the international moral accord

When, in late February 1997, the first cloning of a mammal from a somatic cell made the headlines in all Western countries, the reactions to 'Dolly the cloned sheep' as a media sensation were of a different quality from the public anxieties about soya and maize. Controversy in this case was not related to consumer concerns, fears of environmental disaster or outrage at regulatory misconduct. Instead, this event evoked the 'moral danger' of human cloning. The common feeling was that the transgression of moral boundaries was imminent and had to be prevented. In an interplay between international political actors and the media, within days a consensus developed across the whole industrialised world that human cloning should be prohibited (see Einsiedel et al., chapter 11 in this volume). In Europe, the media event triggered a more or less synchronous mobilisation of the various national publics, but the issue was still discussed at a national level. As with the GM soya case, there was still no common 'European' public, even if debates were now emerging in countries in which the public had hitherto been virtually silent on the issue of biotechnology (for instance, in Greece, Italy, Poland and Ireland).[76]

As an affirmation of this moral consensus, instant policy responses emanated from the Pope, the American President, the British government and the European Commission. The case of Dolly gave authorities an opportunity to demonstrate their responsiveness to public moral sentiment by announcing prohibitions in the form of various pieces of legislation on reproductive medicine and patenting. Their swift action was surely designed to renew trust in regulators and prevent a spillover effect on the image of biological research. However, in the public eye, Dolly became linked to other aspects of biotechnology, again going beyond the borders of a purely scientific understanding. Thus, by evoking human presumptuousness in interfering with (arguably) natural or sacred orders, biotechnology assumed clearly negative moral connotations, in spite of the international moral accord.

National responses to public opposition

Fading consumer trust prompted national publics to exert pressure on their governments. Additionally, in 1996 and 1997 some 'latecomer' countries experienced the trigger events of the first experimental releases of GMOs, and these prompted public debates of a hitherto unknown character. Another factor militating in favour of opposition was that in 1997 and 1998 elections transferred political power to centre–left governments in France, the UK and Germany. The new governments increased the attempts begun under their predecessors to rebuild public accountability in matters of food safety. With governmental changes, critics' arguments became more influential than before, and the balance of power in the EU Council of Ministers also changed.

'Latecomers' catching up

The first country to witness a broad public mobilisation against biotechnology during phase 4 was the new EU member Austria. Although there was no noticeable public or media interest until 1996, the first release of GMOs caused considerable turmoil for the Austrian authorities and the biotechnology industry.[77] A 'people's initiative' (an official petition) calling for a moratorium on agricultural applications and patents received high levels of consent and placed the government and administration under considerable political pressure. In early 1997, the authorities imposed a ban on the import and agricultural use of Bt-maize, despite the Commission's market approval, thus deliberately violating EU regulations. Austria upheld the ban, while the Commission, mainly for procedural reasons, failed to enforce its regulation.

In other small member states, such as Greece and Ireland, with no previous debates on biotechnology, the first GMO releases triggered public opposition in an atmosphere affected by the Dolly story and by controversies over food safety. In Greece the first release took place in spring 1997. This encountered opposition from NGOs, which occupied test fields and succeeded in gaining media attention. As in parts of Austria, Greek agriculture is small in scale and there is an increasing market for organic products. In November 1998, Greece followed the Austrian path in banning GM oilseed rape, which had previously been approved EU-wide.

Because of its close trade relations with the UK, Ireland was particularly affected by the BSE crisis. As in Greece, the first releases took place in spring 1997, meeting NGO resistance that manifested itself in the destruction of fields. In the 1997 election campaign, agricultural biotechnology was a contentious topic among the critical Irish public. During a temporary moratorium, the Irish government organised (public) consultations over the summer of 1998. Subsequently it announced a moderate policy that undertook to consider consumer protection to be a priority, but would aim to secure economic and employment opportunities, revoking its initial promise to prohibit agricultural biotechnology.

'Forerunners' decelerating

Changes in public and governmental attitudes towards biotechnology were not confined to small countries. The most astonishing reversal took place in France, where the conservative government discovered the merits of involving the public in decision-making before the election in spring 1997 (which brought a coalition of the socialist and green parties to power). The scientific-elitist style of biotechnology policy of the previous phases would have rendered such a step almost absurd. Yet, even though it was the French administration that had filed the request for the EU-wide marketing of the US soya/maize imports, it nevertheless decided to suspend distribution of GM maize. Under the new government, environmentalist NGOs, agricultural syndicates and the green Ministry of the Environment took a tough stance against agricultural biotechnology, which in turn had strong vested interests in agriculture, industry and research on its side. In autumn 1997, a 'citizens' conference', designed after the Danish model of consensus conferences, was scheduled for June 1998. Future policies were to be linked with the outcome of this consultation.[78] The French government followed the recommendations that were formulated, and imposed a two-year moratorium on marketing authorisations for certain plants.[79] Remarkably, after the conference the government's position was similar to that in Ireland subsequent to its consultation process: although consumer protection and environmental safety gained high priorities, biotechnology was still officially viewed as a key technology with huge economic potential that had to be supported.

Unexpectedly, the UK too experienced a reversal in its earlier policy. Compared with many other countries, British consumers had never found GM food particularly unacceptable, and biotechnology in agriculture had not been a contentious issue for environmental NGOs. The British system of selectively involving potential dissenters in decision-making had worked well in containing conflicts. The BSE crisis, however, had created a new situation.[80] Before the election, the Conservative government had faced a severe loss of public trust for various reasons. After its victory, the new Labour government was more reluctant than its predecessor to see biotechnology as a predominantly economic issue. Shortly after the election, Britain experienced an intensification in public debate.[81] The British government even considered a temporary moratorium, but instead promised a series of 'public consultations'.[82] In 1999, links with BSE were asserted after allegations were made that experimental data showing negative health effects linked to the consumption of transgenic potatoes by rats had been suppressed. These accusations resulted in a public uproar over regulatory misconduct.

In the early 1990s, Germany had abolished the cautious policy it had adopted during the late 1980s, when it had accepted a number of green caveats. Now it moved in the direction of other European countries and assumed a more positive attitude towards biotechnology. The conservative–liberal government had decided to embrace new technologies, even if the public did not particularly approve of them, in order to regain German competitiveness. Because other issues were at stake, for instance reunification and the rise in unemployment, public opposition had been hardly visible when regions had been applying for special funding to set up new biotechnology facilities. However, surveys showed continuing public unease about agricultural biotechnology. When the Social

Democrat/Green coalition took over in 1998, they acted more cautiously with respect to these hidden public anxieties, although they did not relinquish their industrial modernisation commitments. Agricultural and food biotechnology received less support, while the backing of medical research and the production of drugs and vaccines remained stable.

The Scandinavian countries had always harboured populations that were very critical of biotechnology.[83] The BSE crisis had surprisingly little impact, but the soya/maize episode strengthened the bargaining positions of NGOs opposed to biotechnology, thus affirming already sceptical attitudes. Food retailers and NGOs together lobbied for extensive labelling and a cautiously gradual introduction of GMOs. In Denmark, opposition groups and lobbyists tried to influence regulation policy, or at least its implementation, and they attempted to prohibit the marketing and selling of products that were not properly labelled. Norway continued to hold the most critical position and tried to maintain its commitment to criteria of 'sustainability' and 'societal benefit' in evaluating particular GMOs.

In the Netherlands, the government's obligation to inform the public of both positive and potentially negative effects of the new products had already been a constituent of official policy. This was intended to permit consumers freedom of choice and to provide a basis for public acceptance. The government was, likewise, to guarantee the safety of products and to be responsive to uncertainty and concerns among citizens (van Vugt and Nap, 1997: 31 f.). Thus, in 1996, the Netherlands was very active in informing the public and the media through workshops about market introductions of GM food, and the government promoted participation in EU research programmes for stimulating consumer acceptance of biotechnological products. Differences between consumers, retailers, producers and NGOs could have been divisive (Smink and Hamstra, 1996), but negotiations in line with the Dutch regulatory tradition succeeded in reconciling most of the disagreements. As a result, the Dutch government continued to be in a position to demand significant deregulation, even after the soya conflict.

Exceptions

Two countries did not follow the general European trend. Although a Scandinavian country, Finland's positive and pragmatic attitude towards biotechnology seemed to have been unaffected by general European developments. As in previous phases, biotechnology was still seen as a means to modernise the country and to gain economic advantages, although there were still only a few small biotechnology companies.

In early summer 1998, a referendum on biotechnology was held in Switzerland in order finally to resolve a political conflict that had been ongoing since the late 1980s. At that time, the developments described in other European countries were already under way. The outcome was open, but the referendum represented a clear defeat for the critics of biotechnology. The reasons for this development, which appears to run counter to the new 'critical' European trend, are two-fold. First, though the debate about biotechnology was old, the Swiss political system had reacted very slowly to this issue because of its traditionally time-consuming public participation procedures. Hence, the wording of the referendum did not mirror the recent preoccupation with GM food, but emphasised transgenic animals and, therefore, medical research. Secondly, an estimated 200 pharmaceutical companies and specialised research institutes constituted a major asset to the country's economic well-being. Aware of the importance of biotechnology in maintaining national competitiveness, Swiss industry, researchers, students and government formed a broad alliance and engaged in a coordinated campaign. One reason for their success was that, in the Swiss debate, they concentrated on medical research and pharmaceutical applications rather than on food or agricultural products, in clear contrast to the rest of Europe. Another factor may have been that Swiss pharmaceutical companies are technological world leaders, and are consequently linked to national pride as well as to economic success.

The European consumer

After imported GM crops had reached most European countries, reactions to biotechnology became pan-European, and debates arose in countries in which the public had hitherto barely been aware of biotechnology. In addition, some NGOs became players at the European level. The increased consumer anxiety felt in many European countries indicated a certain harmonisation of beliefs, and the European institutions' actions could be interpreted as reflecting them. This raises the question of whether something like a 'European public', in contrast to phases 1, 2 and 3, had emerged. The preceding periods had seen public controversies remaining strictly confined to national contexts. There had been neither a common European debate nor a European public, since the basic prerequisites of a common language, European mass media or opinion leaders, let alone a common identity, had not existed.

Yet, though the new conflict was no longer confined to single countries, debates were not truly 'European'. The above prerequisites still were lacking, and the publics in the respective countries remained more or less isolated. Instead, from 1996 onwards the GM food controversy triggered parallel, but separate, reactions from the European national publics. The media concentrated on national events, and only occasionally covered controversies in other countries.

The pressure on the national governments and, in consequence, on the European regulatory system led to a redistribution of influence. At the EU level, the European Parliament succeeded in presenting itself as the representative of a fictitious European public opposed to the European government represented by the Commission. This stance was based on an understanding of public interest in terms of consumers' right to choose, notwithstanding issues of risk, but clearly taking account of the recent experiences with BSE.

On the other hand, this consumer orientation conflicted with international agreements, made during phase 3, to liberalise trade. These agreements ruled against non-tariff trade barriers that might prohibit imports. Promoters of free technology flow intended such agreements to make the handling of biotechnology more equal across countries.[84] But these legal requirements left little room for political manoeuvre in response to public opinion or even statutory processes of public participation. Everything relied on scientific risk assessment, and social value judgements were explicitly forbidden. Nor could the precautionary principle override the requirements of the agreements, rendering it more or less toothless (Vogel, 1996).

Consequently, the Commission was more inclined to seek expert advice and to strengthen the role of scientific committees in defining risks in accordance with scientifically defensible, 'hard facts'.[85] The role of scientific experts had become statutorily paramount. Rational as this appeared, it was problematic from a public policy point of view since public trust in experts appeared to be declining. Thus, situations such as that in Austria could occur, where a people's initiative demanded a ban on GM food, whereas EU regulation – in line with international treaties – enforced importation. Such a divisive juxtaposition of the 'people's will' and international trade agreements triggered further public opposition.

Global issues

The USA: a different agenda

The declining public faith in experts and regulators in Europe again highlighted the differences between the US and the European regulatory approaches. Comparative studies of the development of regulation and market introduction of consumer products had already revealed an important division. In general, EU regulation was more paternalistic, involving reliance on experts who were considered to be proactively protecting the citizenry (McKelvey, 1997). This was partly a response to the variety of regulatory styles among the member countries. When, as a reaction to consumer concerns, the frames of the debate on biotechnology in Europe broadened, the debate on protectionism versus the free market intensified and market relations between the USA and the EU came in for serious scrutiny. From the perspective of the USA and Canada, EU policy was protectionist and represented an illegitimate attempt to gain economic advantage (McKelvey, 1997: 135). Considering the virtual absence of public opposition to GM food or releases of GMOs in North America, it was hardly conceivable to the Americans that this represented a substantive political problem in the EU. But, for democratic reasons, the EU could not ignore public opposition. As with previous transatlantic quarrels over food issues, this dispute was fed by differences in public attitudes as well as market interests.[86]

In contrast, in the USA during the mid-1990s the results and implications of the Human Genome Project were causing the furore, rather than GM releases and products. In particular, the public was anxious about the possible use of knowledge about individual risk factors for diseases or behavioural traits by health and life insurance companies, as well as in the labour market. In European countries, these themes attracted little attention, presumably because, in European social security systems, individual risk factors play lesser roles in terms of eligibility and premiums.

Patents: a global asset

Another contested area was that of intellectual property rights. In 1996, the OECD issued a report in which it described different approaches to patenting (OECD, 1996), building on an older report from 1985. Patenting had been a recurrent issue at least since the first patent of an animal was granted in the USA in 1988 and in the EU in 1992. Most disputes, however, arose over two issues. First, the patenting of genetic sequences triggered controversy because it was not clear whether such sequences were indeed true 'inventions', or whether they were just 'found' in nature. Secondly, sequence data were considered 'raw material' for basic research and future development that should not be withdrawn from the public domain. As a result, researchers themselves were involved in a conflict of interests. Some scientists pointed to industry's growing role as a promoter of basic research, and stressed the necessity of securing intellectual property rights as a reward for the companies' spending of research money. Others saw the quick and free flow of scientific data, a prerequisite for basic research, to be in jeopardy. The issue highlighted the differences in understanding between the USA and European countries about what an invention is, and about the proper relations between private enterprise and the public domain.[87]

NGO representatives saw this area as the real challenge for the future. Might genetic information be privatised? Should companies be allowed to acquire the right to do what they wished with such information once a patent had been granted? And should groups such as farmers (especially in the Third World) be denied the right to produce seed? In their view, the issue touched on the more general problem of the ever-increasing power of international capital to command the resources of life.

Yet, though NGOs perceived the patenting question to be crucial, this issue did not have the same power as GM food in mobilising the public. This is because it had no immediate impact on the consumer, but instead concerned such elusive themes as equity, international relations and future development. Obviously, such issues were less contentious to the public than domestic affairs and, especially, risks to human health and the immediate environment. Furthermore, the intricacies of the patenting debates were far too involved and complex to be presented as the catchy and simplistic stories required for effective public communication. Nevertheless, patenting remained an issue of dispute for some time in the European Parliament and elsewhere.[88] It was eventually resolved in 1998, when the Parliament accepted a Commission proposal for a Patent directive largely following industry interests, after long and complicated political negotiations.

In the reservations on patenting, one can trace the impact of the Dolly story. In the aftermath of this media event, the public, governments and EU institutions were unanimous that human cloning had to be rejected. Dolly rendered it easy to insert a prohibition into the directive because, although the economic significance of human cloning remains doubtful, the matter touched upon a common moral code. The prohibition on the patenting of human cloning techniques may be regarded as a high gain, in terms of political publicity, at low cost, since no substantial (industry) interests had to be violated. In contrast, the debate about GM food and the accompanying regulatory turmoil had little influence on the final debate about the Patenting Directive. This was an indication that the general debate on biotechnology had split into diverse conflicts over deliberate release and food products, over cloning and xenotransplantation, over intellectual property rights, and over the 'ownership' of genes.

Life sciences

Intellectual property rights as a basis for the control of genetic material and biotechnological methods had profound implications in two key, previously almost unrelated, areas. The mergers of gigantic companies engaged in the production of pharmaceuticals as well as seeds indicated that the combination of these two areas would indeed secure future gains.[89] It became clear that biotechnology had, finally, achieved a significant status. Even if there were problems concerning the public acceptance of GM food products in Europe, the progress of biotechnology now seemed inexorable. The industry's promises, which had won it backing on the stock market, finally began to come to fruition in the form of multinational companies whose activities centred on biotechnology.

Ironically, this marked the end of the use of the term 'biotechnology' itself. Companies now stopped using the word since it had acquired negative connotations. Instead, they defined the area as 'life sciences', a term with suggestions of more welcome medical applications. This term had been in use for decades to identify the fields of biomolecular and basic medical research, and was employed in the titles of university faculties and scientific journals. It gained a new meaning when industry adopted it to identify companies that apply particular methods of biotechnology, irrespective of the field, to developing new products and acquire patents on genetic sequences as a basic resource. The 'life industry' evolved from the chemical industry through the adoption of biotechnological methods and the development of pertinent products, finally shedding traditional chemical activities in favour of modern biotechnology. This indicated that the use of life (Bud, 1993) was indeed its core field of interest.

Summary

In phase 4, when the first GM crop plants entered the Common Market, concerns about agricultural biotechnology brought about a shift in public debate and policy both in the member countries and at the EU level. The dominating economic frames of previous phases became weaker as public opposition increased, while the directives failed in their objective of harmonising the member countries' interpretations of product assessment. The signals from consumer groups, retailers and NGOs were now explicit: they did not want their food to be produced using biotechnology, and they demanded the right to choose, regardless of safety issues. As a consequence, labelling became paramount. The debate coincided with the advent of new and contentious cloning techniques, re-establishing a link to reproduction that had largely vanished. It seemed as if the debate on biotechnology had returned to a starting point, notwithstanding its preliminary fading in phase 3, although debates diversified and became centred around certain application and problem fields.

In response to public opposition, and supported by government shifts in important member countries, EU policy turned towards a more cautious implementation of GM food products. This, however, caused problems with international and transatlantic trade agreements. Even more far-reaching policy issues centred around intellectual property rights, especially the patenting of genetic sequences. While companies merged to form conglomerates engaged in both medical and agro-biotechnology, the term 'biotechnology' began to disappear from corporate language, being replaced by the even more inclusive, but less contentious, 'life sciences'.

Summary

Regulation to make biotechnology happen

From the very beginning, most national governments, as well as the EU Commission, tried to 'make biotechnology happen'. This motive became dominant in the 1980s, when the economic point of view was advanced, and it has remained paramount ever since. Despite various obstacles, governments adopted a double strategy. On the one hand, they fostered industrial and scientific research even to the extent that they accepted responsibility for the establishment of commercial firms. They succeeded to differing degrees, depending on the national industry's capabilities for engaging in the new technology, and on the willingness of domestic publics to accept its products. They also tried to reassure a critical public – by implementing credible regulation – that risks could be managed. The major questions were: who should be in the position to regulate, and what would such a credible regulation involve? Was self-regulation by the scientific community sufficient, or were there issues that could not be dealt with in such a system? A major problem was to decide what biotechnology could be compared with: is it something entirely new or merely an extension of older techniques? is it like nuclear power or like bread baking? If a solution was found that was generally held to be trustworthy, debates often calmed down.

Over time, however, this strategy became less effective, reflecting a general loss of trust in government and in scientific institutions and experts. Attempts to solve the regulatory problems within an expert system, as in phase 1, failed later on, since governments had to react to the social changes of the 1980s. While some still tried to contain the debate within a strictly scientific frame, others integrated critical actors in the decision-making process as a sign of openness towards the public. It was hoped that this would prevent the spread of the critiques advanced by NGOs without jeopardising the economic potential of biotechnology. However, another reason safety regulation failed to close the debates was that it could no longer cover the issues at stake.

The failure of risk assessment

When biotechnology was introduced, supporters within science, policy and industry expected that it was only a question of time until it became generally accepted by the public. From their points of view, possible risk, to be assessed in a scientifically rational way, was the only obstacle to popular acceptance. Consequently, innovators intended to restrict the debate to classical risk assessment, but they eventually failed to do so because of the lack of universally shared criteria. Studies on risk perception indicate that lay people perceive risks differently from scientific experts (Slovic, 1987; see also Douglas and Wildavsky, 1982: 49–66; Jungermann and Slovic, 1993). Experts tend to keep the problem within what they perceive to be a purely technical frame, whereas lay people implicitly emphasise behavioural, cultural, social and economic aspects. This was hard for scientists to comprehend. Because they followed a rationality that demanded that value considerations be put to one side, they could not understand why risk debates were going beyond the issue of (physical) hazards and involving questions of equity and accountability. Such value choices, however, are the most decisive factor shaping public perceptions (BEP, 1997). When experts kept affirming that risks were negligible or manageable and need not hinder progress, they were reproached by the public for downplaying the potential risks. Because scientists were in a position to make moral judgements, and were stakeholders themselves, they could be accused of being partisan and hiding conflicts of interest. This undermined their credibility.[90]

As a consequence, the question of whether there were significant risks – notwithstanding the apparent lack of accidents – could never be adequately answered. This led to the adoption of the cautious principles of case-by-case and step-by-step assessment in EU regulation, which were themselves predicated on the precautionary principle. Uncertainty was the reason regulation was put in place until ‘more experience’ was acquired; again, this was subject to different interpretations and hence to regulatory uncertainty, leading to delays. Thus, the dispute over risks slowed down the pace of innovation; it even temporarily stopped it or forced it into sidings. Yet, the slow speed of implementation may have contributed to the safety of the new technology. We may speculate that, had it not been for scientific, regulatory and public scrutiny, we might indeed have experienced accidents.

Biotechnology as a sounding board

As biotechnology left the laboratories and entered society, its image varied considerably depending on what was considered technically feasible and what was deemed desirable. The focus on applications meant that the term 'biotechnology' became a symbol with a dual, and contradictory, meaning, depending on whether the user of the term had promoting or preventing interests.[91] Supporters and critics competed in establishing the 'meaning' and definition of biotechnology, and this resulted in terminological shifts. Applications themselves shaped the varying definitions of demands and risks with respect to fields, interests and world-views. Projections of future demand built on promises, sometimes vague, of new products to come that would outperform conventional ones or be entirely novel. According to the particular application, claims of risk were extended beyond human health to a wide variety of issues, from environmental protection to socio-economic factors. Opposition was often (strategically) based on the logical impossibility of proving the absence of any negative long-term outcomes.

Regulators had to find a way through the complexity of the arguments, while uncertainty about risks prevailed. Uncertainty controversies featured two types of argumentation. On the one hand, physical hazards (to health and the environment) were emphasised; on the other, societal and moral issues came to the fore, involving consumer rights, trust in experts and regulation, and the role of agriculture in post-industrial economies.[92] Both types of argument influenced attitudes but, in general, societal and moral arguments turned out to have the greatest impact on consumer behaviour (BEP, 1997; Durant et al., 1998). However, risk issues were the more commonly addressed, since they were (seemingly) independent of individual world-views. Any biotechnology regulation (except the Norwegian and, to some extent, the Swedish and Austrian gene laws) permitted the authorities to address only physical risks, so they had to cover other concerns under risk arguments.[93] In contrast, parliaments were able to take up societal and moral issues as well, even if they had little influence on actual decision-making.[94]

A similar division could be observed at the EU level. While the Commission statutorily focused on risk only, such arguments helped the previously less important European Parliament to realise its potential as the representative of a common but invisible European public. It acquired, temporarily, a role as consumers' advocate.[95] So, during the BSE crisis, the Parliament took up risk issues, but eventually went beyond these and addressed the societal/moral arguments that the Commission could not adequately deal with. Later, in the same vein, the labelling of GM products became an issue not only of paternalistic 'risk prevention', but also of 'consumer democracy' in the Parliament and elsewhere.[96]

National diversity and European integration

In Europe, the new technology encountered a different 'climate' in each country. Its reception varied according to each state's academic tradition, its industry research base and the players that had tended to influence government policy. Obstacles to broad and smooth implementation also often emerged from cultures critical of technology, in the same countries that, during the 1970s and 1980s, had seen social and environmental conflicts over nuclear power, environmental issues, disarmament, women's rights or Third World support. Later, such conflicts found their expression in the presence of green parties in several parliaments. Another factor involved the existence of religious traditions that objected to the 'manipulation of Creation'. This diversity in response to biotechnology made it more likely that national debates would be confined to national boundaries. Thus, in general, early debates throughout Europe focused upon national events. Only when the first transgenic crop plant came on the market was simultaneous media attention provoked in almost all European countries. Similarly, the BSE crisis had a huge transnational impact, and later the advent of Dolly the sheep became an international media event. However, even in these cases, the actual debates remained mostly national.

There was no general rule about how governments dealt with biotechnology conflicts. Their response depended, among other factors, on the way political problems were generally handled and whether there was a generally adversarial or consensual style of resolving them. Government reaction was also affected by whether or not other conflicts allowed biotechnology to appear on the agenda, and by the particular understanding of what biotechnology 'really' means. If regulators understood biotechnology to be similar to other technologies, regulation took into account the varied forms of application, and a strong emphasis was placed upon scientific expertise. In contrast, if biotechnology was seen as something new, entailing unknown risks that had to be broadly debated, regulators tended to enact universal laws affecting the technology itself. Some governments tried to keep issues within the realm of scientific expertise;

others emphasised public accountability. All this contributed to the striking levels of national diversity.

When, in the late 1980s, biotechnology appeared as an economically promising area of technological development, the USA had already taken the lead and Europe was lagging behind. The EU decided to create a common market for biotechnological products in Europe in order to exploit the industrial opportunities. A precondition for this was the implementation of unified regulation, involving a compromise that recognised national differences in regulatory culture and public opinion. After all, Germany appeared as a menacing example of what could happen if public opinion was allowed a free rein. The EU guidelines on contained use and deliberate release met German demands in that the biotechnology itself was regulated, and a kind of precautionary approach was applied, not least in response to public demands.[97]

The directives were intended to serve multiple purposes beyond that of ensuring harmonisation by defining risk assessment requirements. Following international trade agreements, the directives had to be compatible with US regulations, though the USA believed EU regulations to fall short of this.[98] Additionally, the directives turned out to be the master copy for countries outside the EU that lacked regulation but that had close trade relations with Europe, and eventually became EU members. Alas, these aims were set too high. As with any standard resulting from a compromise, the directives suffered from ambiguities. In particular, the wording of Release Directive 90/220 left room for interpretation, in the clear expectation that a common European understanding would develop over time. However, despite intensive negotiations over the years, binding rules on crucial issues such as risk assessment criteria for products could not be achieved. Divergence and convergence did not produce an equilibrium.[99]

Although industry and large NGOs had become European actors, national idiosyncrasies were even more influential when biotechnology entered the agricultural sphere. Since the USA demanded free access to the European market for American GM crops, a veritable trade war was looming. Attempts to 'verticalise' the regulation via directives on Novel Food, Pesticides, Feed, and so on turned out to threaten various other, previously less contentious, areas, rather than to solve the problem. National governments, as well as the European Commission, realised that, after ten years of harmonisation attempts, disharmony among EU member states over biotechnology policy had not been eliminated and a comprehensive regulation would need to be very complicated. Finally, the century ended with a temporary halt to new products in Europe.

Conclusion: no end to conflicts

After a quarter of a century, the future of biotechnology is open. Fields such as basic research and development, the production of pharmaceuticals, food additives and enzymes, forensics and diagnostics have become unthinkable without biotechnological methods. Yet in other fields, such as predictive medicine, gene therapy or environmental remediation, practical implementation has barely begun. Certain areas lag behind for various reasons: seeds commercialised so far have been an agricultural success in the USA but remain contested in Europe; conversely, genetic testing for disease preconditions[100] may entail problems with insurance coverage, and is more debated in the USA than in Europe owing to different health care systems. It has become clear that biotechnology, like other modern technologies, is embedded in society – or, rather, in the societies of various countries – and is subject to their differences. Innovation will continue to accelerate, but one may not take it for granted that a product will be welcomed just because it is new or advantageous for the producer. Public concern brought to the fore (or triggered) by NGOs and attitudes towards the public accountability of regulators and industry will contribute significantly to the success of new products.

Diversity, as it exists at the European level, is also apparent on the global scale. Multinational companies merge and increase their market power in already monopolised fields, such as seeds and pharmaceuticals, and international agreements will enforce homogeneous assessments on a seemingly scientific base. Nevertheless, differences in culture and interests over time and among countries and continents will remain. From a global perspective, the strategies of actors engaged in the promotion or rejection of technology will change, but they will always be influential political factors.

Ten years from now, we may be able to write the history of the debate on biotechnology. At present, however, the controversy is still ongoing, and there are few indications that it will soon come to an end. Whether the rapid cultural and technological changes of our times are too difficult to cope with or the technology is too complex to understand, it has become obvious that the current regulatory scheme in Europe does not fulfil what Sheila Jasanoff (1995: 311) postulated to be the essence of functioning regulation: 'it is a kind of social contract that specifies the terms under which state and society agree to accept the costs, risks and benefits of a given technological enterprise.' To set up such a contract is no less of a challenge today than it was twenty-five years ago.

Acknowledgements

Helge Torgersen is supported by the Österreichischer Fonds zur Förderung der wissenschaftlichen Forschung (project no. P11849), and the Institute of Technology Assessment, Austrian Academy of Sciences. The authors are grateful to Andrea Lenschow and Michael Nentwich for their comments on previous drafts of this chapter.

NOTES

1. According to Rammert, technology comprises 'the inventory of instruments and installations, as well as the repertoire of skills and knowledge, to achieve desired conditions, and to avoid unwanted conditions, in handling the physical, biological and symbolic world' (Rammert, 1993: 10, our translation).
2. Apart from the Showa-Denko case of tryptophane contamination, which was mainly due to a failure of the purification procedure.
3. One could argue that (statutory) regulation is always put in place in order to make possible what it regulates (Majone, 1989).
4. The European Community (EC) changed its name in 1992 to the European Union after the Maastricht Treaty. For the sake of simplicity and consistency, we will refer to both the EC and the EU as the 'EU', irrespective of the actual name used at the time referred to.
5. See, for example, the extensive reviews published earlier by Bud (1993) or Cantley (1995).
6. If no reference is given for a particular claim made in this chapter, then the claim is based on results from the above-mentioned research project. Detailed data may be found in Durant et al. (1998).
7. If we define biotechnology in a literal sense, as an activity putting living organisms to work for humankind, everything from agriculture to fermentation would be included. However, the term is mostly referred to with respect to activities that exceed well-known techniques such as baking or brewing, more in the sense of what has frequently been called modern biotechnology. We may link its onset to the beginning of broader debates arising over recombinant DNA techniques or genetic engineering. Frequently, the term biotechnology was used as a proxy for the combination of the methods for splitting, sequencing, splicing and constructing genes, including methods for their application in various fields such as basic research, industrial production, medicine and agriculture. Genetically modified organisms (GMOs) carry genes modified by these techniques. Consequently, products of biotechnology are the results of production processes applying such methods. In the following, we will stick to this understanding because it seems to be a common denominator, and not because it provides a scientifically exact definition. It should, however, be noted that the term 'biotechnology' has not been restricted exclusively to the process of recombining DNA. Its broader definition includes processes such as protoplast fusion and cell and tissue culture. This is why the notion of, for example, cloning can be considered a biotechnological activity. Although not particularly related from a scientific point of view, this example highlights the links to other fields of modern biological R&D, which gave rise to another set of innovations.

Even if cloning as a technique has nothing to do with genetic engineering, it has a lot to do with how the public perceives modern biotechnology, as the case of the famous sheep Dolly clearly shows (see below). In public debates, these innovations were at times difficult to separate owing to popular representations (Bauer et al., 1998). Therefore we will also have a look at debates and policy responses in such related fields.

8. Obviously different actors conceptualise the public differently but, interestingly, the same actors do too, depending on the context. Hill and Michael, for example, describe what they identified as the decision-maker's concept of 'the ordinary member' of that public, being constructed from ideas of the citizen and of the consumer, respectively. An idealised layperson is thus 'an admixture of (at least) autonomous, thoughtful citizen and concerned, rational decision-making consumers' (Hill and Michael, 1998: 213). This interpretation is in striking contrast to the notion also frequently encountered among scientific and regulatory elites of the 'uninformed public' being basically uninterested and dependent on mass media that keep highlighting certain facts and hiding others.
9. Some of the then Warsaw Pact states oriented themselves towards the US policy.
10. In 1978, DG XII proposed a directive against 'conjectural risks' associated with rDNA work, hence the proposal oriented itself towards the technology. However, various reports from European scientific organisations stressed that the risks had been exaggerated and relaxation was necessary and possible. This argumentation, following the evolution of policy opinion in the USA, prompted DG XII to withdraw the proposal and to opt for non-regulation in the form of a 'recommendation' only.
11. For example, the US Congress's Office of Technology Assessment issued a series of reports emphasising the prospects for medical research, pharmaceutical production, agriculture and environmental remediation (e.g. OTA, 1984).
12. Later, the OECD also assigned to biotechnology the status of a crucial cross-sectional technology, like electricity or microelectronics (OECD, 1988).
13. An annual licence for low-risk experiments replaced notification in advance, and scientific committees checked the safety on their own.
14. In 1980, a British report conveyed the message to government that commercialisation of biotechnology was both possible and desirable; however, competitiveness relied on better technology transfer, which would require more effective support from both government and industry.
15. In France, the early 1980s appeared retrospectively as the 'golden age' of promotional biotechnology policy-making (Gottweis, 1995: 225), since strong government backing led to considerable achievements and a catching-up with leading countries such as the UK.
16. However, firms, especially in the chemical sector, were reluctant to take up this entirely new approach. One reason was that, among chemists only a decade before, the 'old' biotechnology on which the new techniques were based had been deemed outdated compared with the more 'scientific' field of organic chemistry (see note 40).
17. The OECD also issued advice about large-scale industrial production (OECD, 1986), but it took them six more years to come up with recommendations

for deliberate releases for research purposes, which indicated the associated problems (OECD, 1992).

18. Consequently, in 1980 a new British risk assessment scheme served to facilitate large-scale fermentation.
19. There are many interpretations of the precautionary principle, from the rigorous driver's advice for overtaking: 'If in doubt, don't'; to the more cautious environmental protection version: 'If there are serious doubts about the outcome, then the one that is less able to defend itself should be given the benefit of the doubt' (in this case, the environment).
20. Engineered bacteria (the so-called ice-minus strains) would protect strawberry plants from freezing. Opponents made claims of (hypothetical) risks associated with the release. These even included the risk of climatic changes. After extensive review by the NIH, permission was granted to carry out the experiment in 1984. However, NGOs took up the issue and succeeded in placing 'genetic engineering' on the agenda again. Activist Jeremy Rifkin brought a law suit challenging NIH's competence as the first in a series of litigations that significantly contributed to raising awareness or, as others put it, to exaggerating the risks of biotechnology and causing widespread fear. Rifkin's litigation series triggered mostly local opposition against single projects. One of the results was that the US Environmental Protection Agency (EPA) took up biotechnology as an issue and the releases of GMOs became regulated under existing law (Krimsky, 1991).
21. 'As a public controversy, ice minus was history just as soon as the field was sprayed by the moon-suited scientists. A precedent had been set insofar as the first major barrier to the environmental application of GEMs [genetically engineered organisms] was overcome' (Krimsky, 1991: 132).
22. rBST is a bovine hormone for stimulating milk production in cows that is produced in GM bacteria. Its approval for use in dairying triggered debates, legal actions and a Commission of Enquiry of the German parliament in 1989 (Deutscher Bundestag, 1989). After protracted disputes, and contrary to the US government, the EU Commission finally denied its approval for use in milk production, which eventually gave rise to a transatlantic trade conflict.
23. For example, German opponents to biotechnology conceptualised the conflict as one between two different paths of development: 'One path will increase the industrialisation, the technical control and the re-shaping of nature to allow better exploitation. It is feared that this functionalisation . . . of life will not be limited to plants and animals. The other, preferred path of development is described as a path where technological and non-technological solutions to problems are developed that guarantee the protection and sustainability of nature' (Tappeser, 1990: 10 ff.).
24. Later, the ban on rBST was interpreted by the USA as the erection of an unfair trade barrier by the EU.
25. Genetic screening allows the selection of embryos according to their characteristics. Parents obviously want healthy children, but what 'healthy' means depends on cultural definitions. The birth of ill or handicapped children thus becomes the result of a voluntary decision by the parents.

26. For example, the Ökologie-Institut in Freiburg provided counter-expertise, focused on ecological criteria, issuing from an institutionalised base. This was intended to match the expertise of molecular biologists and industry in the biotechnology struggles and litigations during the 1980s in Germany.
27. This had serious practical consequences, because differences in regulatory styles were the basis of frequent quarrels within the EU over the focus of risk assessment (Levidow et al., 1996) and the definition of 'sound science' in their assessment of products.
28. There were different possible reasons for this: they could not find enough support among the public or from the media; their arguments were not heard by the opinion-leaders; they found other means to achieve partial successes; or, given the fact that most NGOs relied heavily on individual activists, they simply compromised on personal grounds.
29. 'The British state either integrates social movements quite well, either directly, in the policy-making process, or through political parties, and gives them some limited influence in exchange for co-operation, or it shuts them out completely, denying them any opportunity to influence policy-making' (Rüdig et al., 1991: 137).
30. 'Through the early 1980s, the strategy of containing regulatory debate within carefully structured expert committees ensured a relatively narrow focus on the physical risks of rDNA research and correspondingly muted attention to the social and political consequences of the new technology' (Jasanoff, 1995: 322).
31. Around 1989, the government tried to counteract the lack of public acceptance by supporting initiatives for multi-actor workshops and setting up an ethics committee on environmental safety. A foundation to support knowledge-based opinion formation about biotechnology by consumer NGOs is still functioning.
32. Its 'inventor', the Danish Board of Technology (founded by the Danish government in the 1980s), is a body created to assess technology, stimulate public debate and advise parliament.
33. The law prohibited releases of GMOs unless the Minister of the Environment found that there were 'special circumstances' for granting an exemption. Eventually, all applications received were exempted from the ban, even without mention of 'special circumstances'.
34. This was also the case in non-member countries joining the EU at a later point in time, for example in Austria and Finland. Finland saw biotechnology as the means to 're-industrialise' R&D in some limited areas. Countries on the other side of the Iron Curtain under Soviet domination, such as Poland and Hungary, slowly tried to establish a research base while orienting themselves towards the US policies, again in the absence of any significant public debate.
35. The OECD is based in Paris; there are twenty-nine members (1999), including the most important industrialised nations and the EU.
36. This was necessary because, if EC regulation was pending, national laws had to be compatible with what was to come.
37. Especially after an unauthorised deliberate release of GMOs in 1987.

38. The ethical debate, which had always been pronounced in France, centred on questions of human medicine and had little influence on the government's position concerning other applications of biotechnology.
39. In particular, the big chemical companies in Germany (with the exception of Hoechst) were reluctant to engage in biotechnology. This was not because of fundamental disadvantages in competitiveness. Ulrich Dolata (Dolata, 1995: 463) located the reason for German reluctance in the different culture of the chemical engineers leading these companies, and their devotion to classic organic chemistry, as opposed to that of the biologists, with their emphasis on biological processes. Meanwhile, US start-up companies had flourished and had partially been incorporated into big established pharmaceutical concerns (Oakey et al., 1990). This had led to a technology transfer, and often entailed their reorientation towards biotechnology.
40. This was in line with an OECD report entitled 'Biotechnology and the Changing role of Government' (OECD, 1988), which emphasised its main role of coordinating R&D in this field and promoting its commercial exploitation.
41. In its report of 1985, the Parliament 'summed up the political situation for biotechnology at the European Community level. It showed a broad awareness for the potential of biotechnology, and the need for a coherent strategy, responding to the need for international competitiveness. But with respect to regulation it saw "special risks with genetic engineering methods"..., demanding a complete ban on field releases "until binding Community safety directives have been drawn up"..., and with similar restrictive views on gene therapy... and animal transgenesis' (Cantley, 1995: 543).
42. The Sixth Amendment (79/831/EEC), shaped by DG XI (Environment) after OECD proposals and the American Toxic Substances Control Act.
43. Cantley (1995: 551), quoting from the summary of a meeting between the Commission and member state representatives.
44. The first Directive proposal on workers' safety covered not only GMOs but 'biological agents' in general.
45. Directives 87/22/EEC for the Production of Pharmaceuticals by New Technologies, 90/679/EEC on Workers' Protection from Risks Related to Biological Agents, or the proposal for a Directive on Intellectual Property Rights in Biotechnology and for a Plant Variety Rights System.
46. On the other hand, the European Parliament demanded more restrictive amendments, put forward by the German rapporteur who favoured an approach oriented to possible risks.
47. Another field of debate was the inclusion of the so-called 'fourth hurdle'. NGOs had demanded that applications of biotechnology should be linked to the demonstration of need, additional to the traditional criteria of purity, safety and efficacy in conventional drug assessment. The Commission strongly opposed such demands since it considered them to be an invitation to raise non-tariff trade barriers. Later, only the Austrian and Norwegian laws took up socio-economic criteria, a move with few practical consequences (see below).
48. Council Directive 90/219/EEC of 23 April 1990 on the Contained Use of Genetically Modified Micro-organisms, *Official Journal*, L117, 08/05/1990, and Council Directive 90/220/EEC of 23 April 1990 on the Deliberate Release

into the Environment of Genetically Modified Organisms, *Official Journal*, L117, 08/05/1990, to be implemented in national regulation within October 1991 (which was not the case in all member countries). The directives were issued by the European Economic Community (after 1993 the European Community) within the European Union, hence it is actually inappropriate to speak about 'EU directives', although they were mandatory throughout the European Union. Nevertheless we will stick to the term for the sake of simplicity.

49. 'According to Technical Progress', an amendment to the Release Directive, was issued in 1994 (Directive 94/15/EC), specifying categories of plants intended for deliberate release and clarifying details of the risk assessment.
50. The EU deliberately did not regulate genetic testing, gene therapy or any other possible applications, fields in which national regulations were implemented in some European countries, but not in all. In particular, there was no reference to ethical considerations other than in relation to risks to health and safety and the environment. Any links to reproduction techniques, cloning and the like were omitted. It even remained questionable how the Release Directive would cover the handling of transgenic animals. Another issue that was not touched upon was the dispute about intellectual property rights. In 1988, a US patent had been granted on the Harvard 'Onco-Mouse', which had raised discussions on the 'patenting of life'. In the aftermath, the issue had also created controversy among scientists. The EU prepared a draft directive that was heavily debated, and it took another ten years before a directive on Intellectual Property Rights could finally be passed (see below).
51. For example, the question of how to deal with deliberate releases of GMOs showed up on European countries' agendas when the release of ice-minus bacteria was debated in the USA. A truly public debate, however, arose only when applications for the release of transgenic plants were made within particular countries.
52. For example, gene therapy, xenotransplantation and genetic testing.
53. Made for the procreation of cows that could produce an anti-microbial protein from human milk in their own milk.
54. The case of Herman the Bull anticipated some of the debates that were to arise later over GM animals as organ donors (xenotransplantation), although medical risks were not at stake.
55. In order to broaden the basis for decision-making, a consensus conference was held on transgenic animals.
56. This step was made possible *inter alia* because the Green Party was no longer in parliament after reunification.
57. A way to reconcile the differences was to assign 'societal benefit' to every release for research purposes, since a gain in knowledge was automatically deemed socially beneficial and capable of promoting sustainable development. This was different for certain product applications, however.
58. Occasional and sudden outbursts of distrust, as in Poland in 1997, were mostly triggered by NGO actions from abroad.
59. The OECD recommendations of 'Good Developmental Principles' (OECD, 1992, part 2) for small-scale field releases did not succeed in playing a similarly unifying role for the US and the EU positions as the 'Recombinant DNA

Safety Considerations' (OECD, 1986) had done six years before. Attempts to link plant biotechnology closely to traditional breeding as a 'baseline for assessing modern biotechnology' (OECD, 1993b) were officially welcomed by both the EU and the USA, but they did not resolve the question of how in practice to assess newly introduced genes conferring traits that had not been seen before in crop plants.

60. There were even more far-reaching differences, since it was by no means clear what an acceptable outcome of the risk assessment was, or what exactly should be prevented when GMOs were released.
61. Although disclosure of such files was prohibited, more data could be obtained in Denmark and the Netherlands than, for example, in France or Germany (Jülich, 1998).
62. Mostly from industry, as well as from the governments of Germany, the UK and the Netherlands.
63. For example, a British company sold 'vegetarian' cheese, which was made with genetically engineered chymosin as an alternative to calf rennin. Most modern washing powders contained recombinant enzymes.
64. In 1994, the Directive on Deliberate Release was updated (Directive 94/15/EC).
65. Environmental remediation with the help of biotechnology was not an issue during this period. On the one hand, there seemed to be too little substantial progress, and on the other hand the release of GMOs in order to eliminate pollution appeared, to many Europeans engaged in environmental protection, as an idea derived from the most cynical technocratic thinking.
66. Sheila Jasanoff (1995: 311) described the study of the political regulation of biotechnology as the 'study of the process by which technological advances overcome public resistance and are incorporated into a receptive social context'. This diagnosis, taken from US experiences, appeared also to hold true for Europe until the mid-1990s.
67. This was acknowledged by an OECD report on 'Biotechnology, Agriculture and Food' in 1993: 'High levels of uncertainty surround the innovation process.' 'Biotechnology innovation involves new forms of co-operation between economic actors situated at different points in the agro-food system.' And: 'Successful innovation demands greater responsiveness to end-users whether they be other firms or the final consumer' (OECD, 1993a: 143).
68. The world-wide acreage covered with transgenic crops exploded between 1996 and 1998, e.g. for soya, from 0.5 to 17.0 million hectares (according to the Austrian Press Agency, 3 December 1998).
69. BSE (Bovine Spongiform Encephalitis) is a nervous tissue disease transmitted by prions (agents consisting of only protein); as it is known as Scrapie disease in sheep. Mostly British cattle acquired the disease probably from being fed food additives derived from sheep carcasses, although experts had deemed any transmission across species borders impossible. There is a very rare and similar condition in humans called Creutzfeld–Jakob disease (CJD). When an unusual number of new cases of this disease occurred in the UK, experts again reassured the public that there were no indications of a trans-species transmission between cattle and humans. In 1995, however, the possibility of a link was officially conceded after the leaking of information from a secret report.

70. The biggest problem for the EU Commission had been the collapse of the common beef market. Apparently giving in to British pressure, the Commission lifted the ban on some beef products in 1996 against heavy opposition. '... a simple appraisal of costs and benefits confirms that in terms of both socio-economic indicators and political legitimacy the policy adopted had disastrous consequences.... the costs of the crisis ... far outweigh the short-term benefits of the approach pursued by the UK government and the European Union institutions' (Baggott, 1998: 64).
71. Although the labelled tomato paste had not evoked any public debates, industry and the EU Commission still tried to prevent mandatory labelling, and food additives were excluded from the regulation's scope. On the other hand, the European Parliament's Economic and Social Select Committee had emphasised the consumer's right to choose and demanded that labelling become mandatory.
72. The labelling debate revealed two fundamentally different approaches. Conventional risk prevention demands a guarantee from the authorities that there are no significant risks for human health with any product. The way the product was generated is considered to be of no interest to the consumer, since only the (physical) properties count. Such an understanding allows labelling only when there is a more or less established risk for the public or for certain persons, e.g. those suffering from allergies. This was basically the position of the US Federal Drug Administration. On the other hand, the consumer choice approach demands labelling in order to indicate the production process. It provides a choice between products that may be substantially identical but produced differently (i.e. a GM versus a non-modified tomato). This is also in line with 'negative' labelling if it can be proven that the product does not contain anything that was genetically modified. Clearly, this approach is much more 'political' and implicitly contradicts the philosophy of international trade agreements (see below). They built on the concept of 'substantial equivalence' proposed by the OECD (OECD, 1993a), which was based on comparison between the modified and unmodified food product. It was acknowledged that uncertainty might play a role, but material differences that should 'matter' had to be established.
73. Although the Commission approved the GM soybeans, it rejected further demands by Parliament for a modification of the Novel Food Regulation. In spring 1996 the EU institutions' stances differed to such an extent that a time-consuming mediation procedure set in.
74. Although the regulation was binding from April 1997 in all member states, labelling requirements were still unclear. The Commission had to specify them for soya and maize that had been permitted before the regulation was enacted. Now labelling was also required if modified DNA or protein were present.
75. For example, the French had always had difficulties in accepting an American dictate in matters of agricultural trade, whereas Austria as a full EU member experienced a new regulatory impotence in the face of binding EU regulations, and the Danes rediscovered an unease over agricultural biotechnology that had been buried under a layer of consensus-oriented deliberations.
76. See the country reports in Durant et al. (1998).

77. The fact that this release was conducted without official permission triggered an immediate response by NGOs and the mass media.
78. During the one-year run up, French political actors kept making reference to this planned event, at the EU level, in international political arenas and even in summit talks of the World Trade Organization (WTO). The conference itself raised consumers' worries about health risks and scepticism about the independence of experts advising the government.
79. In autumn 1998 it decided to keep the harvest of modified maize completely off the market.
80. Like the French, the previous Conservative UK government had serious problems during the BSE crisis, which also had repercussions in the debates over the soya and maize imports pending.
81. During 1998, test fields were destroyed by activists. A court battle took place over contamination by genetically modified plants of an organic farm, among others, and biotechnology companies were 'named and shamed' by government advisers for flouting field trial regulations. The media emphasised consumer concerns about food and choice. The Prince of Wales (amongst others) spoke out against the use of genetically modified foods in favour of 'natural' products from organic farming.
82. A number of other publicly sensitive issues were addressed by the government, including cloning and issues of genetic testing and insurance.
83. Except Finland; see below.
84. The WTO Agreement on Sanitary and Phytosanitary Measures (SPS) demanded transparent, solely science-based risk assessment as the only legitimate basis for even the shallowest trade restrictions. However, 'in the context of the WTO, it is unclear what criteria might justify an import restriction under SPS, short of evidence of significant negative impact following release of a GMO' (Wyndham and Evans, 1998: 2 ff.)
85. Within DG XXIV, an expert committee was set up to assess genetically modified products in order to circumvent differences between evaluations by national governments. The evaluations had proven to be hard to reconcile with each other.
86. For example, the American stance towards artificially processed products that were 'nutritionally enhanced' (e.g. de-cholesteroled or sugarless) usually was more relaxed than European perceptions of an 'adulteration' of such food (Hoban, 1997). On the other hand, the perceived European preoccupation with 'naturalness' as well as the frequently rather lukewarm attitude, compared with American enthusiasm, to competition and economies of scale appeared irrational on the other side of the Atlantic.
87. It forced some revisions also in traditional plant variety protection, including such issues as the right to propagate an organism, e.g. in order to produce seed for own purposes. A different aspect of this issue was the exploitation of genetic data from developing countries' indigenous species or crop varieties by companies from the industrialised world. Issues of intellectual property rights were linked to attempts to secure biological diversity through environmental protection. When this was negotiated at the 1992 UN Conference on Environment and Development in Rio de Janeiro, the issue of sustainable

development was emphasised. In the end, 171 parties signed the Convention on Biological Diversity, including the EU, but the USA did not agree to the negotiated results. This too had consequences, especially in EU countries, for public opinion of the role of US companies and their attempts to commercialise GMOs.

88. The neglect of ethical concerns eventually led to a turning down of the EU Commission's proposal for a Patenting Directive in the European Parliament in 1995, which in turn flagged up the issue for industry (Galloux et al., 1998: 182).
89. For example, the mergers between Hoechst and Rhône-Poulenc and between Zeneca and Astra in late 1998 resulted in conglomerates that covered exactly these two areas, after they had reduced most other activities they had previously been engaged in.
90. Another problem arose from the time perspective of risk assessment. Classical technical risk assessment is in general retrospective: the risk of future hazards can be determined by applying statistical methods to past experiences. Many opponents of biotechnology held a different concept of risk, which was oriented not to past experiences but to possible futures (see Krohn and Krücken, 1993). This change in time perspective had severe consequences for the debates on risk. Because the future cannot be controlled, it is logically impossible to exclude future hazards and debates on risk cannot be closed.
91. For example, when Dolly the sheep appeared, the question of whether this was biotechnology or not split even the experts' community.
92. This distinction can also be made for example in the case of BSE: arguments about the hazard of acquiring Creutzfeld–Jacob disease concern a physical risk; those about the behaviour of governments (e.g. whether the British government was right in not disclosing ambiguous data) are societal or moral arguments.
93. A good example is the different reasons why labelling was demanded. On the one hand, labelling should serve to indicate health risks for those consumers who are, for example, allergic to certain ingredients. On the other hand, labelling of a product with respect to the technology that was used to produce it was seen as a genuine consumers' right. It was perceived as the only means by which a consumer decision on the market could give signals to producers about the acceptability of their products, rather than legal prohibition.
94. Both the German and the Austrian Commissions of Enquiry were considered to have had little impact on actual biotechnology policy (Grabner and Torgersen, 1998).
95. This role shed some light on the presently prevailing understanding of democracy in the economy-centred common market: impotent citizens (or, for paternalising state authorities, subjects) should possibly turn into powerful consumers in order to pursue a moral matter. See also note 9 above.
96. This was also reflected in new alliances between NGOs, consumer organisations and retailers.
97. The EU did not follow the European Biotechnology Council's recommendation to adopt the British model, which built much more on scientific evidence.
98. This led to the conflict over 'product versus process'.

99. Among the forces towards convergence were the pressure exerted by international competition and the globalisation of trade and industry; the necessity to adhere to guidelines issued by international organisations such as the OECD or the World Trade Organisation; and concerns about Europe as an area with high wages and standard of living that demands the production of goods with high added value. Among the forces towards divergence were the increasing importance of NGOs, acting mostly nationally, which could command high public trust, as compared with the erosion of trust in expert knowledge after food scandals; developments triggered by the election of governments formed by more leftist or green parties; a strengthening of the importance of national parliaments by governments built on weak majorities; and the acknowledgement of the importance of dealing with public concerns by state institutions and companies.
100. Apart from testing in cases of severe inherited diseases, which is generally welcomed.

REFERENCES

Baggott, R. (1998) 'The BSE Crisis. Public Health and the "Risk Society"', in P. Gray and P. 'tHart (eds.), *Public Policy Disasters in Western Europe*, London and New York: Routledge, pp. 61–78.

Bandelow, N. (1997) 'Ausweitung politischer Strategien im Mehrebenensystem, Schutz vor Risiken der Gentechnologie als Aushandelsmaterie zwischen Bundsländern, Bund und EU', in R. Martinsen (ed.), *Politik und Biotechnologie. Die Zumutung der Zukunft*, Baden-Baden: Nomos, pp. 153–68.

Bauer, M. (1995) 'Resistance to New Technology and Its Effects on Nuclear Power, Information Technology and Biotechnology', in M. Bauer (ed.), *Resistance to New Technology: Nuclear Power, Information Technology and Biotechnology*, Cambridge: Cambridge University Press, pp. 1–41.

Bauer, M., J. Durant and G. Gaskell (1998) 'Biology in the Public Sphere: a Comparative Review', in J. Durant, M. Bauer and G. Gaskell (eds.), *Biotechnology in the Public Sphere*, London: Science Museum, pp. 217–27.

Beck, U. (1986) *Die Risikogesellschaft. Auf dem Weg in eine andere Moderne*, Frankfurt am Main: Suhrkamp.

(1988) *Gegengifte. Die organisierte Unverantwortlichkeit*, Frankfurt am Main: Suhrkamp.

Behrens, M., S. Meyer-Stumborg and G. Simonis (1997) *Genfood. Einführung und Verbreitung, Konflikte und Gestaltungsmöglichkeiten*, Berlin: Sigma.

BEP (Biotechnology and the European Public Concerted Action Group) (1997) 'Europe Ambivalent on Biotechnology', *Nature* 387: 845–7.

Bongert, E. (1997) 'Towards a "European Bio-Society"? Zur Europäisierung der neuen Biotechnologie', in R. Martinsen (ed.), *Politik und Biotechnologie. Die Zumutung der Zukunft*, Baden-Baden: Nomos, pp. 117–34.

Bud, R. (1993) *The Uses of Life. A History of Biotechnology*, Cambridge: Cambridge University Press.

Cantley, M. (1995) 'The Regulation of Modern Biotechnology: A Historical and European Perspective', in D. Brauer (ed.), *Biotechnology*, vol. 12, New York: VCH, pp. 505–681.

Deutscher Bundestag (1989) *Zum gentechnologisch hergestellten Rinderwachstumshormon*. Enquete-Kommission 'Gestaltung der technischen Entwicklung; Technikfolgen–Abschätzung und-Bewertung' des 11. Deutschen Bundestages, BT-Drucksache 11/4607, Bonn.

Dolata, U. (1995) 'Nachholende Modernisierung und internationales Innovationsmanagement. Strategien der deutschen Chemie- und Pharmakonzerne', in T. von Schell and H. Mohr (eds.), *Biotechnologie–Gentechnik. Eine Chance für neue Industrien*, Berlin: Springer, pp. 456–80.

Douglas, M. and A. Wildavsky (1982) *Risk and Culture. An Essay on the Selection of Technological and Environmental Danger*, Berkeley: University of California Press.

Durant, J., M. Bauer and G. Gaskell (eds.) (1998) *Biotechnology in the Public Sphere*, London: Science Museum.

Ellul, J. (1964) *The Technological Society*, New York: Alfred A. Knopf.

European Council (1990), Directive 90/219/EEC of 23 April 1990 on the Contained Use of Genetically Modified Micro-organisms; and Directive 90/220/EEC of 23 April 1990 on the Deliberate Release into the Environment of Genetically Modified Organisms, *Official Journal of the European Communities* L117/1–27, Brussels.

(1994), Directive 94/15/EC of 15 April 1994 for the First Adaptation of the Directive 90/220/EWG on the Deliberate Release of GM Organisms Into the Environment to Technical Progress. Abl./L 103/20, Brussels.

Galloux, J.-C., H. Prat Gaumont and E. Stevers (1998) 'Europe', in J. Durant, M. Bauer and G. Gaskell (eds.), *Biotechnology in the Public Sphere*, London: Science Museum, pp. 177–85.

Gaskell, G., M. Bauer and J. Durant (1998) 'The Representation of Biotechnology: Policy, Media and Public Perception', in J. Durant, M. Bauer and G. Gaskell (eds.), *Biotechnology in the Public Sphere*, London: Science Museum, pp. 3–12.

Gill, B. (1991) *Gentechnik ohne Politik. Wie die Brisanz der Synthethischen Biologie von wissenschaftlichen Institutionen, Ethik- und anderen Kommissionen systematisch verdrängt wird*, New York and Frankfurt am Main: Campus.

Gill, B., J. Bizer and G. Roller (1998) *Riskante Forschung. Zum Umgang mit Ungewißheit am Beispiel der Genforschung in Deutschland. Eine sozial- und rechtswissenschaftliche Untersuchung*, Berlin: Edition Sigma.

Gottweis, H. (1995) 'Governing Molecules. The Politics of Genetic Engineering in Britain, France, Germany, and in the European Union', habilitation paper, Faculty of Humanities, University of Salzburg.

Grabner, P. and H. Torgersen (1998) 'Österreichs Gentechnikpolitik – Technikkritische Vorreiterrolle oder Modernisierungsverweigerung?' *Österreichische Zeitschrift für Politikwissenschaft* 1: 5–27.

Gundelach, P. (1991) 'Research on Social Movements in Denmark', in D. Rucht (ed.), *Research on Social Movements. The State of the Art in Western Europe and the USA*, Frankfurt am Main: Campus, pp. 262–94.

Hall, P.A. (1993) 'Policy Paradigms, Social Learning and the State. The Case of Economic Policymaking in Britain', *Comparative Politics*, 25(3): 275–96.

Hill, A. and M. Michael (1998) 'Engineering Acceptance: Representations of "The Public" in Debates on Biotechnology', in P. Wheale, R. von Schomberg and P. Glasner (eds.), *The Social Management of Genetic Engineering*, Aldershot: Ashgate, pp. 201–18.

Hoban, T.J. (1997) 'Consumer Acceptance of Biotechnology: an International Perspective', *Nature Biotechnology* 15: 232–4.

INRA (1993) 'Biotechnology and Genetic Engineering. What Europeans Think about It in 1993', survey conducted in the context of Eurobarometer 39.1.

Jasanoff, S. (1995) 'Product, Process or Programme: Three Cultures and the Regulation of Biotechnology', in M. Bauer (ed.), *Resistance to New Technology. Nuclear Power, Information Technology and Biotechnology*, Cambridge: Cambridge University Press, pp. 311–31.

Jülich, R. (1998) 'Öffentlichkeitsbeteiligung im Geltungsbereich der EG-Richtlinien 90/219 und 90/220 im internationalen Vergleich. Die Ausgestaltung von Informations- und Partizipationsrechten in den EU-Mitgliedstaaten, der Schweiz und Norwegen', Öko-Institut, Freiburg, Darmstadt, Berlin.

Jungermann, H. and P. Slovic (1993) 'Characteristics of Individual Risk Perception', in Rück Bayerische (ed.), *Risk Is a Construct*, Munich: Knesebeck, pp. 85–101.

Kliment, T., O. Renn and J. Hampel (1994) 'Die Wahrnehmung von Chancen und Risiken der Gentechnik aus der Sicht der Öffentlichkeit', in T. von Schell and H. Mohr (eds.), *Biotechnologie–Gentechnik. Eine Chance für neue Industrien*, Berlin/Heidelberg: Springer, pp. 558–83.

Krimsky, S. (1991) *Biotechnics and Society. The Rise of Industrial Genetics*, New York: Praeger.

Krohn, W. and G. Krücken (1993) 'Risiko als Konstruktion und Wirklichkeit. Eine Einführung in die sozialwissenschaftliche Risikoforschung', in W. Krohn and G. Krücken (eds.), *Riskante Technologien: Reflexion und Regulation. Eine Einführung in die sozialwissenschaftliche Risikoforschung*, Frankfurt am Main: Suhrkamp, pp. 9–44.

Levidow, L., S. Carr, R. von Schomberg and D. Wield (1996) 'Regulating Agricultural Biotechnology in Europe: Harmonisation Difficulties, Opportunities, Dilemmas', *Science and Public Policy* 23: 135–7.

Macer, D. (1994) 'Bioethics for the People by the People', Christchurch, New Zealand/Tsukuba, Japan: Eubios Ethics Institute.

McKelvey, M. (1997) 'Moving to Commercialisation: Invited Response', in *Transgenic Animals and Food Production, Kungelik Skogs- och Lantbruksakademiens Tidskrift* 20: 133–8.

Majone, D. (1989) *Evidence, Argument and Persuasion in the Policy Process*, New Haven, CT: Yale University Press.

Nielsen, T.H. (1999) 'Bioteknologi og biopolitik, 1976–1999', manuscript, Oslo.

Oakey, R., W. Faulkner, S. Cooper and V. Walsh (1990) *New Firms in the Biotechnology Industry*, London: Pinter.

OECD (1986) 'Recombinant DNA Safety Considerations', Paris.

(1988) 'Biotechnology and the Changing Role of Government', Paris.

(1992) 'Safety Considerations for Biotechnology', Paris.

(1993a) 'Biotechnology, Agriculture and Food', Paris.

(1993b) 'Traditional Crop Breeding Practices: an Historical Review to Serve as a Baseline for Assessing the Role of Modern Biotechnology', Paris.
(1996) 'Intellectual Property. Technology Transfer and Genetic Resources. An OECD Survey of Current Practices and Policies', Paris.
Ogburn, W.F. (1973) 'The Hypothesis of Cultural Lag', in Eva Etzioni-Halevy and Amitai Etzioni (eds.), *Social Change: Sources, Patterns, and Consequences*, 2nd edn, New York: Basic, pp. 477–80.
OTA (Office of Technology Assessment, US Congress) (1984) *New Developments in Biotechnology. 4: US Investment in Biotechnology*, OTA-BA-360, Washington D.C.: US Government Printing Office.
Pline, C. (1991) 'Popularizing Biotechnology: the Influence of Issue Definition', *Science, Technology and Human Values* 16(4): 474.
Praag, P. van (1991) 'The Netherlands: Action and Protest in a Depillarized Society', in D. Rucht (ed.), *Research on Social Movements. The State of the Art in Western Europe and the USA*, Frankfurt am Main: Campus, pp. 295–320.
Rammert, W. (1993) *Technik aus soziologischer Perspektive*, Opladen: Westdeutscher Verlag.
Rucht, D. (1991) 'The Study of Social Movements in Western Germany: between Activism and Social Science', in D. Rucht (ed.), *Research on Social Movements. The State of the Art in Western Europe and the USA*, Frankfurt am Main: Campus, pp. 175–202.
(1995) 'Impact of Anti-Nuclear Power Movements', in M. Bauer (ed.), *Resistance to New Technology. Nuclear Power, Information Technology and Biotechnology*, Cambridge: Cambridge University Press, pp. 277–91.
Rüdig, W., J. Mitchell, J. Chapman, P.D. Lowe (1991) 'Social Movements and Social Sciences in Britain', in D. Rucht (ed.), *Research on Social Movements. The State of the Art in Western Europe and the USA*, Frankfurt am Main: Campus, pp. 121–48.
Schomberg, R. von (1998) 'An Appraisal of the Working in Practice of the Directive 90/220/EEC on the Deliberate Release of GM Organisms', STOA, European Parliament, Luxembourg.
Seifert, F. and H. Torgersen (1997) 'How to Keep out What We Don't Want. An Assessment of "Sozialverträglichkeit" under the Austrian Genetic Engineering Act', *Public Understanding of Science* 6: 301–27.
Sharp, M. (1985) 'The New Biotechnology – European Governments in Search of a Strategy', *Industrial Adjustment and Policy IV Series*, Sussex European Papers No. 15.
Slovic, P. (1987) 'Perceptions of Risk', *Science* 236: 280–5.
Smink, G.C.J. and A.M. Hamstra (1996) *Informing Consumers about Foodstuffs Made with Genetic Engineering*, Leiden: SWOKA.
Tappeser, B. (1990) in 'Kurzcommentar für die Arbeitsgemeinschaft ökologischer Forschungsinstitut', K. Grosch, P. Hampe and J. Schmidt (eds.), *Herstellung der Natur? Stellungnahmen zum Bericht der Enquete-Kommission 'Chancen und Risiken der Gentechnologie' des 10. Dentschen Bundestags*, Frankfurt/New York: Campus.
Thielemann, H. (1998) 'Kommunikation im Konflikt um die gentechnische Insulinherstellung bei der Hoechst AG', in O. Renn and J. Hampel

(eds.), *Kommunikation und Konflikt. Fallbeispiele aus der Chemie*, Würzburg: Königshausen & Neumann, pp. 153–81.
Tichy, G. (1995) 'Sozialverträglichkeit – ein neuer Standard zur Technikbewertung?', *Österreichische Zeitschrift für Soziologie* 19(4): 50–61.
Vogel, D. (1996) *Trading up. Consumer and Environmental Regulation in a Global Economy*, Cambridge, MA: Harvard University Press.
Vugt, F. van and A.M.P. Nap (1997) 'Regulatory and Policy Issues as Viewed within a Cultural Framework: a European Perspective', in *Transgenic Animals and Food Production, Kungl. Skogs- och Lantbruksakademiens Tidskrift* 20: 31–6.
Wyndham, A. and G. Evans (1998) 'National Biosafety Legislation and Trade in Agricultural Commodities', *BINAS News* 4(2 & 3): 2–6.

3 Media coverage 1973–1996: trends and dynamics

Jan M. Gutteling, Anna Olofsson, Björn Fjæstad, Matthias Kohring, Alexander Goerke, Martin W. Bauer and Timo Rusanen. With the further cooperation of: Agnes Allansdottir, Anne Berthomier, Suzanne de Cheveigné, Helle Frederiksen, George Gaskell, Martina Leonarz, Miltos Liakopoulos, Arne Thing Mortensen, Andrzej Przestalski, Georg Ruhrmann, Maria Rusanen, Michael Schanne, Franz Seifert, Angeliki Stathopoulou and Wolfgang Wagner

Introduction

Owing to the complexity and vulnerability of their technical and social systems, modern societies are sometimes characterised as 'risk societies' (Beck, 1986). A dynamic and transnational exchange of potentially conflicting information occurs between actors participating in the discourse about the consequences of risk-related decisions (Hilgartner and Bosk, 1988). In these complex societies, in which more and more immediate information is required in coping with everyday life, individuals have become highly dependent on the mass media. In most European countries, the media have assumed additional functions to informing the public: for instance, serving as society's watchdog, signalling injustice or unwanted developments, as well as acting as the main source of entertainment. As the developments and applications of modern biotechnology become more apparent, the mass media will also begin to provide these services for this particular domain of technological innovation. Furthermore, the level of individual dependency on the media will be especially high in the case of modern biotechnology because it is virtually impossible to gather information on this subject through direct personal experiences. Genetically modified soya, gene therapy, the square tomato, Dolly the sheep and Herman the bull have become well-known public 'icons' of modern biotechnology, not through personal acquaintance but through the information provided by the electronic and mass print media. This also implies that the public can construct an image of the 'biotechnology reality' only on the basis of what the media themselves decide to convey.

Thus, it is reasonable to identify the media as important modifiers in the processes of shaping popular perceptions and influencing the public's decision to either accept or reject new developments such as biotechnology.

Many scholars of mass communication have formulated theories of the media's role in society, in which realistic and constructionist approaches can be distinguished. The realist notion is based on the assumption of an existing set of events, which can be reported in an objective and balanced manner by competent, fair and unbiased journalists. According to this realistic approach, journalism is like a mirror on reality (see, for example, Kepplinger, Ehmig and Ahlheim, 1991). In keeping with this perspective, many (biotechnology) scientists will view the task of the press as involving the accurate and unbiased reporting of technical facts. They will argue that the complexity of the technology lends authority to the expert point of view and compels the press to highlight precisely this aspect (see, for example, Hagedorn and Allender-Hagedorn, 1997).

In contrast to this perspective, other scholars have stated that the media construct meaning by presenting a mediated world, rather than mirroring a more or less 'objective' reality (see, for example, DeFleur and Ball-Rokeach, 1989). In many conceptions of the construction of mediated reality, processes of news selection play a crucial role, particularly in the sense that a variety of social factors determine the press's notions of what is newsworthy and how issues ought to be framed. Clearly, on any day, many more events occur than the media are able to report, owing to limitations of time, space and technology. Therefore, at diverse stages in the process of news production, information is being processed: it is traced, collected, translated, edited, shortened or expanded, and transferred. After the final editorial process the public receives information about news events with little conception of these processes. In their theory of news value, Galtung and Ruge (1965) postulated that news value is mainly determined by the actuality of the event, in a temporal, geographical or psychological sense, and by its significance. A radical event, with long-lasting consequences for a large group of people, has a high news value. However, many studies in this area indicate that initially establishing a particular event as newsworthy depends on the journalist's highly personal and subjective assessments of its news value (see, for an overview, Servaes and Tonnaer, 1992).

The framing of news can be understood as the process through which complex issues are reduced to journalistically manageable dimensions, resulting in a particular focus on a certain issue. This reasoning implies that journalistic framing may also lead to journalistic selection, for example by placing heavy reliance on the information from very particular sources. With regard to modern biotechnology, US newspaper coverage is characterised by an inordinate dependence upon industrial and scientific sources, most of which emphasise economic considerations and potential benefits (Hornig, 1994). Nonetheless, if journalists are aware of differences in perception among experts and laypeople, whether owing to selective outlooks or to conflicting value orientations, they may be more inclined to emphasise their role as watchdogs. This is especially likely in the context of discussions of the potential risks and benefits of modern biotechnology, in which journalists may define their role as that of seeking out and framing the issue in terms of danger, controversy, and so on (Hornig, 1992; Frewer, Howard and Shepherd, 1997). These dynamic processes may encourage the media to highlight some aspects of biotechnological innovation at particular times, but they may also encourage the media to entirely neglect other issues related to biotechnology (Kitzinger and Reilly, 1997).

Research questions

In this chapter, we will describe the dynamics of the coverage of modern biotechnology in the opinion-leading press across Europe, from the perspective of journalistic selection and framing processes. Our study of media coverage spans a twenty-four-year period, starting in 1973 with the invention of recombinant DNA (rDNA) technology, a year considered to represent the inception of a new era of modern biotechnology. Our study ends in December 1996, a few months before the breaking of the Dolly story, which may be seen as another landmark in the development of modern biotechnology. (In chapter 11 of this book more will be said about Dolly and the media.) It is clear that the public's perception of modern biotechnology may be greatly influenced by the press's earlier coverage of modern biotechnology in its broadest sense. So the present study sheds light on the media reactions, and to some extent also the public reactions, to modern biotechnology.

As we have seen, many studies of the functioning of the mass media focus on journalistic selection and framing processes. The selection of events to be reported and the way events are portrayed may have a profound impact on the public's perceptions of these events, and may even influence policy formulation. Consequently, by studying journalistic products we may be able to gain insight into these influential processes. Thus, the basic questions we ask are: 'What is reported?' and 'How is it reported?', referring to selection and framing processes respectively. Our study of the development of the coverage of modern biotechnology proceeds from these fundamental questions. We will present data from the opinion-leading press of twelve European countries, enabling us to look at differences and similarities across Europe. Furthermore, we will look at developments across time.

First, we will describe the media coverage of modern biotechnology in general terms, analysing the main themes, actors and locations that recur in the coverage across Europe. This analysis will provide the basis for our exploration of journalistic selection processes. Next, we will focus on framing: we will analyse the negative or positive tone of the coverage, the framing of modern biotechnology in terms of hope or doom, and the distribution of risk and benefit arguments in the coverage. Indeed, the focus throughout our study will be on the press's treatment of issues related to the potential risks and benefits of modern biotechnology. The journalistic framing of modern biotechnology as either a salutary development or one in which risks are believed to outweigh all foreseeable benefits is likely to have a major impact on readers (see also Hornig, 1992; Frewer, Howard and Shepherd, 1997). In other words, we will analyse the contents of articles in which risks or benefits, or both, are described. Finally, we will look at the relationship between the media coverage of modern biotechnology and public perceptions across Europe.

Method

Content analysis may be described as a systematic preparatory and data-reducing method that seeks to achieve the objectivity of other social scientific research methodologies (Holsti, 1969; Krippendorff, 1980). In this study, the unit of analysis comprises any article containing references to modern biotechnology found in any part of the opinion-leading newspapers of the twelve European countries (with the exclusion of advertisements). These articles were further subdivided and coded according to whether they were describing conditions for, or the procedures and consequences of, research, analysis, practical handling and intervention into the genome of humans, animals and/or plants. Different examples include genetic engineering, gene therapy, modern biotechnology as a business enterprise, the genetic modification of organisms, novel foods, genetic fingerprinting, prenatal genetic diagnostics, the ethics of modern biotechnology, the statutory and non-statutory regulation of genetic technology, and initiatives supporting or opposing genetic technology. Articles addressing these diverse themes could be found in virtually all sections of the newspapers: national and international news, business pages, science pages, editorials and debate, and so on. Each relevant article was coded and the information incorporated in a transnational database.

We adopted this broad conception of modern biotechnology for several reasons. First, we wanted to ensure a thorough coverage that did not occlude the variation among different nations. Secondly, we wanted our unit of analysis to be concordant with the two other parts of the international research project, namely the analysis of the development of policy processes in the field of modern biotechnology and the analysis of public perceptions of modern biotechnology across Europe (as measured by Eurobarometer Biotechnology 1996, see 'Biotechnology and the European Public', 1997, and chapters 6, 7, 8 and 12 of this book). Finally, modern biotechnology is, in itself, a very broad issue with applications in many different fields, each of which demands consideration.

The assumption underlying our definition of what constitutes the 'opinion-leading press' is that they may be considered to be exponents of the 'media arena' in each country. Across Europe, certain newspapers and news magazines are identifiable as opinion-leading sources of information, and have assumed this status in relation to other media, to important decision-makers (such as politicians, civil servants, experts and industrialists), as well as to the general public. So, by analysing the opinion-leading press, we can realistically expect to gain an accurate impression of the social dynamics of information processing relating to modern biotechnology. In addition, we will gain insight into the development of these information flows over time. This method of choosing the opinion-leading press of the participating countries automatically leads to the inclusion of a very diverse set of media: in some countries large-circulation dailies, in others small-circulation elite newspapers, and, in still others, news weeklies. Other news media (for example, television) may also have an important opinion-leading function. However, we were able to analyse only print material because, unlike television broadcasts, these are systematically preserved – in paper, microfiche or electronic form – for the entire twenty-four-year period covered by our study. Appendix table 3A.1 contains some key data on the media in our study. This table also incorporates information on the manual or on-line selection procedures that were employed in the various countries. It should be noted that on-line searching with national equivalents for each keyword allows one to locate every article on the subject, whereas manual searching would probably lead to an underestimation of the actual number of articles in that particular period.

Coding frame (variables)

The three types of variables we studied were incorporated in a common coding frame, identical for all countries. Native speakers performed all of the coding tasks. In order to improve the comparability of the coding across countries, the coding frame was discussed thoroughly and revised after a trial coding of substantial numbers of articles in the various countries. Inter-coder reliability was also achieved so as to ensure consistent coding across and within countries. However, this process could not resolve all transnational coding issues. The more substantial differences in the characteristics of the research material in the various countries could not be obviated. All comparisons between countries should be interpreted with this caveat in mind.

The coding frame consisted of registration variables, variables relating to journalistic processes of selection and framing and, finally, variables reflecting an evaluation or judgement of the content of the article. We registered each article that qualified as a unit of analysis according to country and year of publication. Year of publication was then recoded and allotted to one of three equal-size eight-year periods, which we call octades: octade I – 1973–80; octade II – 1981–8; and octade III – 1989–96. These time periods were used to study developments, or trends, in the coverage of modern biotechnology. Because of a lack of clear internal criteria that could be used over all countries simultaneously, this periodisation was based on a criterion external to developments in modern biotechnology (see the twelve country profiles of media coverage of modern biotechnology in Durant, Bauer and Gaskell, 1998).

With respect to the variables relating to journalistic processes of selection and framing, we identified the appearance of manifest descriptions of the risks and benefits of modern biotechnology. These articles then underwent five coding stages. First, their contents were coded according to their inclusion, or otherwise, of thirty-five categories or sub-themes. Then, the coding of article content was further reduced to nine themes: 'Medical', 'Basic research', 'Animal and agricultural', 'Economic', 'Regulatory and policy', 'Ethical', 'Identification', 'Safety and risk' and 'Other themes'. In each article three different thematic codings were permitted. A second coding stage involved classifying articles according to the sorts of actors mentioned. Thirty-five classifications were identified, which were then consolidated to form nine actor groups: 'Scientists', 'NGOs', 'Politicians', 'Industry', 'Ethical committees', 'Media and public voice', 'International actors', 'EU' and 'Other actors'. The category 'Media and public voice' comprises references to modern biotechnology in other media or accounts of public opinion. In each article two actor codings were allowed. The third criterion for coding was the geographical location of the biotechnological activities identified in the articles. Locations (two codes allowed per article) were categorised according to a coding frame that comprised over fifty countries, including all participating European countries and the USA. Next these locations were collapsed into three location categories: 'Own country', 'Other European countries' and 'USA'. Further, the framing of modern biotechnology was coded in terms of the presentation of either progress/utility scenarios (the belief that biotechnology will have positive benefits) or doom scenarios (a pessimistic world-view in which biotechnology is conceived in terms of runaway technology or likened to Pandora's box). The final coding stage was termed 'coder impression of modern biotechnology'; the coder's impressions were rated on two separate five-point scales relating to positive and negative assessments, respectively. These scales were combined and recoded to provide a single score.

Results

Our database of the opinion-leading press in the twelve European countries contains 5,404 articles. In the following sections we will describe these articles in terms of our basic research questions. First, we briefly examine the database (table 3.1). Then we address the 'What?' and 'How?' questions. In doing so, we will describe the major themes relating to modern biotechnology covered in Europe's opinion-leading press (figure 3.1 and appendix table 3A.2), followed by a consideration of the main actors (figure 3.2 and appendix table 3A.3) and the main locations of activity in the European coverage of modern biotechnology (figure 3.3 and appendix table 3A.4). Then we switch to the analysis of the overall negative or positive characteristics of the coverage (table 3.2). Finally, the data from this media study are compared with data on the public perception and public knowledge of modern biotechnology from one of the other modules of this international project (table 3.3).

A first glance at the data

Table 3.1 presents, per country, the number of articles in the database and the relative distribution of these articles over the three octades. Looking at the table, we notice that, for most countries, more articles were found in octade II (1981–8) than in octade I (1973–80), and more in octade III (1989–96) than in octade II. This may suggest an increase in intensity in the coverage of the subject of biotechnology across Europe in the twenty-four-year period under study. Of course, we should bear in mind that a degree of sampling bias may have distorted these figures (see also appendix table 3A.1). However, if we take into account the different sampling strategies and then calculate – for each country and each octade – the average number of relevant articles in a random issue of each national opinion-leading daily or weekly, indications of an increase in coverage over the octades is reinforced. The exception here is Greece, where the probability of finding an article on biotechnology remains very small throughout the period studied.

Table 3.1 *Number of articles and distribution across Europe by octade, 1973–1996*

Country	Number of articles and type of medium	Number of articles selected per octade and estimated number of articles for a random issue					
		Octade I (1973–80)		Octade II (1981–88)		Octade III (1989–96)	
		Selected	Random issue	Selected	Random issue	Selected	Random issue
Austria	191 (daily)	36	0.05	66	0.14	89	0.32
	111 (weekly)	17	0.04	31	0.09	63	0.30
Denmark[a]	300 (daily)	13	0.01	127	0.05	160	–
Finland	375 (daily)	26	0.01	97	0.04	252	0.10
France[b]	623 (daily)	17	0.05	216	0.13	390	0.63
Germany	418 (daily)	42	0.00	132	0.16	244	0.29
	170 (weekly)	13	0.00	44	0.21	113	0.54
Greece[a,c]	65 (daily)	11	0.00	13	0.01	41	0.00
Italy	340 (daily)	51	0.00	128	0.07	161	–
Netherlands[c]	1,119 (daily)	5	0.01	184	0.13	930	0.37
Poland	132 (daily)	14	0.03	53	0.04	65	0.16
	76 (weekly)	31	0.07	16	0.04	29	0.07
Sweden	734 (daily)	99	0.04	254	0.10	381	0.15
Switzerland	211(daily)	17	0.05	61	0.18	133	0.39
United Kingdom	539 (daily)	107	0.67	156	1.01	276	2.21
Total	5,404						

Notes: (–) cannot be computed, or insufficient data.
[a] Two dailies; no correction applied.
[b] In octade 1973–80 only 1975 studied.
[c] In octade 1973–80 only four years.

Main themes in the press coverage of modern biotechnology across Europe

Figure 3.1 and appendix table 3A.2 present the distribution of the main themes in the coverage of modern biotechnology across Europe. Figure 3.1 shows, for each country, the frequency with which the eight basic themes arise ('Medical', 'Basic research', 'Economics', 'Animal and agricultural', 'Ethics', 'Regulatory and policy', 'Identification', and 'Safety and risk'). Appendix table 3A.2 contains information on the six most commonly mentioned themes, and has been subdivided into 'risk-only', 'benefit-only' and 'risk and benefit' articles (articles in which neither risks nor benefits are mentioned are ignored in this table).

As we can see from figure 3.1, in most countries 'Medical' and 'Basic research' issues are the most frequently cited themes in the press coverage. 'Regulatory' and 'Animal and agricultural' issues follow, and then 'Economic' and 'Ethical' issues. 'Animal and agricultural' issues are relatively important in Finland, the Netherlands and Poland, and are accorded considerable attention in Denmark, but they were rarely mentioned in Greece. 'Regulatory' issues are discussed relatively frequently

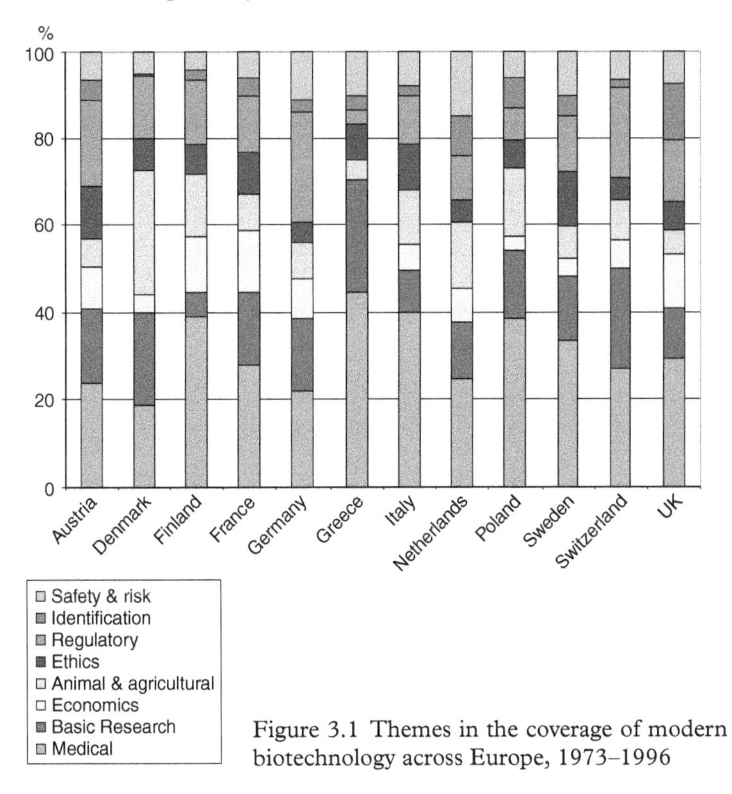

Figure 3.1 Themes in the coverage of modern biotechnology across Europe, 1973–1996

in Switzerland and Germany, but, again, the Greek press seldom broaches the issue. 'Economic' issues are comparatively important in the opinion-leading newspapers of France and Finland, but this issue is barely addressed in Poland and Denmark and is entirely neglected in Greece. 'Ethical' issues receive relatively higher coverage in the Austrian, Swedish, French, Greek and British opinion-leading press.

Table 3A.2 identifies the six most frequently encountered themes (in order of frequency, these are 'Medical', 'Basic research', 'Regulatory and policy', 'Animal and agricultural', 'Economics' and 'Ethics'), which collectively comprise more than 87 per cent of the total. Overall, these six themes are the most frequently observed in all octades. (One exception to this general rule is that the theme of 'Safety and risk' ranks third in the first octade, with 14 per cent, but only seventh in the cumulative ranking, with 7 per cent). In addition, this table shows the relative emphasis of press coverage on the issue of risks versus benefits (coded into three categories: 'risk only', 'benefit only' or 'mixed risk and benefit').

As one might expect, another general rule is that 'Medical' issues and 'Basic research' are presented with greatest frequency in 'benefit-only' articles. The relationship between risk–benefit assessment and the theme of 'Animal and agricultural' issues is more complex. In some countries (Finland, France and Italy), 'benefit-only' articles predominate in this context; in others, 'Animal and agricultural' issues are found almost as much in other types of article as well (Sweden, the Netherlands and, most clearly, Denmark). In eight of the countries 'Regulatory and policy' issues are reported most frequently as 'risk only', but in France, the UK, Italy and Poland these issues appear most frequently in 'mixed risk and benefit' articles. In general, 'Ethical' issues are found most commonly in 'risk-only' articles. Throughout Europe, 'Economic' issues appear most frequently in 'benefit-only' articles, though in Denmark they can also appear in 'mixed risk articles'.

Dynamics in the coverage of themes Looking at the dynamics in the reporting of the six main biotechnological themes across Europe, a fairly consistent picture emerges. The theme of 'Basic research' declined in importance over the entire twenty-four-year period. The only – partial – exceptions are Poland, which accorded it most attention in octade II, and Greece, where 'Basic research' was ascribed most significance in octade III. The situation for the treatment of 'Medical' issues is a little more complex. In several countries (Austria, Finland, Germany, Switzerland and the UK) there was a decrease in the relative treatment of 'Medical' issues over the whole twenty-four-year period. In other countries, however, there was an overall increase (Denmark, France, Poland and Sweden). In Greece and the Netherlands, 'Medical' issues received a similar proportion of the biotechnology coverage, but with strong increases and decreases, respectively, in octade II. In most countries, there was a relative increase in the attention accorded to 'Animal and agricultural' issues between octade I and octade II. In Poland and the UK this was followed by a decrease during octade III. Changes in the treatment of 'Regulatory' issues may be separated into three clusters. In some countries, the proportion of coverage was relatively constant over the three octades (Finland, Germany, Italy, Poland, Sweden and the UK); whereas in others the attention devoted to these issues either diminished (France, Greece and the Netherlands) or increased (Austria, Denmark and Switzerland). The same can be said for the coverage of 'Ethical' issues. In some countries interest in this aspect of the biotechnology debate did not appreciably change during the three octades (Austria, Germany, Greece, Italy and Switzerland); in other countries the attention devoted to these issues either diminished (Denmark and Poland) or increased (Finland, France, the Netherlands, Sweden and the UK). 'Economic' issues received relatively more attention during our research period in Finland, France, Germany, the Netherlands and the UK, but less so in Austria.

For the third octade we studied the intercorrelations among the various themes across the participating countries. The results of this analysis indicate that there is no significant correlation between the occurrence of the themes 'Basic research', 'Economic' issues, 'Ethical' issues or 'Animal and agricultural' issues. There is, however, a negative correlation between 'Medical' and 'Regulatory' issues (Spearman rank order correlation $-.69$, $p < .005$). This indicates that, in countries with a relatively high coverage of 'Medical' issues, relatively few articles were found on 'Regulatory' issues.

Main actors

Figure 3.2 and appendix table 3A.3 show which actors or spokespersons played a role in the coverage of modern biotechnology across Europe. Figure 3.2 presents for each country the percentages of the observations made by the nine actor groups: 'Scientists', 'Industry', 'Politicians', 'NGOs', 'Media and public voice', 'Ethical committees', 'International actors', 'EU' and 'Others'. Table 3A.3 contains the data for only the four most frequently mentioned actor groups, subdivided into 'risk-only', 'benefit-only' and 'mixed risk and benefit' articles (articles in which neither risks nor benefits are mentioned were not included in this table).

In most countries, 'Scientists' and 'Industry', groups that focus primarily on the beneficial aspects of modern biotechnology, constitute more than half the references to actors. In Poland, Finland, and Greece, they make up at least 75 per cent of the references. Yet, in Denmark, Switzerland, Austria, the Netherlands and Sweden, these two actor groups enjoy considerably less prominence. In Denmark and Switzerland, the

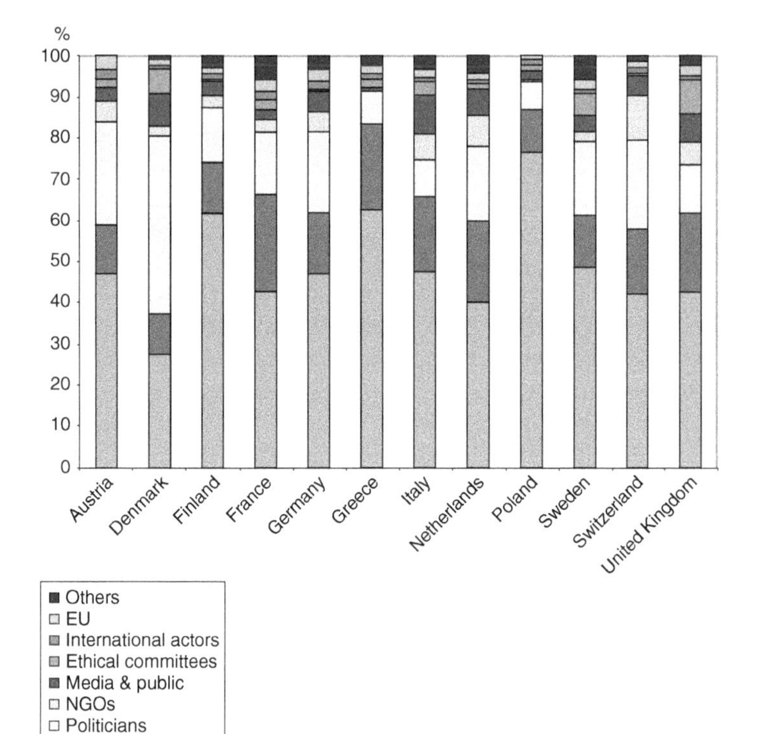

Figure 3.2 Actors in the coverage of modern biotechnology across Europe, 1973–1996

views of 'Politicians' and 'NGOs' rank highest, whereas in Finland, Greece and Poland these two actor groups receive considerably less attention. 'Media and public voice' is relatively important in Italy (10 per cent), Denmark (8 per cent) and the UK (7 per cent), but is hardly discernible in Poland, France and Greece (2 per cent or less). 'Ethical committees' gained a relatively high degree of coverage in the UK (9 per cent), Denmark (6 per cent) and Sweden (5 per cent); but in other countries, coverage of the views of such groups is insubstantial (3 per cent or less). Similarly, 'International actors' and the 'EU' receive little attention as actor groups, with the views of the EU achieving most attention in Austria (3 per cent).

Table 3A.3 identifies four frequently encountered actor groups ('Scientists', 'Politicians', 'Industry' and 'NGOs'), which together comprised more than 85 per cent of the total coverage. 'Scientists' take up more than 46 per cent of all codings, followed by 'Politicians' (18 per cent), 'Industry' (17 per cent) and 'NGOs' (about 5 per cent). In general, these four actor groups are commonly observed in all octades. The remaining, less prominent, actor groups have not been included in this table. In most countries, 'Scientists' are the most frequently cited actor group in 'benefit-only' and 'mixed risk and benefit' articles. The views of industry spokespersons and 'Politicians' are also, though less commonly, found in this context. In Poland, more than three-quarters of all identified actors are 'Scientists', and in many other countries 'Scientists' represent by far the largest single group of newspaper commentators. In contrast, in Denmark the 'Politicians' have assumed this status irrespective of the type of article.

In 'risk-only' articles the relative importance of the different actor groups is slightly different. In a number of countries, 'Scientists' again comprise the main actor group (Finland, France, the UK, Italy and Poland), but in other countries 'Politicians' are the largest group (Sweden, Switzerland and Denmark) or share this rank with 'Scientists' (Greece, Austria, the Netherlands and Germany). The journalistic selection and framing processes regarding 'NGOs' are also quite complex. In some countries, 'NGOs' are important actors in 'risk-only' and 'mixed risk and benefit' articles (most clearly so in France, Austria, the Netherlands, Germany, the UK and, perhaps, Italy), whereas in others they are most strongly associated with 'risk-only' stories (Sweden and Switzerland). In other countries (Finland, Greece, Poland and, perhaps surprisingly, Denmark), NGOs do not seem to play a very significant role (in terms of numbers involved).

Dynamics in references to actor groups Looking at the dynamics of actor-group coverage across Europe, the following picture emerges. The relative importance of 'Scientists' diminished in almost all countries (their status was stable in Poland, and there was an increase in Sweden). Over the entire twenty-four-year period, the relative prominence of 'Politicians' increased in many countries (Austria, Denmark, Finland, France, Italy, Switzerland and the UK), but stayed reasonably stable in Germany, the Netherlands, Poland and Sweden. The role of 'Industry' spokespersons in the coverage of modern biotechnology increased in almost all countries from the first to the second octade (Denmark, Finland, France, Germany, Greece, Italy, Poland, Switzerland and the UK); in many countries this was followed by a decrease during octade III. In Sweden, however, an incremental decrease in the attention accorded to 'Industry' spokespersons is discernible throughout the twenty-four-year period.

Relations between main actors and themes To look for correlations between the coverage of the main actors and the main themes in the coverage of modern biotechnology across Europe we calculated rank order correlations between these variables. We confined ourselves here to octade III, which, for most countries, contained sufficient numbers of observations of each variable to permit analysis.

A negative correlation between the occurrence of 'Politicians' and 'Industry' actor groups (Spearman rank order correlation −.79, $p < .001$) indicates that, in countries in which 'Politicians' were regular commentators in the coverage of modern biotechnology, fewer articles were to be found articulating the views of 'Industry' spokespersons. A similar relation pertains between 'Scientists' and 'NGOs' (−.77, $p < .005$), and there is a positive correlation between 'Politicians' and 'NGOs' (.61, $p < .05$). This indicates that, in countries in which 'NGOs' enjoyed little coverage, 'Politicians' also received fewer mentions, but references to 'Scientists' were proportionately greater. Regarding themes and actors, significant relationships were found between 'Medical' issues and 'Scientists' (.64, $p < .05$), 'Politicians' (−.72, $p < .005$) and 'NGOs' (−.69, $p < .01$). In countries characterised by many articles on 'Medical' issues, more 'Scientists' tended to serve as actors, but relatively few 'Politicians' or 'NGOs'. There are also significant correlations between 'Regulatory' issues and particular actor groups: 'Industry' (−.50, $p < .05$), 'Politicians' (.71, $p < .005$) and 'NGOs' (.54, $p < .05$). These correlations indicate that, in countries that accorded a relatively high degree of attention to 'Regulatory' issues, 'Industry' spokespersons received relatively little attention and, conversely, 'Politicians' and 'NGOs' enjoyed a higher proportion of the coverage.

Main locations of activity of modern biotechnology across Europe

Figure 3.3 and appendix table 3A.4 display the results of our assessment of the 'locations of activity' for modern biotechnology identified in the European press (the categories were 'Own country', 'Rest of Europe', 'USA', 'Other' and 'None mentioned'). Overall, 94 per cent of the articles contained at least one reference to a location of activity for modern biotechnological activity, and many articles also included a second reference. Of these, 46 per cent were made to 'Own country', which is by far the largest category. 'USA' was cited in 22 per cent of all observations and 'Rest of Europe' in 18 per cent; 'other locations' comprised 8 per cent of references. This pattern is observable throughout Europe, with the exception of Greece, in which only 13 per cent of references concerned 'Own country', with 'Rest of Europe' (40 per cent) and 'USA' (34 per cent) exceeding this proportion. Italy follows the general trend, but less conspicuously: only 32 per cent of references were made to 'Own country' whereas slightly more than 30 per cent of references were made to 'USA'.

Table 3A.4 identifies the three most important 'locations of activity' in terms of the appearance of 'risk-only', 'benefit-only' and 'mixed risk and benefit' articles. In many countries, reference to 'Own country' tended to be made in the context of 'risk-only' and 'mixed risk and benefit' articles. The exception here is again Greece. Concerning 'benefit-only' articles,

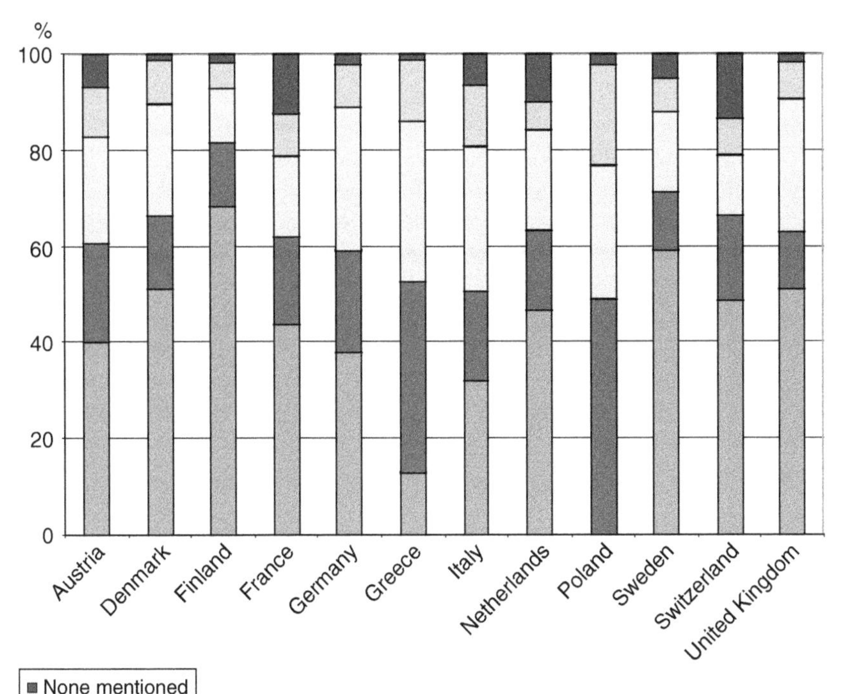

Figure 3.3 Reported locations of modern biotechnology activity across Europe, 1973–1996

'Own country' is, again, the most frequently cited 'location of activity'; however, in several countries 'USA' is the most frequently mentioned location (Germany, Greece and Italy).

Dynamics in references to locations of activity Over the twenty-four years studied, references to 'Own country' increased in many countries (Austria, Denmark, Finland, Germany, the Netherlands, Switzerland and the UK) and decreased in a few others (Greece and Sweden, the latter of which had a relatively high proportion of references to 'Own country' in octade I). In the first set of countries, during this period one can also observe a decrease in the proportion of citations of 'USA' as 'location of activity' for modern biotechnological research (with the exception of Finland). Relatively more attention was accorded to 'USA' in France, Italy, Poland and Sweden. The proportion of references to 'Rest of Europe' varies, with more references being made to this location in Denmark, Greece, the Netherlands and Sweden than in Austria, France, Italy, Poland, Switzerland and the UK.

Relations between locations of activity, actors and themes In looking for relationships among the coverage of 'locations of activity', the main actor groups and biotechnological themes, we calculated rank order correlations. Again, we confined our analysis to the third octade.

This analysis indicates that references to 'Own country' as the 'location of activity' correlate negatively with references to 'Rest of Europe' (−.98, $p < .001$) and 'USA' (−.74, $p < .005$). References to 'Rest of Europe', however, correlate positively with mentions of 'USA' (.67, $p < .05$). This indicates that, in countries in which references to 'Own country' were frequently made, few references to other countries were found. There is also a significant relationship between the mention of 'Economic issues' and references to 'Own country' (.65, $p < .05$). Among the remaining themes and actor groups, no significant correlations with 'location of activity' were found.

The overall negative or positive characteristics of modern biotechnology

Table 3.2 contains the data on journalistic framing in terms of: progress versus doom scenarios; the coder's overall impression of the coverage; and the proportions of 'risk-only' and 'benefit-only' articles per country. These data give an impression of the framing of modern biotechnology in the European opinion-leading press.

Overall, the framing of modern biotechnology in terms of 'progress/utility' is much more commonplace than framing in terms of doom scenarios (runaway technology, Pandora's box, etc.). In 42 per cent of articles, 'progress/utility' was emphasised (reversed ranking in table 3.2), whereas in only 5 per cent of articles was doom predicted. In Italy, Greece and Poland the 'progress/utility' scenario was especially dominant; though in Denmark, Austria, Sweden and (somewhat surprisingly) Greece, we observed a relatively high proportion of articles anticipating disaster.

Table 3.2 also displays the combined measure of the coder's impression of the positive versus negative treatment of modern biotechnology. In most countries this combined measure results in a positive overall impression, with Finland, Greece, Italy and Poland having the most positive overall evaluations. In Denmark and Sweden, however, the general assessment is negative.

With respect to the proportions of 'risk-only' and 'benefit-only' articles (reversed ranking in table 3.2) a fairly consistent picture emerges. Across Europe, 11 per cent of the articles are coded as 'risk only' and 42 per cent

Table 3.2 *Journalistic framing of modern biotechnology in the opinion-leading press across Europe, 1973–1996*

	Doom scenario		Progress/utility scenario		Coder impression		Risk-only articles		Benefit-only articles		
Country	%	Rank	%	Rank (reversed)	Mean	Rank	%	Rank	%	Rank (reversed)	Overall rank
Austria	8	3[a]	46	6	0.8	6	13	5	51	7	6
Denmark	12	1	25	2	−0.1	2	19	1[a]	21	1	1
Finland	4	9	51	8[a]	1.3	9[a]	5	11	53	8	10
France	6	6	37	4	0.6	5	11	6	43	6	5
Germany	5	7[a]	49	7	0.9	7	6	9[a]	56	9[a]	8
Greece	9	2	72	11	1.3	9[a]	6	9[a]	63	11[a]	11
Italy	7	5	64	10	1.9	11	18	3	42	5	7
Netherlands	2	10[a]	22	1	0.3	4	9	7	29	3	4
Poland	5	7[a]	74	12	2.7	12	2	12	63	11[a]	12
Sweden	8	3[a]	42	5	−0.2	1	16	4	36	4	3
Switzerland	2	10[a]	29	3	0.0	3	19	1[a]	27	2	2
United Kingdom	2	10[a]	51	8[a]	1.2	8	6	9[a]	56	9[a]	9
Overall	5		42		0.7		11		42		

[a] Tied ranks.

as 'benefit only', indicating a generally positive attitude towards modern biotechnology. Denmark and Switzerland fail to conform to this trend, both countries exhibiting a relatively high proportion of 'risk-only' articles and a relatively low proportion of 'benefit-only' articles. In Finland and Poland this situation is reversed; here we observe a relatively low proportion of 'risk-only' articles and a relatively high proportion of 'benefit-only' articles. The proportion of 'risk-only' articles varies from 2 per cent (in Poland) to 19 per cent (in Switzerland and Denmark); the incidence of 'benefit-only' articles varies from 21 per cent (in Denmark) to 63 per cent (in Poland and Greece).

If the national data from table 3.2 are assigned to different groups according to their overall level of negative/positive coverage of modern biotechnology, one finds a clear affinity among four of the five indicators: 'progress/utility scenarios', 'coder impression', 'risk-only' articles and 'benefit-only' articles. The Spearman rank order correlations among these variables are all significant and are between .62 ($p < 0.05$) and .89 ($p < .001$). None of the correlations between the four variables and 'framing of doom' is, however, significant. Most of the twelve countries display a similar pattern with respect to these four indicators (only the Italian ranking on coder impression is inconsistent with the other rankings). Looking at the overall ranking based on these indicators, we conclude that in Denmark the coverage was coded as most negative, followed by Switzerland and Sweden. Overall, the most positive coverage was found in Finland, Greece and Poland.

Dynamics in negative or positive coverage of modern biotechnology

Looking at the dynamics in the framing of risk and benefit information, we observe that, for the transition from octade I to octade II, five countries show an increase in 'risk-only' articles and four countries a stationary situation (not shown in table). In three countries a decline in the proportion of 'risk-only' articles can be observed over these years. Regarding the 'benefit-only' articles, an increase in ten countries can be observed, but only in six of these countries (Sweden, France, Austria, Germany, the UK and Poland) was this trend accompanied by a decrease, or no change, in the number of 'risk-only' articles over the same period. This result is in accordance with the coder impression data, which suggest that, in the transition from octade I to octade II, more positive coverage appeared in the UK, France, Germany and Sweden, but it remained more or less the same as before in Austria and Poland. Combined, these results suggest that, at least in these six countries, the overall presentation of modern biotechnology became more positive between 1973 and 1988.

However, looking at the framing of risk and benefit in the transition from octade II to octade III, a different picture emerges. At this stage, many countries show either an increase in the proportion of 'risk-only' articles accompanied by a relative decrease in 'benefit-only' articles (Switzerland, Austria, Germany, the UK and Denmark), or no change in the proportion of 'risk-only' articles accompanied by a decrease in the proportion of 'benefit-only' articles (France and Italy). In six of these seven countries, this picture is reinforced by evidence of a change to a more negative coder impression of modern biotechnology from octade II to octade III (the exception being Germany, where a change to a more positive evaluation is observed). Overall, then, these results suggest the emergence of a more negative framing of modern biotechnology in newspaper reports over this period. In only three countries do we observe the reverse process: a decrease or absence of change in the proportion of 'risk-only' articles and an increase in 'benefit-only' articles (Finland, Greece and Sweden). For Greece and Sweden this is supported by a more positive coder impression. In these two countries alone, it would seem that a more positive general impression of modern biotechnology appeared between 1981 and 1996 in the opinion-leading press.

Media coverage and public perception and knowledge of modern biotechnology

The connections between the media coverage of modern biotechnology and the study of public perceptions of this technology can, of course, be conceived in a variety of ways. It is reasonable to expect the way biotechnology is portrayed in the media to have an effect on public perceptions, especially since few other sources of information are available. But it is also clear that the media tend to adapt their coverage to the perceived attitudes of their readership. Indeed, one may argue that causality flows both ways. So far in this chapter, we have examined the media coverage alone, but in the project as a whole we have also collected detailed information on the public perception and knowledge of modern biotechnology in the various countries covered in this media study (see chapters 4, 5, 11, 12 and 13).

For octade III, taking our data on the relationships among themes, actor groups, 'locations of activity' and evaluations of risks versus benefits (see above and table 3.2), we performed an additional analysis. For all these variables, rank order correlations were computed with data relating to the public perception of modern biotechnology across Europe (combined attitude score for encouragement of six applications of biotechnology)

Table 3.3 *Spearman rank order correlations across countries between public perception and knowledge, with positive or negative overall characteristics of the coverage of modern biotechnology, and themes, actor groups and location of activity*

	Public perception	Public knowledge	Characteristics of media coverage
Public perception	–	n.s.	.53*
Public knowledge	n.s.	–	−.63*
Main themes			
Basic research	n.s.	n.s.	n.s.
Medical issues	.69*	n.s.	.51*
Economic issues	n.s.	n.s.	n.s.
Ethical issues	n.s.	n.s.	n.s.
Regulatory & policy issues	−.78**	n.s.	n.s.
Animal & agricultural issues	n.s.	n.s.	n.s.
Main actor groups			
Scientists	n.s.	n.s.	.61*
Industry	.72**	n.s.	n.s.
Politicians	−.66**	n.s.	−.71**
NGOs	n.s.	n.s.	n.s.
Main location of activity			
Own country	n.s.	.53*	n.s.
Rest of Europe	n.s.	−.56*	n.s.
USA	n.s.	n.s.	n.s.

* $p < .05$; ** $p < .01$
n.s. = not significant.

and public knowledge (based on a nine-item knowledge scale from Eurobarometer 1996). Table 3.3 summarises these correlations. The first observation from this analysis is that a positive correlation exists between the public's attitude towards modern biotechnology and the overall positive or negative characteristics of the media coverage (Spearman rank order .53, $p < .05$). This suggests that countries in which the public is positive about biotechnology also tend to have a positive media coverage, and vice versa. In addition, there is a relationship between the public's knowledge of biotechnology and the media's relative emphasis upon risks and benefits: in countries with the more negatively framed coverage, the public has more knowledge about modern biotechnology (−.63, $p < .05$); and in countries with a positive media framing, the public has less knowledge (see chapters 7 and 8 for more detailed information). Figure 3.4 summarises the relationships between attitude and knowledge across Europe and positive/negative coverage (countries in figure 3.4 are ordered on the

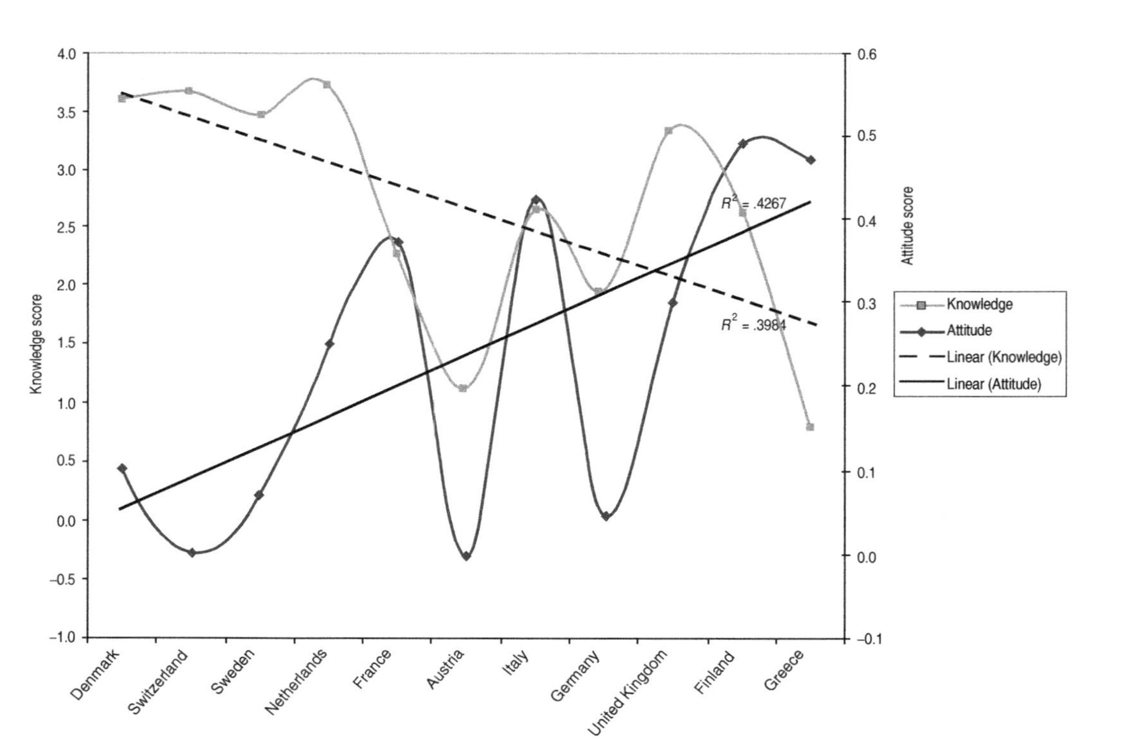

Figure 3.4 Relation between negative or positive coverage of modern biotechnology and public attitude and knowledge across Europe

Note: The countries are ordered from left to right according to their overall negative or positive coverage (Denmark has the most negative coverage; Greece has the most positive coverage).

x-axis from left to right according to their overall ranking of negative versus positive coverage; see table 3.2). Figure 3.4 also contains linear regressions of these relations indicating moderately strong relations ($R^2 = .43$ for the relation between attitude and negative/positive coverage, and $R^2 = .40$ for knowledge).

Table 3.3 indicates that the public perception of modern biotechnology in a country is more positive when the opinion-leading media pay a considerable amount of attention to 'Medical' issues, and relatively little to 'Regulatory' issues. In addition, these data suggest that public attitudes towards biotechnology are more positive when media coverage accords the views of 'Industry' spokespersons a high profile, and when it is rather less inclined to quote the views of 'Politicians'. Yet there are no significant correlations between the frequency of references to particular actor groups and public knowledge about modern biotechnology; and there is no obvious relationship between public attitudes and the cited 'location of activity'. The overall positive or negative characteristics of the coverage of modern biotechnology appear not to correlate systematically with the occurrence of the various major themes or the location of activity. Only attention to 'Medical' issues is an exception here. There is also a significant positive correlation with references to 'Scientists' and a negative correlation with 'Politicians', indicating that, in countries where the overall characteristic of the coverage of modern biotechnology is more positive, relatively more references to 'Medical' issues or 'Scientists' appear in the opinion-leading press, and relatively fewer references to 'Politicians'.

Discussion and conclusions

This study started with several goals in mind. In the first place we wanted to analyse systematically the dynamics involved in the European opinion-leading press coverage of modern biotechnology over the past twenty-four years. A second goal of our media study was to attune our analyses with other modules of the European Biotechnology project, namely the analysis of developments in modern biotechnology policy across Europe and studies of European public opinion towards modern biotechnology as assessed by the 1996 Eurobarometer Biotechnology survey. For two main reasons modern biotechnology has been conceptualised rather broadly in this media study: the technology domain itself has potential applications in many different fields; and we wanted national issues and particulars to be included within the study. Our final database consists of more than 5,400 articles from twelve European countries, which were selected from the original print material or more sophisticated storage media, and then coded and assessed according to a common coding frame. The participating countries provide an excellent coverage of the entirety of Europe, from the north to the south, and from the east to the west.

This procedure allowed us to apprehend the changing content and tone of the coverage of modern biotechnology within the various European countries, while being sensitive to the differences and similarities among these countries. As far as we are aware, this is the most substantial study of the media coverage of modern biotechnology so far undertaken. But, having said this, some additional remarks about the research, and its limitations, are due. Electing to study only the opinion-leading press in the various European countries means that our data sets contain partial incompatibilities, because in some countries major daily newspapers were analysed, whereas in others newspapers as well as news magazines were included, and in some countries only elite newspapers were selected. It should also be repeated that, for practical reasons, no attempt was made to incorporate material from other opinion-leading media, such as television.

Examining the results of this study, it is immediately obvious that the entire twenty-four-year period saw a clear increase in the attention devoted to modern biotechnology in most European countries (most of the analyses were performed at a country level). A second important conclusion is that, in general, the coverage of modern biotechnology is fairly positive. On the other hand, the countries studied vary – sometimes sharply – in terms of negative versus positive journalistic framing. In this investigation, we used several indicators to evaluate journalistic framing: whether articles invoke 'doom' or 'progress/utility' scenarios; their relative emphasis upon risks and benefits; and a subjective measure of coder impression. With the exception of the invocation of 'doom scenarios', each of these indicators has a high rank order correlation, indicating that these countries can be placed on a single scale from the positive (with Poland, Greece and Finland as exemplars) to the more negative framing of modern biotechnology (typified by Denmark, Switzerland and Sweden). Because 'doom scenarios' did not correlate with the other indicators, they were not used in ranking the various countries.

Another conclusion is that the European opinion-leading press pays considerable attention to 'Medical' issues related to modern biotechnology as well as to 'Basic research'. Indeed, the relatively positive framing of the coverage in some countries is correlated with a high proportion of press articles dealing with 'Medical' issues. This indicates that a major part of the coverage this technology receives emphasises the potential benefits for human health and progress that its further development may provide. In another study, it was found that US newspapers also tend to give much attention to the salutary aspects of modern biotechnology (Hornig, 1994). So, in this respect, the overall European and US media approaches exhibit clear similarities.

Over the whole twenty-four-year period, the attention devoted to 'Basic research' diminished. However, the dynamics involved in the coverage of 'Medical' issues are more complicated, with some countries displaying an increase in the press coverage of this aspect, and others a decrease. The European opinion-leading press also pays attention to 'Regulatory and policy' issues, but these were accorded less attention than either 'Medical' issues or 'Basic research'. 'Regulation and policy' is a more prominent issue in those countries in which the press coverage of biotechnology is mostly negatively framed; these questions were usually addressed in the context of elucidating the potentially disadvantageous implications of modern biotechnology, and of conveying to the public that biotechnology may encompass unwanted developments and that prudence is imperative. Over the whole research period we also observed an increase in the attention devoted to 'Animal and agricultural' issues in many countries, presumably owing to the fact that this area of biotechnology has experienced significant growth in the twenty-four years under study. Considerably less attention was given to either 'Ethical' or 'Economic' issues, and the dynamics of the coverage of these issues do not present a very clear picture.

So, in conclusion, the hypothesis put forward by some biotechnology proponents – that the European opinion-leading press has painted a dire picture of this industry – is at odds with the facts. Overall, coverage has been positive and varied in its themes. On the other hand, the press seems to have failed to communicate the importance that policy-makers attach to economic issues, and has neglected the ethical issues that are very important to the general public throughout Europe.

In the European opinion-leading press, 'Scientists' and 'Industry' spokespersons are the most frequent sources of information and opinions, followed at a certain distance by actors from the political or societal domain, such as 'Politicians' or 'NGOs'. Again, here there is a parallel between the European situation and the results from an earlier study carried out in the USA (Hornig, 1994), as well as from the more recent study by Kohring, Goerke and Ruhrmann (1999), which incorporated both US and European data. Among the European countries, however, differences in the relative prominence of the various actors' groups can be found. Particularly noteworthy is the finding that, in countries in which the coverage of modern biotechnology is framed in a positive manner, 'Scientists' enjoyed more coverage than 'Politicians'. Conversely, in countries with a more negative framing, we observed more references to 'Politicians' than to 'Scientists'. In addition, over the entire twenty-four-year period most countries witnessed a decrease in the input of 'Scientists' as opinion-leading press sources, in favour of 'Industry' and 'Politicians'. Most references to the locations of activity of modern biotechnology were to 'Own country', followed by 'USA' and other European countries.

The 'Biotechnology and the European Public' project offers the opportunity of analysing the relationship between, on the one hand, public perception and public knowledge about modern biotechnology and, on the other, the results of our media analysis. These analyses are informed by the notion that media coverage of a particular issue, through processes of selection or framing, may have an impact upon what the public thinks or knows about that issue. In this study we did observe a clear relation between the positive/negative framing in the various countries and the public perception of modern biotechnology, as assessed by the Eurobarometer Biotechnology 1996 (see 'Biotechnology and the European Public', 1997). In countries with a relatively negative media framing of biotechnology, the public perception of this new technology was also more negative; whereas in countries with a more positively framed coverage the public's attitude was more positive. Negative or positive framing also correlates with the level of public knowledge about modern biotechnology. However, in this case, greater public knowledge was found in those countries in which the opinion-leading press negatively framed the coverage of modern biotechnology; and lower levels of knowledge were found in countries in which a more positive framing occurred. Further studies will have to establish the determinants of these relationships.

A final conclusion of our study is that journalistic selection and framing processes play an important role in determining how the European opinion-leading press covers the many facets of modern biotechnology, and that these processes may reflect varying national sensibilities. Studying the media coverage of biotechnology, we frequently encountered differences in the national treatment of this subject. The level of the coverage itself, the emphasis upon the different themes, the reliance placed upon various actors and stakeholders, and the positive/negative framing of this technology all exhibit considerable inter-national heterogeneity. Moreover, in this respect, our finding that there is a correlation between the nature of framing and the inclusion of certain issues in discussions of biotechnology, on the one hand, and the public's perception of, and knowledge about, modern biotechnology, on the other, is especially significant. This finding emphasises the importance of the media coverage of this technology. Although the existence, or otherwise, of a causal relationship between the media coverage of biotechnology and the public's knowledge and opinions of it cannot be deduced from this correlation analysis, finding evidence for causality will be an interesting task for future studies.

Appendix

Table 3A.1 *Main characteristics of the research material*

Country	Newspaper name (and type)	Circulation: population	Estimation of total coverage	Size of sample in study	Sampling procedures	Selection procedures
Austria	*Die Presse* (daily) *Profil* (weekly)	1:100 (*Presse*)	1,396	302 (*Presse* 191; *Profil* 111)	*Presse*: 1973–9, 28% *Presse*: 1980–5, 20% *Presse*: 1986–96, 10% *Profil*: 1973–85, 100% *Profil*: 1986–96, 50%	1973–93 manual 1994–6 on-line
Denmark	*Information* (daily) *Politiken* (daily)	1:34 (*Politiken*)	–	300 (*Information* 168; *Politiken* 132)	1973–89, 1996, 100% 1990–5 (only large articles)	1973–89 manual, index 1990–6 on-line
Finland	*Savon sanomat* (daily)	1:55	375	375	1973–96, 100%	Manual
France	*Le Monde* (daily)	1:179	1,483	549	1975, 1982–6, 100% 1987–96, 25%	1973–86 manual, index 1987–96 on-line
Germany	*Frankfurter Algemeine Zeitung* (daily) *Der Spiegel* (weekly)	1:162 (*FAZ*)	1,594 (*FAZ* 1,254; *Spiegel* 340)	588 (*FAZ* 418; *Spiegel* 170)	*FAZ*, 33% *Spiegel*, 50%	Manual
Greece	*Kathemerini* (daily) *Eleftherotypia* (daily)	1:283 1:89	88	65 (*Kathemerini* 16; *Eleftherotypia* 49)	1973–85, every 2nd yr, 100 days/year 1986–96, 100%	Manual

Italy	*Corriere della sera* (daily)	–	936	340	1973–86 (est.), 100% 1987–96, 20 days/year	1973–84 manual 1984–91 on-line
Netherlands	*Volkskrant* (daily)	1:41	1,185	1,119	1973–85, every 2nd yr, 50% 1986–96, 100%	1973–92 manual 1993–6 on-line
Poland	*Polityka* (weekly) *Rzeozpospolita* (daily) *Trybuna ludu* (daily)	1:200	113 (*Polityka*)	254 (*Polityka* 76; *Rzeozpospolita* 160; *Trybuna ludu* 18)	*Polityka*, 100% Others, 1 day/week	Manual *Trybuna ludu* 1973–82 *Rzeozpospolita* 1983–96
Sweden	*Dagens Nyheter* (daily)	1:21	734	734	1973–96, 100%	1973–92 manual 1992–6 on-line
Switzerland	*Neue Zürcher Zeitung* (daily)	1:45	1,537	211	1973–96, 14%	Manual
United Kingdom	*Times* (daily) *Independent* (daily)	1:83 1:226	5,471	539 (*Times* 256; *Independent* 283)	1973–87, 1996, 20 day/year 1988–95, 15 day/year	*Times* 1973–87, *Independent* 1988–96 1973–80 manual, index, 1981–96 on-line

Table 3A.2 *Main themes in the coverage of modern biotechnology across Europe: trends, 1973–1996*

		Theme																	
		Basic research			Medical			Animal/Agricultural			Regulatory			Ethical			Economic		
Country	Type of article	%	I–II	II–III	%	I–II	II–III	%	I–II	II–III	%	I–II	II–III	%	I–II	II–III	%	I–II	II–III
Austria	Risk only	3	↓	↔	17	|	↓	4	)	↔	32	↑	↑	30	↔	↓	0	↔	↔
	Benefit only	28	|	)	32	↔	↔	5	↔	↔	13	↔	↓	2	↔	↔	14	↔	↓
	Risk & benefit	6	↓	↔	14	↑	↓	9	)	↔	17	)	↑	25	|	↓	9	↔	|
	Observations	*98*	*21%*		*110*	*24%*		*30*	*7%*		*90*	*20%*		*56*	*12%*		*43*	*9%*	
Denmark	Risk only	14	↑	↔	25	↓	↔	28	↓	↔	17	↑	↔	9	)	↔	0	↔	↔
	Benefit only	30	↑	↔	20	↑	↑	30	↑	↔	10	↑	↓	1	↔	↔	5	)	↔
	Risk & benefit	20	↓	|	16	↔	↑	29	)	↔	15	↑	↔	9	|	↔	5	↔	↔
	Observations	*173*	*21%*		*155*	*19%*		*238*	*29%*		*118*	*14%*		*59*	*7%*		*33*	*4%*	
Finland	Risk only	7	)	↔	31	↓	)	7	)	↔	22	↑	↑	13	↑	↓	2	↔	↔
	Benefit only	4	)	|	42	↔	|	17	|	↔	10	↑	↔	2	↔	↔	17	↔	↑
	Risk & benefit	6	|	|	38	↓	↔	13	↑	↔	18	)	↔	13	)	↔	9	↔	↔
	Observations	*44*	*5%*		*320*	*39%*		*117*	*14%*		*122*	*15%*		*56*	*7%*		*105*	*13%*	
France	Risk only	14	↓	↔	20	↑	)	3	)	↔	12	↓	↔	24	↑	|	5	)	↔
	Benefit only	21	↔	|	28	↔	)	10	↑	↔	11	↓	↔	2	↔	↔	24	↑	|
	Risk & benefit	9	↓	↔	33	↑	|	5	)	↔	16	↓	↔	18	↑	↔	10	)	↔
	Observations	*145*	*17%*		*248*	*29%*		*63*	*7%*		*115*	*13%*		*87*	*10%*		*134*	*15%*	
Germany	Risk only	7	|	↔	13	↔	↔	3	↔	↔	33	↑	↔	11	↔	↑	2	↔	)
	Benefit only	28	)	↓	28	↓	↔	10	)	↔	15	↔	↑	1	↔	↔	11	)	↔
	Risk & benefit	6	|	↔	18	↔	↓	9	↔	)	31	↔	↔	8	)	↔	10	↔	↔
	Observations	*206*	*17%*		*267*	*22%*		*102*	*8%*		*308*	*25%*		*58*	*5%*		*113*	*9%*	
Greece	Risk only	0	↔	↔	33	↑	↑	0	↔	↔	17	↑	↓	33	↓	↓	0	↔	↔
	Benefit only	33	|	↑	52	↑	↓	3	↓	↔	0	↔	↔	2	↔	↔	0	↔	↔
	Risk & benefit	16	↔	↑	32	↑	↓	8	↔	↑	8	↓	↔	20	↑	↓	0	↔	↔
	Observations	*25*	*26%*		*42*	*44%*		*4*	*4%*		*3*	*3%*		*8*	*8%*		*0*	*0%*	

<table>
<tr><td>Italy</td><td>Risk only</td><td>7</td><td>|</td><td>↔</td><td>40</td><td>↔</td><td>↔</td><td>11</td><td>↔</td><td>|</td><td>13</td><td>↑</td><td>↔</td><td>20</td><td>↔</td><td>↔</td><td>4</td><td>↔</td><td>⟩</td></tr>
<tr><td></td><td>Benefit only</td><td>14</td><td>|</td><td>↓</td><td>45</td><td>↔</td><td>↑</td><td>19</td><td>↔</td><td>↔</td><td>5</td><td>↔</td><td>↔</td><td>1</td><td>↔</td><td>↔</td><td>7</td><td>⟩</td><td>|</td></tr>
<tr><td></td><td>Risk & benefit</td><td>7</td><td>|</td><td>|</td><td>36</td><td>↓</td><td>⟩</td><td>8</td><td>⟩</td><td>↔</td><td>14</td><td>⟩</td><td>↔</td><td>15</td><td>↔</td><td>↔</td><td>6</td><td>⟩</td><td>↔</td></tr>
<tr><td></td><td>Observations</td><td>76</td><td>10%</td><td></td><td>307</td><td>40%</td><td></td><td>96</td><td>12%</td><td></td><td>80</td><td>10%</td><td></td><td>8</td><td>1%</td><td></td><td>48</td><td>6%</td><td></td></tr>
<tr><td>Netherlands</td><td>Risk only</td><td>7</td><td>⟩</td><td>↔</td><td>15</td><td>↓</td><td>⟩</td><td>20</td><td>↑</td><td>⟩</td><td>19</td><td>↓</td><td>↔</td><td>19</td><td>↑</td><td>↔</td><td>3</td><td>↔</td><td>↔</td></tr>
<tr><td></td><td>Benefit only</td><td>15</td><td>↓</td><td>|</td><td>36</td><td>↔</td><td>↑</td><td>19</td><td>↑</td><td>↔</td><td>7</td><td>|</td><td>↔</td><td>1</td><td>↔</td><td>↔</td><td>12</td><td>↑</td><td>|</td></tr>
<tr><td></td><td>Risk & benefit</td><td>8</td><td>↑</td><td>↔</td><td>18</td><td>↑</td><td>↔</td><td>25</td><td>↑</td><td>⟩</td><td>14</td><td>↑</td><td>↔</td><td>10</td><td>⟩</td><td>↔</td><td>7</td><td>⟩</td><td>↔</td></tr>
<tr><td></td><td>Observations</td><td>260</td><td>14%</td><td></td><td>498</td><td>27%</td><td></td><td>307</td><td>17%</td><td></td><td>204</td><td>11%</td><td></td><td>103</td><td>6%</td><td></td><td>163</td><td>9%</td><td></td></tr>
<tr><td>Poland</td><td>Risk only</td><td>2</td><td>↑</td><td>↓</td><td>6</td><td>↓</td><td>↔</td><td>0</td><td>↔</td><td>↔</td><td>0</td><td>↔</td><td>↔</td><td>3</td><td>↓</td><td>↔</td><td>0</td><td>↔</td><td>↔</td></tr>
<tr><td></td><td>Benefit only</td><td>43</td><td>↔</td><td>|</td><td>60</td><td>↑</td><td>↑</td><td>42</td><td>|</td><td>↓</td><td>9</td><td>⟩</td><td>↔</td><td>4</td><td>↔</td><td>↔</td><td>16</td><td>↔</td><td>|</td></tr>
<tr><td></td><td>Risk & benefit</td><td>21</td><td>⟩</td><td>↓</td><td>48</td><td>↔</td><td>↑</td><td>30</td><td>↑</td><td>↔</td><td>6</td><td>|</td><td>↔</td><td>18</td><td>↓</td><td>⟩</td><td>2</td><td>↔</td><td>↔</td></tr>
<tr><td></td><td>Observations</td><td>82</td><td>17%</td><td></td><td>185</td><td>39%</td><td></td><td>74</td><td>15%</td><td></td><td>15</td><td>3%</td><td></td><td>25</td><td>5%</td><td></td><td>18</td><td>4%</td><td></td></tr>
<tr><td>Sweden</td><td>Risk only</td><td>8</td><td>↓</td><td>↔</td><td>18</td><td>↑</td><td>|</td><td>6</td><td>⟩</td><td>↔</td><td>20</td><td>↔</td><td>↑</td><td>27</td><td>↑</td><td>↓</td><td>0</td><td>↔</td><td>↔</td></tr>
<tr><td></td><td>Benefit only</td><td>17</td><td>↓</td><td>↓</td><td>55</td><td>↑</td><td>↑</td><td>8</td><td>↔</td><td>↔</td><td>4</td><td>|</td><td>↔</td><td>1</td><td>↔</td><td>↔</td><td>8</td><td>⟩</td><td>↓</td></tr>
<tr><td></td><td>Risk & benefit</td><td>16</td><td>↓</td><td>↓</td><td>28</td><td>↑</td><td>↔</td><td>8</td><td>⟩</td><td>↔</td><td>9</td><td>↔</td><td>⟩</td><td>18</td><td>↑</td><td>↔</td><td>2</td><td>↔</td><td>↔</td></tr>
<tr><td></td><td>Observations</td><td>178</td><td>15%</td><td></td><td>403</td><td>34%</td><td></td><td>90</td><td>8%</td><td></td><td>154</td><td>13%</td><td></td><td>150</td><td>13%</td><td></td><td>45</td><td>4%</td><td></td></tr>
<tr><td>Switzerland</td><td>Risk only</td><td>3</td><td>⟩</td><td>↔</td><td>17</td><td>↑</td><td>↓</td><td>10</td><td>↔</td><td>↑</td><td>41</td><td>↑</td><td>↑</td><td>14</td><td>++</td><td>↓</td><td>1</td><td>↔</td><td>↔</td></tr>
<tr><td></td><td>Benefit only</td><td>32</td><td>↔</td><td>|</td><td>29</td><td>↓</td><td>|</td><td>14</td><td>↔</td><td>|</td><td>8</td><td>↔</td><td>↑</td><td>1</td><td>↔</td><td>↔</td><td>12</td><td>⟩</td><td>↔</td></tr>
<tr><td></td><td>Risk & benefit</td><td>16</td><td>↓</td><td>↑</td><td>25</td><td>↑</td><td>↓</td><td>9</td><td>↑</td><td>↔</td><td>22</td><td>↑</td><td>↔</td><td>12</td><td>↓</td><td>↔</td><td>8</td><td>↑</td><td>↔</td></tr>
<tr><td></td><td>Observations</td><td>111</td><td>23%</td><td></td><td>130</td><td>27%</td><td></td><td>44</td><td>9%</td><td></td><td>130</td><td>27%</td><td></td><td>25</td><td>5%</td><td></td><td>31</td><td>6%</td><td></td></tr>
<tr><td>United Kingdom</td><td>Risk only</td><td>6</td><td>|</td><td>⟩</td><td>21</td><td>↓</td><td>|</td><td>2</td><td>↔</td><td>↔</td><td>22</td><td>↑</td><td>↓</td><td>18</td><td>↓</td><td>↑</td><td>5</td><td>↑</td><td>↓</td></tr>
<tr><td></td><td>Benefit only</td><td>15</td><td>|</td><td>↓</td><td>33</td><td>↓</td><td>↑</td><td>6</td><td>⟩</td><td>|</td><td>9</td><td>↔</td><td>↔</td><td>1</td><td>↔</td><td>↔</td><td>13</td><td>⟩</td><td>↔</td></tr>
<tr><td></td><td>Risk & benefit</td><td>5</td><td>↓</td><td>↔</td><td>20</td><td>|</td><td>↔</td><td>8</td><td>↔</td><td>↔</td><td>23</td><td>↔</td><td>↔</td><td>13</td><td>↔</td><td>⟩</td><td>12</td><td>⟩</td><td>|</td></tr>
<tr><td></td><td>Observations</td><td>106</td><td>12%</td><td></td><td>267</td><td>29%</td><td></td><td>53</td><td>6%</td><td></td><td>127</td><td>14%</td><td></td><td>58</td><td>6%</td><td></td><td>109</td><td>12%</td><td></td></tr>
</table>

Notes: 1. Trends in the coverage of modern biotechnology are indicated in the columns headed I–II and II–III (representing the three octades I:1973–80, II:1981–8, and III:1989–96). The symbols in the columns indicate relative increases (⟩ = 5–10%; ↑ > 10%), relative decreases (| = 5–10%; ↓ > 10%), or an absence of change (↔, < 5%).

2. We present articles containing information only about the risks of modern biotechnology, only about the benefits, or about both risks and benefits. Articles containing neither risk nor benefit information are not included in this table. The percentages indicate, for each country, the proportion of observations of different themes of modern biotechnology for each type of article (three observations of a theme per article allowed). The six themes mentioned most frequently in the overall coverage are presented in this table; the other themes are not included.

3. *Observations* rows indicate the number of observations per theme and their proportion compared with all themes.

Table 3A.3 *Main actors in the coverage of modern biotechnology across Europe: trends, 1973–1996*

<table>
<tr><th></th><th></th><th colspan="9">Type of coverage</th></tr>
<tr><th></th><th></th><th colspan="3">Risk only</th><th colspan="3">Benefit only</th><th colspan="3">Risk & benefit</th></tr>
<tr><th>Country</th><th>Actors</th><th>%</th><th>I–II</th><th>II–III</th><th>%</th><th>I–II</th><th>II–III</th><th>%</th><th>I–II</th><th>II–III</th></tr>
<tr><td>Austria</td><td>Scientists</td><td>32</td><td>↓</td><td>↓</td><td>60</td><td>|</td><td>↑</td><td>40</td><td>|</td><td>)</td></tr>
<tr><td></td><td>Politicians</td><td>32</td><td>↑</td><td>↑</td><td>20</td><td>|</td><td>↔</td><td>23</td><td>↑</td><td>↔</td></tr>
<tr><td></td><td>Industry</td><td>4</td><td>↔</td><td>)</td><td>14</td><td>)</td><td>↓</td><td>15</td><td>↔</td><td>↔</td></tr>
<tr><td></td><td>NGOs</td><td>13</td><td>)</td><td>)</td><td>1</td><td>↔</td><td>↔</td><td>10</td><td>↔</td><td>↔</td></tr>
<tr><td></td><td>Observations</td><td>54</td><td></td><td></td><td>188</td><td></td><td></td><td>83</td><td></td><td></td></tr>
<tr><td>Denmark</td><td>Scientists</td><td>21</td><td>↓</td><td>|</td><td>32</td><td>↑</td><td>↔</td><td>27</td><td>↓</td><td>↔</td></tr>
<tr><td></td><td>Politicians</td><td>43</td><td>↑</td><td>↔</td><td>35</td><td>↑</td><td>↓</td><td>47</td><td>↑</td><td>↓</td></tr>
<tr><td></td><td>Industry</td><td>7</td><td>↑</td><td>↓</td><td>16</td><td>↑</td><td>↔</td><td>9</td><td>↑</td><td>|</td></tr>
<tr><td></td><td>NGOs</td><td>3</td><td>↔</td><td>)</td><td>3</td><td>↔</td><td>↔</td><td>2</td><td>↔</td><td>↔</td></tr>
<tr><td></td><td>Observations</td><td>103</td><td></td><td></td><td>116</td><td></td><td></td><td>335</td><td></td><td></td></tr>
<tr><td>Finland</td><td>Scientists</td><td>46</td><td>↑</td><td>↓</td><td>66</td><td>)</td><td>|</td><td>59</td><td>↓</td><td>↓</td></tr>
<tr><td></td><td>Politicians</td><td>29</td><td>↑</td><td>↔</td><td>11</td><td>↔</td><td>)</td><td>14</td><td>↑</td><td>↔</td></tr>
<tr><td></td><td>Industry</td><td>9</td><td>)</td><td>↔</td><td>15</td><td>↔</td><td>)</td><td>10</td><td>↔</td><td>↔</td></tr>
<tr><td></td><td>NGOs</td><td>3</td><td>↓</td><td>↔</td><td>2</td><td>↔</td><td>↔</td><td>5</td><td>)</td><td>↔</td></tr>
<tr><td></td><td>Observations</td><td>35</td><td></td><td></td><td>332</td><td></td><td></td><td>222</td><td></td><td></td></tr>
<tr><td>France</td><td>Scientists</td><td>39</td><td>↑</td><td>↓</td><td>46</td><td>↓</td><td>↔</td><td>43</td><td>|</td><td>↓</td></tr>
<tr><td></td><td>Politicians</td><td>13</td><td>↔</td><td>|</td><td>14</td><td>↑</td><td>↔</td><td>14</td><td>|</td><td>↑</td></tr>
<tr><td></td><td>Industry</td><td>13</td><td>↑</td><td>↔</td><td>29</td><td>)</td><td>↓</td><td>18</td><td>↑</td><td>)</td></tr>
<tr><td></td><td>NGOs</td><td>8</td><td>↓</td><td>↔</td><td>2</td><td>↔</td><td>↔</td><td>6</td><td>)</td><td>↔</td></tr>
<tr><td></td><td>Observations</td><td>72</td><td></td><td></td><td>326</td><td></td><td></td><td>201</td><td></td><td></td></tr>
<tr><td>Germany</td><td>Scientists</td><td>31</td><td>|</td><td>↓</td><td>64</td><td>↔</td><td>↓</td><td>30</td><td>↓</td><td>|</td></tr>
<tr><td></td><td>Politicians</td><td>31</td><td>↔</td><td>↔</td><td>12</td><td>|</td><td>)</td><td>27</td><td>↑</td><td>↓</td></tr>
<tr><td></td><td>Industry</td><td>7</td><td>)</td><td>↔</td><td>16</td><td>↔</td><td>)</td><td>19</td><td>|</td><td>↑</td></tr>
<tr><td></td><td>NGOs</td><td>8</td><td>↑</td><td>↔</td><td>1</td><td>↔</td><td>↔</td><td>9</td><td>)</td><td>↑</td></tr>
<tr><td></td><td>Observations</td><td>86</td><td></td><td></td><td>469</td><td></td><td></td><td>341</td><td></td><td></td></tr>
<tr><td>Greece</td><td>Scientists</td><td>33</td><td>↑</td><td>↑</td><td>64</td><td>)</td><td>|</td><td>68</td><td>↑</td><td>↓</td></tr>
<tr><td></td><td>Politicians</td><td>33</td><td>↑</td><td>↓</td><td>5</td><td>↓</td><td>)</td><td>5</td><td>↓</td><td>|</td></tr>
<tr><td></td><td>Industry</td><td>0</td><td>↔</td><td>↔</td><td>24</td><td>↔</td><td>↑</td><td>18</td><td>↔</td><td>↑</td></tr>
<tr><td></td><td>NGOs</td><td>0</td><td>↔</td><td>↔</td><td>0</td><td>↔</td><td>↔</td><td>0</td><td>↔</td><td>↔</td></tr>
<tr><td></td><td>Observations</td><td>6</td><td></td><td></td><td>59</td><td></td><td></td><td>22</td><td></td><td></td></tr>
<tr><td>Italy</td><td>Scientists</td><td>39</td><td>|</td><td>↓</td><td>54</td><td>↓</td><td>)</td><td>45</td><td>|</td><td>↓</td></tr>
<tr><td></td><td>Politicians</td><td>15</td><td>↑</td><td>↔</td><td>2</td><td>↑</td><td>↓</td><td>13</td><td>↔</td><td>↔</td></tr>
<tr><td></td><td>Industry</td><td>11</td><td>↑</td><td>↔</td><td>26</td><td>↓</td><td>↑</td><td>14</td><td>)</td><td>|</td></tr>
<tr><td></td><td>NGOs</td><td>9</td><td>↑</td><td>↔</td><td>7</td><td>↓</td><td>)</td><td>5</td><td>)</td><td>↑</td></tr>
<tr><td></td><td>Observations</td><td>54</td><td></td><td></td><td>199</td><td></td><td></td><td>255</td><td></td><td></td></tr>
</table>

Table 3A.3 (*cont.*)

<table>
<tr><th></th><th></th><th colspan="9">Type of coverage</th></tr>
<tr><th></th><th></th><th colspan="3">Risk only</th><th colspan="3">Benefit only</th><th colspan="3">Risk & benefit</th></tr>
<tr><th>Country</th><th>Actors</th><th>%</th><th>I–II</th><th>II–III</th><th>%</th><th>I–II</th><th>II–III</th><th>%</th><th>I–II</th><th>II–III</th></tr>
<tr><td>Netherlands</td><td>Scientists</td><td>22</td><td>↑</td><td>↑</td><td>51</td><td>↓</td><td>↔</td><td>29</td><td>↑</td><td>↓</td></tr>
<tr><td></td><td>Politicians</td><td>22</td><td>↑</td><td>↓</td><td>11</td><td>|</td><td>↔</td><td>24</td><td>↑</td><td>↔</td></tr>
<tr><td></td><td>Industry</td><td>12</td><td>↑</td><td>|</td><td>25</td><td>)</td><td>↔</td><td>27</td><td>↑</td><td>↑</td></tr>
<tr><td></td><td>NGOs</td><td>17</td><td>↑</td><td>↓</td><td>5</td><td>↔</td><td>↔</td><td>12</td><td>)</td><td>)</td></tr>
<tr><td></td><td>Observations</td><td>161</td><td></td><td></td><td>506</td><td></td><td></td><td>304</td><td></td><td></td></tr>
<tr><td>Poland</td><td>Scientists</td><td>50</td><td>↓</td><td>↑</td><td>33</td><td>↔</td><td>↔</td><td>72</td><td>|</td><td>↔</td></tr>
<tr><td></td><td>Politicians</td><td>13</td><td>↑</td><td>↓</td><td>7</td><td>↓</td><td>)</td><td>3</td><td>)</td><td>|</td></tr>
<tr><td></td><td>Industry</td><td>0</td><td>↔</td><td>↔</td><td>11</td><td>↑</td><td>↓</td><td>12</td><td>↑</td><td>↔</td></tr>
<tr><td></td><td>NGOs</td><td>0</td><td>↔</td><td>↔</td><td>0</td><td>↔</td><td>↔</td><td>2</td><td>↔</td><td>↔</td></tr>
<tr><td></td><td>Observations</td><td>8</td><td></td><td></td><td>232</td><td></td><td></td><td>117</td><td></td><td></td></tr>
<tr><td>Sweden</td><td>Scientists</td><td>26</td><td>↔</td><td>↓</td><td>67</td><td>↑</td><td>↑</td><td>51</td><td>|</td><td>↔</td></tr>
<tr><td></td><td>Politicians</td><td>33</td><td>↓</td><td>↑</td><td>10</td><td>↓</td><td>↔</td><td>17</td><td>↔</td><td>↔</td></tr>
<tr><td></td><td>Industry</td><td>7</td><td>|</td><td>↔</td><td>16</td><td>↔</td><td>↓</td><td>11</td><td>|</td><td>↔</td></tr>
<tr><td></td><td>NGOs</td><td>8</td><td>↔</td><td>↔</td><td>0</td><td>↔</td><td>↔</td><td>2</td><td>↔</td><td>↔</td></tr>
<tr><td></td><td>Observations</td><td>151</td><td></td><td></td><td>362</td><td></td><td></td><td>252</td><td></td><td></td></tr>
<tr><td>Switzerland</td><td>Scientists</td><td>10</td><td>↑</td><td>↓</td><td>53</td><td>↓</td><td>↓</td><td>31</td><td>↓</td><td>↔</td></tr>
<tr><td></td><td>Politicians</td><td>38</td><td>↑</td><td>↔</td><td>13</td><td>)</td><td>↑</td><td>29</td><td>↑</td><td>↓</td></tr>
<tr><td></td><td>Industry</td><td>8</td><td>)</td><td>↔</td><td>24</td><td>↑</td><td>↔</td><td>21</td><td>↓</td><td>↔</td></tr>
<tr><td></td><td>NGOs</td><td>34</td><td>↔</td><td>↑</td><td>3</td><td>↔</td><td>↔</td><td>6</td><td>↔</td><td>↑</td></tr>
<tr><td></td><td>Observations</td><td>77</td><td></td><td></td><td>95</td><td></td><td></td><td>48</td><td></td><td></td></tr>
<tr><td>United Kingdom</td><td>Scientists</td><td>40</td><td>↓</td><td>↔</td><td>56</td><td>↓</td><td>↓</td><td>32</td><td>↓</td><td>↔</td></tr>
<tr><td></td><td>Politicians</td><td>16</td><td>↑</td><td>↓</td><td>10</td><td>↔</td><td>↑</td><td>18</td><td>)</td><td>↔</td></tr>
<tr><td></td><td>Industry</td><td>9</td><td>↔</td><td>↑</td><td>25</td><td>↑</td><td>↔</td><td>21</td><td>↑</td><td>|</td></tr>
<tr><td></td><td>NGOs</td><td>7</td><td>↑</td><td>↓</td><td>4</td><td>↔</td><td>)</td><td>10</td><td>)</td><td>↔</td></tr>
<tr><td></td><td>Observations</td><td>45</td><td></td><td></td><td>391</td><td></td><td></td><td>196</td><td></td><td></td></tr>
</table>

Notes: 1. See note 1 to table 3A.2.

2. We present articles containing information only about the risks of modern biotechnology, only about the benefits, or about both risks and benefits. Articles containing neither risk nor benefit information are not included in this table. The percentages indicate the proportion of observations of different actor groups per type of article (two observations of actors per article allowed). The four actor groups mentioned most frequently in the overall coverage are presented in this table; the other actor groups are not included.

3. *Observations* rows indicate the number of observations of actors in risk-only, benefit-only and risk and benefit articles, respectively.

Table 3A.4 *Main location of reported activity in the coverage of modern biotechnology across Europe: trends, 1973–1996*

		Type of coverage								
		Risk only			Benefit only			Risk & benefit		
Country	Main location of activity	%	I–II	II–III	%	I–II	II–III	%	I–II	II–III
Austria	Own country	45	↑	⌉	33	↔	↓	36	↑	⌉
	Other Europe	23	⌉	|	24	↓	⌉	19	↓	↔
	USA	15	↓	↔	25	↔	↔	29	↓	↔
	Observations (N)	*53*			*219*			*80*		
Denmark	Own country	61	↑	↓	55	↑	⌉	47	↑	↓
	Other Europe	15	⌉	↑	9	⌉	⌉	18	⌉	↑
	USA	16	⌉	↔	18	↑	↔	28	↓	↔
	Observations (N)	*86*			*99*			*284*		
Finland	Own country	51	↑	↓	72	|	↓	65	↔	|
	Other Europe	17	⌉	↑	13	↓	↑	13	|	↔
	USA	14	↑	|	11	↑	↑	14	|	⌉
	Observations (N)	*35*			*308*			*215*		
France	Own country	28	↓	↑	46	↔	↔	48	⌉	|
	Other Europe	21	↑	↓	18	↓	↔	19	↓	⌉
	USA	12	↑	|	19	↑	↔	19	↑	|
	Observations (N)	*67*			*301*			*168*		
Germany	Own country	52	↑	↓	31	⌉	↔	41	↑	↓
	Other Europe	11	|	↑	24	↓	⌉	19	↓	⌉
	USA	22	↔	↔	35	↔	|	28	↓	↔
	Observations (N)	*64*			*450*			*277*		
Greece	Own country	20	↓	↔	11	↓	↑	14	↓	↑
	Other Europe	20	↔	↑	38	⌉	↔	55	↑	|
	USA	20	↑	↓	40	⌉	↔	23	↓	↑
	Observations (N)	*5*			*55*			*22*		
Italy	Own country	41	↑	|	29	↑	⌉	31	⌉	↓
	Other Europe	19	↓	↑	18	↓	⌉	21	↓	⌉
	USA	16	⌉	↔	35	↓	↔	31	↑	⌉
	Observations (N)	*37*			*153*			*156*		
Netherlands	Own country	61	↓	↑	45	↑	⌉	54	↑	↑
	Other Europe	9	↑	↔	17	↑	↔	9	⌉	↔
	USA	6	↑	↓	28	↓	↔	23	↑	↓
	Observations (N)	*111*			*382*			*220*		
Poland	Own country	?	?	?	?	?	?	?	?	?
	Other Europe	16	↔	↑	41	↓	↔	49	↔	↑
	USA	33	↓	0	36	↑	↑	20	|	|
	Observations (N)	*6*			*160*			*75*		
Sweden	Own country	62	↓	↓	53	↑	|	69	|	↓
	Other Europe	14	↔	↑	13	|	↔	11	↔	↑
	USA	9	↔	|	23	⌉	↔	13	↔	↔
	Observations (N)	*139*			*337*			*205*		

Table 3A.4 (*cont.*)

Country	Main location of activity	Type of coverage: Risk only %	I–II	II–III	Benefit only %	I–II	II–III	Risk & benefit %	I–II	II–III
Switzerland	**Own country**	88	↑	↔	37	↘	↑	63	↑	↓
	Other Europe	5	↑	|	24	↔	↔	6	↔	↑
	USA	0	↔	↔	16	↑	↘	9	↓	↑
	Observations (N)	*42*			*70*			*32*		
United Kingdom	**Own country**	62	↑	↓	50	↘	↑	55	↓	↑
	Other Europe	10	↔	↑	13	↔	|	11	↘	↔
	USA	21	↓	↑	29	|	↔	26	↘	↓
	Observations (N)	*39*			*377*			*149*		

Notes: 1. See note 1 to table 3A.2.

2. We present articles containing information only about the risks of modern biotechnology, only about the benefits, or about both risks and benefits. Articles containing neither risk nor benefit information are not included in this table. The percentages indicate, for each country, the proportion of observations of different locations of reported activity per type of article (two observations of location of activity per article allowed). The locations are 'Own country', 'Other European countries', and 'USA', respectively. Other locations are not included in this table.

3. *Observations* rows indicate the number of observations of location of activity in risk-only, benefit-only and risk and benefit articles, respectively.

REFERENCES

Beck, U. (1986) *Die Risikogesellschaft*, Frankfurt am Main: Suhrkamp.

Biotechnology and the European Public (1997) 'Europe Ambivalent on Biotechnology', *Nature* 387: 845–7.

DeFleur, M. and S. Ball-Rokeach (1989) *Theories of Mass Communication*, White Plains, NY: Longman.

Durant, J., M.W. Bauer and G. Gaskell (1998) *Biotechnology in the Public Sphere: a European Sourcebook*, London: Science Museum.

Frewer, L., C. Howard and R. Shepherd (1997) 'Public Concerns in the United Kingdom about General and Specific Applications of Genetic Engineering: Risk, Benefit, and Ethics', *Science, Technology and Human Values* 22(1): 98–124.

Galtung, J. and Ruge, M.H. (1965). 'The Structure of Foreign News: The Presentation of the Congo, Cuba and Cyprus Crises in Four Norwegian Newspapers'. *Journal of Peace Research*, 2, 64–91.

Hagedorn, C. and S. Allender-Hagedorn (1997) 'Issues in Agricultural and Environmental Biotechnology: Identifying and Comparing Biotechnology Issues from Public Opinion Surveys, the Popular Press and Technical/Regulatory Sources', *Public Understanding of Science* 6: 233–45.

Hilgartner, S. and C.L. Bosk (1988) 'The Rise and Fall of Social Problems: a Public Arenas Model', *American Journal of Sociology* 94(1): 53–78.

Holsti, O.R. (1969) *Content Analysis for the Social Sciences and Humanities*, Reading, MA: Addison Wesley.
Hornig, S. (1992) 'Framing Risk: Audience and Reader Factors', *Journalism Quarterly* 69(3): 679–90.
(1994) 'Structuring Public Debate on Biotechnology: Media Frames and Public Response', *Science Communication* 16(2): 166–79.
Kepplinger, H.M., S. Ehmig and C. Ahlheim (1991) *Gentechnik im Widerstreit. Zum Verhältnis von Wissenschaft und Journalismus*, Frankfurt am Main: Campus.
Kitzinger, J. and Reilly, J. (1997). 'The Rise and Fall of Risk Reporting: Media Coverage of Human Genetics Research, "False memory syndrome" and "Mad cow disease".' *European Journal of Communication.*
Kohring, M., A. Goerke and G. Ruhrmann (1999) 'Das Bild der Gentechnologie in den internationalen Medien. Eine Inhaltsanalyse meinungsführender Zeitschriften', in O. Renn and J. Haempel (eds.), *Gentechnik aus der Sicht der Öffentlichkeit*, Frankfurt am Main: Campus.
Krippendorff, K. (1980) *Content Analysis: an Introduction to Its Methodology*, Newbury Park, CA: Sage Publications.
Servaes, J. and C. Tonnaer (1992) *De nieuwsmarkt: vorm en inhoud van de internationale berichtgeving*, Groningen: Wolters-Noordhoff.

4 The institutions of bioethics

Jean-Christophe Galloux, Arne Thing Mortensen, Suzanne de Cheveigné, Agnes Allansdottir, Aigli Chatjouli and George Sakellaris

The discovery of genetic recombination techniques, at the beginning of the 1970s, opened up extraordinary new possibilities for the biological modification of organisms; and the potential for these interventions immediately prompted profound ethical concerns. As the call for a moratorium before the Asilomar Conference (held in California in February 1975) showed, microbiologists were well aware of these issues. The risks associated with unintended genetic combinations, particularly the danger of disturbing ecological balances and, ultimately, threatening human welfare, were seriously contemplated. Such concerns resulted in a voluntary moratorium during which the acceptability of certain experiments could be discussed. Since then, public references to 'bioethics' (the term was coined in the USA in 1971 by the oncologist Van R. Potter, 1971) have grown consistently. Roughly speaking, the term 'bioethics' relates to a body of reflections, involving multidisciplinary approaches, that deal with all aspects of the responsible management of human life in the face of complex and rapid developments in biomedicine. In the present chapter, we will provide a sketch of the history of the development of bioethics, including the media background, we will clarify the philosophical notions behind the term and, finally, we will trace the process of institutionalisation that bioethics has undergone over the past twenty to thirty years. In doing so, we will concentrate on four European countries: Denmark, France, Italy and Greece. We will also consider these issues at the level of Europe as a whole.

A short history of biotechnology and ethics

The development of genetic recombination techniques inspired speculation about the possibility of designing living organisms, and provoked ethical discussions concerning the proper boundaries for the application of microbiological techniques. However, during the 1970s, these dangers did not seem sufficiently realistic, or imminent, to demand political initiatives. The discussions were, therefore, largely confined to professional circles, with the first ethical research committees appearing during the mid-1960s in the USA, following the 1964 Helsinki declaration on medical ethics. The birth of bioethics was, in fact, the result of diverse and sometimes paradoxical anxieties within a scientific community trying to come to terms with the enormity of the power it was acquiring; and it represented a search for an international consensus on values in the life sciences domain. But it also symbolised a desire to evade formal state regulation. Largely inspired by liberal Anglo-Saxon ideologies, life scientists were striving to achieve a – preferably institutionalised – professional self-regulation.

In this period, however, the public debate developed slowly. References were regularly made to ethical issues in the European elite press when addressing the subject of biotechnology (see Gutteling et al., chapter 3 in this volume). Yet, it is illuminating to compare the four countries examined in this study in terms of the prominence of ethical concerns in discussions of biotechnology.[1] In Italy, ethics represent, on average, 10.4 per cent of all themes considered in the context of biotechnology, in France 9.0 per cent, in Greece 7.8 per cent and in Denmark only 6.7 per cent. The European average is 7.1 per cent of biotechnology discussions.

We can regard this variation in the prominence of the ethics theme as an indicator of the relative attention given by the national elite press to ethical issues in the context of biotechnology. Figure 4.1 illustrates the development of the ethics theme in the newspaper coverage of biotechnology in the four countries under study. There are clearly considerable variations over time in the relative salience of the ethics theme in the media debate. However, a general trend is discernible in which the media's interest in this issue peaked in the 1970s, declined in importance during the more optimistic 1980s, only to resurface during the 1990s. There are also strong local variations that reflect national differences in the form of this debate. The data from both the Danish and the Italian media show that a substantial amount of attention was accorded to ethical issues in the years immediately before the institution of formal bioethics bodies. At the moment of institutionalisation (in 1988 in Denmark and 1990 in Italy), however, the salience of the ethics theme diminished somewhat. A similar phenomenon is observable in France at the time of the creation of the national ethics committee in 1983 and coinciding with the passage of the Bioethics Law in 1994. This would tend to support the sociological hypothesis associated with 'public arena' models, which stipulate that, once an issue has been taken up by a formal policy body, public interest in the issue is diminished or diverted towards other concerns (Hilgartner and Bock, 1988).

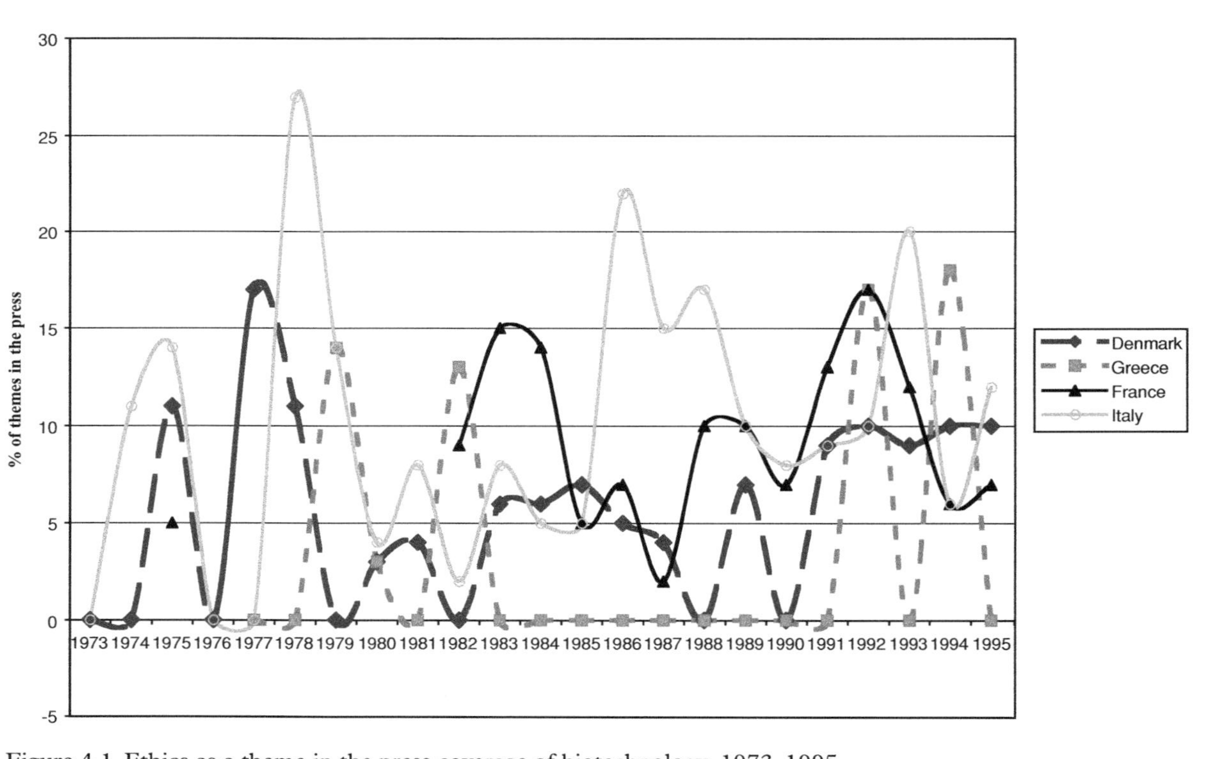

Figure 4.1 Ethics as a theme in the press coverage of biotechnology, 1973–1995

In the early days of biotechnological development, scientists were generally considered trustworthy enough to be able to handle the new risks and ethical problems. This changed during the 1980s, primarily as a result of industrial pressure. As a research field, modern biotechnology was then growing rapidly, and it was producing techniques and ideas that were obviously commercially promising. As a result, the medical industry – as well as the agrochemical and other industries – entered the public scene as powerful agents. In most European countries, perceptions regarding modern biotechnology in the early 1980s were distinctly positive. Generally speaking, biotechnology was treated in the media as another striking example of scientific progress, and even in Denmark (of the four countries compared in this study, Denmark has the most influential tradition of criticising new technologies) the attitude was positive (Durant et al., 1998). Indicative of this, very few people rejected the claim that modern biotechnology is one of the most important fields for future scientific development. Nor did they criticise the argument that industry and agriculture should exploit the opportunities biotechnology presents in the interests of economic growth and for the sake of progress itself.

In Denmark, more than in the other three countries, 'counter-experts' and public movements were able to influence the political agenda; although the critics insisted on the implementation of safety measures and the imperative for political control, they did not demand absolute prohibitions. As a result of public discussions in the early 1980s, the political majority in Denmark became convinced that the common goal – that of achieving a congenial climate for biotechnological research and development – could best be accomplished by creating a legal and administrative framework that would prevent the risks of the new technology getting out of control. The Gene Technology Act, passed in 1986, answered this need. Yet bioethics did not become an essential political issue before the establishment, in 1987, of the Danish Council of Ethics. Similar councils on bioethics were created in France (1983), in Italy (1990) and in Greece (1992). Before we look at them in detail, a comment on the primary arguments invoked in bioethical discussion is necessary.

Ethical discourse

As a philosophical realm, 'ethics' covers all efforts to found normative positions upon reasoned argument rather than upon intuitions or emotive reactions. Appealing to ethics is to demand reflection and a rational approach to the study of values. In popular usage, however, the concept of ethics is more narrowly applied. Here, reasoning based on almost universally accepted normative principles is not classified as ethical; instead, the concept tends to define only normative reasoning based on controversial principles. Another difference between the philosophical and the popular use of the word is that 'ethics' in everyday language tends to be equivalent to what philosophers call 'normative ethics' or 'meta-ethics'. For this study, these terminological variations may cause problems in both interpreting the raw data and discussions thereof. For instance, in the media study carried out in the Biotechnology and the European Public project, newspaper articles were coded as treating biotechnology in a 'progress' frame or in an 'ethical' frame, 'ethical' here denoting that biotechnology is discussed as an issue of debatable moral acceptability, while notions of 'progress' seem to be inherent in a range of unquestioned values (Durant et al., 1988). Yet, in fact, the notion of 'progress' is in itself essentially ethical, since it implies something along the lines of 'a change for the better'. Moreover, what constitutes 'better' is certainly open to dispute. Thus, the borderline between these two frames is very hard to delineate. Nevertheless, it seems obvious that there is an important everyday distinction to be made here. It is for this reason that we seek below to clarify the different ethical frameworks that are called upon in bioethics discussions, as well as the types of argument that are employed.

Ethical frames

Seeing 'ethics' as a rational enterprise – an effort to clarify arguments rather than just to sustain attitudes – we need to examine current ethical theories and determine whether the main differences between the philosophers' and non-philosophers' (scientists, professionals, politicians, laypeople, etc.) conceptions of ethics reflect important differences in the way ethical questions are conceived by these two groups.[2] In this sense, looking for logically elaborated theories is less important than elucidating arguments that belong to the same family. Such similarities can, indeed, be found by looking for evidence of shared axioms. Our hypothesis is that three principles are essential and relevant to characterising the different ways in which bioethics became a political issue during the 1980s and 1990s: the principle of utility, the principle of democracy and the principle of veneration.

The principle of utility According to the principle of utility, moral questions are questions about the optimal balance between happiness or well-being and suffering or discomfort. According to utilitarians, our actions should be judged by the consequences they have for the 'sum of happiness' in the world. Happiness is measured in terms of individuals' feelings or preferences, which should in themselves be rational since a moral agent must be able to evaluate his/her actions in relation to rational criteria. For instance, to argue that soya plants should be genetically modified because this is a way to increase crop yields while reducing expenditure on herbicides is to apply the principle of utility as an underlying premise. We call this kind of reasoning Utility-argumentation or, for short, U-argumentation.

For three main reasons, U-argumentation is closely associated with science. First, it sees human activity in terms of finding means of reaching predetermined ends, and this is exactly what science seeks to accomplish. Secondly, in U-argumentation, human ends are considered to be 'natural', in the sense that they are related to biological, social and psychological needs. Moral action would clearly need to be based upon a good understanding of these human needs and science offers the most efficacious means of acquiring this understanding. Third, utilitarians presuppose that choices between alternative courses of action would, ideally, be based upon some kind of methodical evaluation, and this might well entail the development of scientific procedures.

In view of the above, the 'welfare state' can be seen as a product of utilitarian thinking. According to utilitarian thinking, the most important relation between one person and another is one of mutual care and it is everybody's duty to maximise the sum of human happiness. Therefore, the stronger are responsible for the welfare of the weaker, with respect to both humans and animals. As for the relationship between humans and Nature, the general position is that Nature exists to be tamed and used by humans. Thus all values in Nature – aesthetic, religious or otherwise – derive from humans' usage and needs.

The principle of democracy According to the principle of democracy, every person has the right and a duty to make personal judgements about what constitutes right and wrong, so long as they accept the equal right of other members of society to do the same. As a rational moral agent, each person is considered autonomous, and in principle one is obliged (and entitled) to consider oneself as having exactly the same status as everybody else within a given society. For instance, to argue that food containing genetically modified ingredients should be labelled because we have the right to know what we eat is to apply the principle of democracy as a presupposed premise. We call this approach D-argumentation.

D-argumentation is a way of thinking that has developed in European philosophy from the seventeenth- and eighteenth-century contract theories of society up until, among others, Habermas's studies of the communicative conditions for democracy. Central to this development is Kant's definition of ethics as a study of the forms of ethical reasoning rather than of their content. Decisions about right and wrong, good or bad, must in principle be left to the autonomous person. This autonomous individual is considered to be his or her own – universally obliged – master and law-maker. According to this approach, in politics ethical positions should be institutionalised as rules of procedure rather than as specific prohibitions and allowances, leaving it up to the agents themselves to make their own concrete judgements.

The principle of veneration According to the principle of veneration, moral questions are questions about the relation between the moral agent (an individual or some collective 'we', such as 'humankind') and an 'other', possessing value, integrity and power of its own and, as such, demanding respect or even deference. Duties and values are, essentially, founded in this relation to something or somebody in possession of value and power independent of the agent. For instance, from this perspective one might argue that experiments involving the genetic modification of human reproductive cells should not be allowed because this constitutes interfering with Nature. Here the presupposed premise would be the principle of veneration. We term this kind of approach V-argumentation.

The most obvious example of ethical veneration concerns ethical thinking based on religious assumptions: God is the universal source of values and the human *raison d'être* is, essentially, to worship divine perfection. But it should be remembered that religious ethics is not the only example of V-argumentation. Philosophers and non-philosophers have created refined positions based on veneration for 'the other' person involving all living beings, or Nature defined as something more than an entity existing for human exploitation. In this conception, Nature is deemed infinite in its complexity; therefore, because humans have little understanding of it, the idea of their 'taming' Nature is fundamentally absurd. This position demands the establishment of ethical principles that are in accordance with Nature, not opposed to it.

Ethical argumentation

These are three general categories of ethical thinking, each with its distinct profile, and although they compete with each other they are not necessarily contradictory. In current social and political practice, 'ethics' is a battleground of disagreements, but conflict is magnified in areas, such as modern biotechnology, that raise new problems and dilemmas. To define these disagreements as 'ethical' – and not as merely 'political' or 'economic' – is to stress the presuppositions of rationality upon which the debate is conducted. However, before we examine the contents of bioethical discourse and, in particular, the activities of the different bioethics bodies, some preliminary remarks are required.

- The themes covered by ethical committees (especially the central ones) are similar among different countries.
- The 'solutions' adopted, on the other hand, are not always the same. To give an example, the Group of Advisers on the Ethical Implications of Biotechnology (GAEIB) did not reject requests for patenting human genes in its opinion of 1996, whereas the French national committee did so in December 1991. The reasoning and the arguments involved are not always employed in the same way. However, on the main 'bioethical' questions, such as human cloning, there is an obvious consensus. The differences that exist are indicative of the cultural diversity within Europe.
- The 'values' referred to, in the opinions expressed by bioethical committees, are often drawn from traditional legal frameworks. This is particularly the case at the level of the GAEIB/European Group of Ethics and of the French bioethics committee. The GAEIB, for instance, refers formally, in its opinions, to legal texts (directives or international conventions). Though some legal principles are of moral origin, this is certainly not always the case. The borders between ethical norms and legal norms are obvious and must be preserved. In some cases, however, this border has been crossed by bioethical committees. There is a danger of confusion here that is not always perceived by the actors concerned.

This raises more general questions: what are the sources of bioethics? if ethical norms have to be established, who is entitled to do so? if ethical norms have to be found among pre-existing values in a given society, is a new discipline such as bioethics required? These questions remain unanswered at the theoretical level. Yet, in practice, they were resolved in most European countries by the establishment of official bioethical committees and by the institutionalisation of bioethics.

The process of institutionalisation of bioethics

The purpose of this section is to describe how ethics has become an explicit issue in politics, and to point out some of the similarities and differences between countries in handling this issue. In most European countries, bioethics has become institutionalised by the creation of political bodies with consultative and administrative functions. In principle, this common development seems to indicate a need for, and a will to respect, fairly high standards of rationality in making ethical decisions.

Nonetheless, the ways in which these decisions are arrived at differ considerably.

Science has been a dominant factor in European culture for over a century. In all countries, the support scientific culture now confers to U-argumentation allows it to displace V- and D-argumentation. One would, however, expect some degree of compromise to be reached. In addition, it is reasonable to expect variation in the nature of conflicts depending upon the way in which they are handled. Thus, in different countries, the debate may be fought around, variously, Utilitarian versus Democratic positions, or around Utilitarian versus Veneration positions. In our comparisons, this sort of variation is visible at a national level, with Denmark differing from the other three countries and, to a lesser degree, France diverging from the other countries.

The institutionalisation of bioethics indicates that ethics has become an explicit issue in politics. The discussion of ethics at an institutional level is now considered an acceptable means of resolving conflicts concerning biotechnology and an appropriate way to deal with more basic issues related to the nature of ethics. The establishment of special bodies dedicated to bioethics has been a pivotal event in this field. But, as we have noted, the dangers of institutionalisation are two-fold: first, it can have the effect of stifling public debate; second, committees can become part of the intellectual Establishment.

The establishment of the ethics committees

The 1980s in Europe witnessed the inception of central bioethics committees and the beginning of the regulation of biotechnological activities. In all four countries examined here, and at the European level, bioethics committees were established in the nine years between 1983 and 1992:

- 21 February 1983, French National Ethics Committee (CCNE: Comité Consultatif National d'Ethique pour la santé et les sciences de la vie);
- 3 June 1987, Danish Council of Ethics (DER: Det Etiske Raat);
- 28 March 1990, Italian National Committee for Bioethics (CNB: Comitato Nazionale di Bioetica);
- 20 November 1991, Group of Advisers on the Ethical Implications of Biotechnology to the European Commission (GAEIB);
- 1992, the creation of the Greek Bioethics Committee was approved, although the committee was only actually set up in 1997.

At this stage, state interventions were largely directed at encouraging self-regulation, and, as such, few administrative constraints were imposed. But at the European level, this emphasis upon 'soft' modes of regulation began to change into 'hard' regulation, prescriptive and prohibitive, from the beginning of the 1990s. This shift followed the lead of Denmark, which had passed a Gene Technology Act in 1986, and Germany, which had passed a Genetic Engineering Law in 1990.[3]

Considering the fact that reproductive technologies had been under development in most European countries since the mid-1970s, there was something of a time-lag before governments took serious action. Indeed, the debates this technology almost immediately raised show that the life sciences were already having an impact upon society. However, bioethics forums were already in existence some time before the emergence of the central bodies mentioned above; and it is clear that institutionalisation is a protracted process. The local committees, hospital committees or research committees established in the USA towards the end of the 1960s appeared in Europe only at the beginning of the 1970s. In this period there is also evidence of formalisation: the activities of local bodies were coordinated and new bodies were created at the national level.

Thus, in 1974 the main state medical research institution in France, INSERM (Institut National de la Santé et de la Recherche Médicale), set up its own bioethics committee (Comité Consultatif d'Ethique Médicale). In Denmark, a research ethics committee was created in 1973 in the Council of Medical Research. But this movement did not spread to Italy until the 1980s (Spagnolo and Sgreccia 1987–8). In Greece, a ministerial order of 1978 demanded the creation of ethics committees in every hospital, yet the plan was rarely implemented (Dalla-Vorgia et al., 1993).

Institutions The process of institutionalisation is characterised by the centralisation of existing local and specialised bodies. But, in addition to this, the sanction of the state must be obtained, involving the legitimisation of existing bodies by the legal authorities. This stage generally involves an administrative or legal act and the provision of central funding.

With regard to bioethics, this first took place in Denmark, where the Danish Council of Ethics was created by law (3 June 1987) and the Central Scientific Ethical Committee of Denmark, established in 1978, was confirmed by another law in 1992. The Council is now attached to the parliament and to the Ministry of the Interior, and the Committee is attached to the Ministry of Health. In France, the National Ethics Committee (CCNE) was introduced by a decree of 21 February 1983 issued by the President of the French Republic himself. More recently, the existence of the Committee was confirmed by a law of 29 July 1994. The CCNE is attached to two ministries: Public Health and Research. The Greek Bioethics Committee was also instituted by law (No. 2071/92), and is attached to the General Secretary of Research and Technology in the Ministry of Development. In Italy, the National Committee for Bioethics was established by a decree issued by the President of the Council on 28 March 1990; this committee is attached directly to the Presidency of the Council of Ministries. We shall consider each of these cases in more detail in the next section, but first let us examine the institutionalisation of bioethics at the European level.

At the European institutional level, interest in bioethics emerged in the mid-1980s. From the point of view of the Commission, this was stimulated by its participation at several international meetings, including that in Hakone, Japan, in 1985, and the 1988 Rome meeting on the ethical implications of human genome sequencing. In 1989, the Commission hosted its own meeting on environmental ethics. In the wake of these, at the end of the 1980s the Commission took the initiative of adding bioethics to its existing concerns. Then, in 1989, it launched its first European Bioethics Conference, concerning the human embryo in medical biological research. At the beginning of the 1990s, it went on to establish ad hoc groups dealing with bioethics. The Commission's involvement in the field deepened during the following years. In November 1991, it set up the Group of Advisers on the Ethical Implications of Biotechnology (GAEIB) and imposed ethical standards on all research funded within the Fourth Framework Programme. It could be argued that these initiatives were attempts to address the concerns expressed to the Commission by the Parliament and other bodies. Yet this is not entirely the case. These initiatives (which included the Eurobarometer) were also involved with the broader questions of public acceptance of biotechnology, which the Commission was addressing at the same time.

A new body replaced the GAEIB in February 1998 – the European Group on Ethics in Sciences and New Technologies (EGE). Its existence was recognised by article 7 of Directive 98/44/EC on the legal protection of biotechnological inventions, dated 6 July 1998. The EGE is attached to the Secretariat General of the Commission. It should be noted that, before any debate on the moral aspects of patenting in biotechnology developed, the Commission was already aware of the ethical dimensions of this question.[4] With the creation of the EGE, the Commission once again took the initiative by widening the scope of ethics; now it deals not only with biotechnology but also with science and new technologies in general.

The Council of Europe followed, in parallel, the same path and held its first meeting on bioethics in December 1989. At the opening conference, the Secretary General, Catherine Lalumière, asked for the creation of a European bioethics committee. A preliminary body, the CAHBI (Ad Hoc Committee on Bioethics), had already been established in 1985, and was renamed in 1989. Gradually, from the 1980s onwards, bioethics became a part of the administrative process of regulation at the European level.

Who pushed for institutionalisation and for what purpose? The European Group on Ethics was clearly established for political reasons: it was the Commission's answer to the demands made by the European Parliament. It also enables the Commission to legitimate its policies in fields in which ethical issues are at stake. But at the time of its inception, discussion of a directive had made the problem of the patentability of biological products a subject for vigorous debate in the Parliament. The GAEIB twice issued an opinion on this question (30 September 1993 and 25 September 1996), and on each occasion GAEIB's opinion was invoked by the Commission in attempting to buttress its position during negotiations with the Parliament. In its Proposal on the competitiveness of European industry, the Commission stated:

> It is desirable that the Community have an advisory structure on ethics and biotechnology which is capable of dealing with ethical issues where they arise in the course of Community activities. Such a structure should permit dialogue to take place where ethical issues which Member States or other interested parties consider to require resolution could be openly discussed. . . . The Commission considers that this would be a positive step towards increasing the acceptance of biotechnology and towards ensuring the achievement of a single market for these products. (European Commission, 1996)

In Italy, the CNB was created in the context of national debates concerning reproductive technologies, and its launch saw confrontations between religious and secular ideas on bioethics. The creation of the CNB had followed a Christian Democrat deputy's proposal, which had been advanced in order to settle a heated parliamentary debate. A multi-party coalition had proposed a bill that would stimulate discussions about abortion and the genetic manipulation of life. At the same time, communist groups proposed the launch of a parliamentary commission to study the ethical problems arising from genetic engineering and reproductive technologies. From the outset, this was an attempt to reconcile religious and secular bioethics, and sought to alleviate the tensions by means of institutionalisation. Bioethics is still a contested domain in Italy; but this is unsurprising in a country with a strong Catholic tradition, relative political instability and scientific lobbies able to exert considerable pressure.

France has a long tradition of 'administrative consultancy'. Since the absolutist monarchies of the mid-seventeenth century, the French state has been replete with advisory councils in every sector of its competence. The activity of these consultative bodies tends to be proportional to the activity of the administration, and they often constitute an important part of the decision-making process. The CCNE belongs to this tradition. President Mitterrand involved himself personally in the creation of the CCNE; as such, its inception can be seen as a political act. But this involved only the formalisation, at a national level, of the pre-existing INSERM bioethics committee. It retained the same function, the same president (Professor J. Bernard) and a similar composition. The establishment of the CCNE appears to have satisfied two principal requirements: first, the French scientific community's need for clear guidelines for the conduct of its research in the developing field of biotechnology; secondly, the state's need for moral legitimacy in formulating legislative initiatives in regulating this new field. Neither in France nor in Italy were bioethics bodies set up in response to the demands of public, practitioners or lawyers. The chief source of appeals was the scientific community.

In Denmark, however, the system differs from those of France and Italy. There are two central bodies: the first, the Central Scientific Ethical Committee (CSEC), deals with research in the field of biomedicine; the second, the Danish Council of Ethics (DER), tackles general questions raised by the development of the life sciences. The CSEC presides over seven regional committees; it is in charge of the harmonisation of the solutions these committees suggest, and also functions as a sort of appeal court (Solbakk, 1991). The CSEC system of committees was organised, in accordance with the Second Helsinki Declaration, solely for the benefit of the research sector, and not the public. The DER, in contrast, was established in response to the political will of the Danish parliament, and was designed as an instrument by which public debate on the controversial aspects of biotechnology could be fostered.

The Greek Bioethics Committee styles itself as an intermediary between the scientific community and the public, disseminating information and facilitating the expression of opinions from a variety of different parties. For increasing the level of public dialogue, it also presents itself as a 'catalyst' for bioethics discussions. This body was initiated by a law, passed by the Greek parliament in 1992 (No. 2071/92), which provided a general mandate for the creation of advisory committees; but the Bioethics Committee was formed only on 11 April 1997 by an act of the Minister of Development. The Committee is attached to the General Secretary of Research and Development, which in turn belongs to the Ministry of Development. On 12 September 1998, the Committee underwent a transition within the ministry, and its focus changed from medical ethics to bioethics in general.

To conclude this section, several general observations about these bodies may be made. Most are consultative and multidisciplinary. They are not formally dependent upon governments, as is the case with other administrative bodies; however, this legal autonomy does not preclude constraints of a less formal political or sociological nature. All of these bodies participate at a national level in helping governments address the various ethical issues arising out of progress in science and technology. Nevertheless, it must be recognised that 'this ethical dimension is gradually being institutionalised' (Lenoir, 1998: 2). Central ethics committees have become the tools of central policy-makers; they are established by state authority and come to function as part of the administrative system. With some reservations vis-à-vis the Danish Council of Ethics, their role can be seen as that of providing governments with technical expertise. Furthermore, although the legitimacy of these bodies derives from the state, in many instances they in turn provide the state with scientific legitimacy.

Ethics committee for the establishment?

This generalised movement towards the institutionalisation of bioethics raises important questions. For whose benefit were ethics committees created? For instance, does the institutionalisation of bioethics in governmental committees mean that the bioethical discourse has been appropriated by a small minority? Are these bodies really efficient in influencing state policy related to the field of bioethics? With respect to the first question (excepting Denmark's DER), it is clear that the demand for official bioethics committees came, in the first instance, from the scientific community and, secondarily, from politicians.

The composition of the committees provides an immediate indication of the purpose for which they were established. The Danish Council of Ethics comprises seventeen members, the majority of whom are from a non-scientific background, with men and women being equally represented. This body works for the administration of the Ministry of the Interior and the parliament and enjoys considerable fiscal support. Eight members are designated by the Parliament and nine by the Ministry.

The French National Ethics Committee is a typical administrative and expert body, and it contrasts sharply with the more public-oriented Danish model. It is composed of forty-one members who are designated by a wide range of authorities: the President of the Republic, the ministries of Health, Research, Communication, Education, Industry and Justice, as well as some of the 'grands corps d'Etat' such as the Conseil d'Etat and the Cour de Cassation. The most prestigious state research bodies or academies can also make appointments, such as INSERM, Institut National de la Recherche Agronomique, Centre National de la Recherche Scientifique, the Academy of Sciences, the Collège de France, the Institut

Pasteur and the Conference of University Presidents, as can the presidents of the National Assembly and the Senate. The chief objective of this complex system of designation is to achieve an accurate and balanced representation of the scientific community. But, as a consequence, between 1996 and 1998 members of the CCNE were predominantly male (more than two-thirds) and university professors involved in the life sciences (twenty-seven out of forty-one). There were only four lawyers, none of whom were academics, and, even more surprisingly, there were only four theologians, philosophers or ethicists. In addition, more than two-thirds of the committee is Parisian. The appointment procedures are rather obscure and the turnover is low, suggested by the fact that eight out of the forty-one members have been in place for the full fifteen years.

The European Group on Ethics has twelve members appointed for their technical or scientific competence or on the more ambiguous basis of 'personal qualities'. These members are appointed by the Commission. Presently they are five women and seven men; seven of the twelve are university professors; and five of the twelve are there purely for their experience in, and knowledge of, the life sciences. In contrast to the French national bioethical body, life sciences and social sciences are fairly evenly represented (five for the first, seven for the second). However, analogous to the French CCNE, the nomination process is largely uncontrolled; further, like the CCNE, the EGE has clearly been conceived as an administrative expert body in bioethics. In view of the limited number of its members, not all EU countries are represented in the EGE; however, there is only one member per member state. North/south origin is carefully balanced.

The Italian CNB consists of forty members, who are nominated on a multidisciplinary basis. This body is comparable to the French CCNE in the sense that it is designed to act as an administrative body of experts. This committee basically comprises individuals from a scientific or political background, and their opinions reflect the divergence in views characteristic of the Italian public. Thus, discussions on major issues, such as the status of the embryo, generate no greater consensus inside the CNB than they do in the public sphere.

The Greek Bioethics Committee (GBC) is composed of seven members. Appointed by the General Secretary of Research and Technology, they are all men and all university professors, with the notable exception of a journalist. In terms of background, there are roughly equal numbers of social scientists and biologists. All of them are specialists in the field of biomedicine and bioethics. Therefore, the GBC can also be seen as an administrative expert body.

In conclusion, with the exception of the Danish Council of Ethics, the central committees are designed as expert bodies, whose members come predominantly from the scientific community or have had some involvement with scientific research. Overall, it can be seen that the national bioethics committees mostly represent what might be called a 'scientific establishment', even if, in the legal texts, the desirability of members being drawn from a plurality of backgrounds is generally endorsed. Does this mean that this minority has managed to appropriate the ethical discourse on biotechnology? In other words, has the institutionalisation of bioethics led to an impoverishment of the debate on bioethics? Two additional factors may help in addressing this issue: first, the existence and role of ethical bodies in 'competition' with the aforementioned committees; second, the power of the central bioethics bodies in relation to the media.

The European Group on Ethics has – at its level in the category of European central bodies – only one competitor: the Advisory Group on Ethics (AGE) set up by Europabio, the European Association of Bioindustries, at the beginning of 1998. AGE was created to advise Europabio and its members on ethical questions raised by their research into, and the exploitation of, biotechnology. To date, it has not yet released an opinion and, therefore, it is extremely difficult to evaluate its activities. But the EGE also has connections with other national, central bodies, with which it meets, and to which some of the EGE's members belong. But some of the 'younger' national central bodies, those that have not yet clearly defined their positions, may be susceptible to influence by the prominent and well-established EGE. To a certain extent this does indeed seem to have been the case for the Greek Bioethics Committee. This is certainly the impression given by the title and contents of the first volume of a booklet edited by the GBC: 'Views of the European Bioethics Committee about the Ethical Consequences of Biotechnology'. However, the other groups do tend to act independently of the EGE. This situation helps to maintain a minimum of diversity among bioethical forums. Nevertheless, it must be recognised that the EGE is the only expert body that is consulted and listened to by the Commission. This factor obviously considerably strengthens its authority.

In Denmark, the danger of limiting the participation in, and diversity of, bioethics debates is largely obviated by the coexistence of two separate bodies: the Central Scientific Ethical Committee, which comprises technical experts and is devoted to research, and the Danish Council of Ethics, a body open to the public and geared to stimulating public debate. In a similar vein, Italy hosts numerous unofficial institutions dealing with bioethics, such as the Fondazione Centro San Raffaele del Monte

Tabor, the Centro di Bioetica dell' Università Cattolica del Sacro Cuore in Rome, the Centro di Bioetica della Fondazione Instituto Gramsci of Rome, and the Instituto Giano di Bioetica di Genova. Their number and their vitality help to maintain a genuine public debate on bioethics, and they do not hesitate to criticise the advice proffered by the National Committee for Bioethics. In France, the CCNE retains a leading position that is unchallenged by any other body. Unlike the CCNE, the local research committees set up since 1988 are unable to deal with general questions, and there are no private bodies that can do so. In Greece, the Orthodox Church established its own Bioethics Committee. And, although it does not have a formal relationship with the state, its opinion is taken seriously by state authorities. From 1997 onwards, a number of other small-scale advisory committees have also been set up to deal with bioethical issues, and these have assumed advisory roles within the different ministries.

Relationships among these central bodies, and between their members and the media, also have a profound impact upon the public debate on bioethics. Some institutions, and members, have established very close and fruitful relationships with the mass media. The cases of Noëlle Lenoir, head of the EGE, and Axel Kahn who were both members of the French CCNE, provided good examples of this. Their close ties with the media maximised the prominence and influence enjoyed by their organisations. Nonetheless, generally speaking, the central ethics committees are extremely careful with regard to communicating their positions to the media and publicly providing advice. This is in accordance with the principles on which most of these bodies were founded: in most cases, they received a mandate from the state to educate the public (this is the case in Denmark, involving the DER, and in France and Greece; in Italy this task is not specified for the national committee). As a result, close ties between the media and bioethics central committees are deemed acceptable only when they can be seen to be fostering public debate on questions of bioethics. On the other hand, if the public debate does not really exist, media coverage may have the effect of enhancing the institutional authority of state bodies, and conferring disproportionate influence on them. This is, by and large, the case in France. In Italy, on the other hand, the number of publications on the subject of bioethics prepared by 'competitors' (for example, *Politeia*, *Problemi di Bioetica*, *Quaderni di Etica e Medicina*) ensures a far greater plurality in public debates.

We may now turn to the second main question: What is the real impact of these bodies on the public policies adopted in the field of biotechnology? If these bodies, because of their institutionalisation and their role in the regulatory process, are asked for advice by governments, they will almost certainly exert some influence. However, this does not mean that their advice is always adhered to by governments and legislators. Several reasons for this may be identified. First, the problems the committees deal with do not always need a legal solution. Second, governments and legislators have to take into account political as well as ethical considerations when formulating policy. Political factors may stymie the adoption of new laws (a good example of this is Italy concerning the problem of reproductive technologies), or lead to the acceptance of alternative legislative solutions. Third, advice offered may conflict with existing laws. A recent example is provided by the Commission's request for the EGE to decide whether or not it is ethical to fund research on embryos. The EGE's advice was released in November 1998 and, despite being fairly balanced, it was not accepted by Jacques Santer (president of the European Commission at the time) and Edith Cresson (head of DG XII). For the first time, advice requested by the Commission had been subordinated to political considerations.

Conclusions

To conclude, the institutionalisation of bioethics raises serious problems concerning the development of public debate in the field of bioethics. In particular, there is the serious danger of suppressing the diversity of ethical opinions traditionally expressed within our societies, and, instead, imposing upon society the 'ethics of the scientific establishment'. The different ethical positions we considered in the first section are not always adequately represented. Moreover, with the establishment of a central European bioethics committee, a related risk is that of losing the diversity of ethical approaches among the different European countries.

These outcomes are not especially surprising. Most of the national bodies are designed to advise governments, in their capacities as expert bodies, and to educate the public as teachers. In other words, their mission is not to host public discussions or to stimulate public debate. At any rate, these bodies have imposed their views less than their institutional presence in the regulatory process or in the decision-making process regarding biotechnology. These central bioethical committees have become a tool to legitimate the regulatory process in the field of biotechnology. Yet, in seeking to enshrine values within the law, there is always the risk of creating confusion between legal and ethical norms, as well as between morals and the law.

A final danger is that of limiting the public debate on bioethics. The demand for discussions of bioethics is part of a general phenomenon in modern Western societies, reflecting a public desire for the establishment of clearly stated, fundamental norms. One can also discern this prevalent trend in the multiplication of declarations of rights and an increasing emphasis on basic human rights. As such, it is appropriate for debates on bioethics to encourage reflection upon values and the public's discussion of moral choices. But these values should not be treated as concepts that can be adopted as and when required in order to help legitimise political or scientific interests. Avoiding this sort of scenario requires public participation in bioethical bodies to be substantially increased. Bioethics bodies must follow the example of Denmark, and come down, as it were, from their ivory towers. The lack of consensus conferences illustrates just how insufficient public involvement has been, and points to an obvious means by which it could be enhanced.

NOTES

1. In this paper we analyse only the data sets from the four countries under study. In Denmark, the newspapers analysed were *Information* and *Politiken*, and the data set is a sample from total press coverage ($n = 300$). In France, the newspaper analysed was *Le Monde*; the data set is a sample from total press coverage ($n = 622$). The data set from the French press does not include coverage before 1982 apart from a single year, 1975. In Greece, the newspapers analysed were *Kathimerini* and *Eleftheropia*, and the data set was constructed by a fixed sampling ratio ($n = 65$). In Italy, the newspaper analysed was *Il Corriere della Sera* and the data set was constructed through a variable sampling ratio from the total press coverage ($n = 340$). Each article was coded for up to three themes (see chapter 3 in this volume). This analysis makes no distinction among the order of themes coded.
2. Readers unfamiliar with philosophical ethics may find good introductions and discussions in, for instance, Singer (1991), Rachels (1993), or Dyson and Harris (1994).
3. For details, see Durant et al. (1998).
4. *Directive Proposal on Patentability of Biotechnological Inventions*, Doc COM (88) 496 final/SYN 169 (21/10/88).

REFERENCES

Dalla-Vorgia, P., V. Kalapothaki and A. Kalandidi (1993) 'Bioethics in Greece', *Quaderni di Bioetica e Cultura* January: 101.

Durant, J., M.W. Bauer and G. Gaskell (eds.) (1998) *Biotechnology in the Public Sphere. A European Sourcebook*, London: Science Museum.

Dyson, A. and J. Harris (eds.) (1994) *Ethics and Biotechnology*, London: Routledge.

European Commission (1996) 'Proposal for a Council Decision Implementing a Programme of Community Action to Promote the Competitiveness of European Industry', *Bulletin* EU 1/2-1996, point 1.3: 77.

Hilgartner, S. and C.L. Bock (1988) 'The Rise and Fall of Social Problems: a Public Arenas Model', *American Journal of Sociology* 94: 53–78.
Lenoir, N. (1998) 'Introduction to the EGE' at http://europa.eu.int/comm/european_group_ethics/gee1_en.htm#introd.
Potter, Van R. (1971) *Bioethics: Bridge to the Future*, Englewood Cliffs, NJ: Prentice Hall.
Rachels, J. (1993) *The Elements of Moral Philosophy*, 2nd edn, New York: McGraw-Hill.
Singer, P. (ed.) (1991) *A Companion to Ethics*, Cambridge, MA: Blackwell Publishers.
Solbakk, H. (1991) 'Ethics Review Committees in Biomedical Research in the Nordic Countries', in *Arsmelding (1990)*, Oslo: NAVF, p. 19.
Spagnolo, A. and E. Sgreccia (1987–8) 'I comitati e commissioni di bioética in Italia e nel mondo', *Vita e Pensiero*, 500–14.

5 Controversy, media coverage and public knowledge

Martin W. Bauer and Heinz Bonfadelli

During their life cycles, some new inventions leave the inventor's laboratory and enter the public sphere. However, before the new technology has materialised into new services and products, it first assumes a largely symbolic existence. Thus, its 'real' appearance is preceded by requests for venture capital, from the state, banks or the stock market, and by attempts to persuade investors, producers and consumers to endorse the new technology in imagination. In this way, new technologies enter the public sphere to confront existing expectations, concerns and value. The further development of a technology is thus determined by many variables, and the contingency of the trajectory it assumes will often obviate the possibility of predicting the outcomes.

In this chapter we explore the emerging representation of modern biotechnology in Western Europe during the early 1990s. Our focus is on the intensity of representation rather than on its quality. We assume that public attention and awareness are a precondition for the formation of clear images, opinions and attitudes about an emergent technology. We also recognise that biotechnology is represented in a variety of different modes: it exists as knowledge in the individual 'mind'; as an informal topic of conversation at work or over coffee, beer or dinner; and as a theme in formalised mass media coverage. The relations among these different modes of representation are of key theoretical interest (Bauer and Gaskell, 1999). It is crucial to ask how important an issue biotechnology is in the various modes of individual cognition and informal/formal communication. Further, how do these modes of representation interact to create a public image of biotechnology that has the power to influence its future development? And do these modes of representation interact with each other to increase or decrease the salience of biotechnology and, in this way, affect its development?

The 'knowledge gap hypothesis' is a mid-range theory of mass media effects that explores the intensity of representations of non-local issues by examining the relations among the intensity of media coverage, the level of controversy and the level of pluralism characterising a particular public sphere. Representation is, therefore, defined in terms of the distribution of knowledge on a given topic in the public sphere. The theory predicts that, over a number of years of media coverage of an issue, this distribution will change. First, more information will not necessarily lead to more equal distribution of knowledge; on the contrary, the differential between the knowledge of the elite and that of the remainder of the public will widen (the knowledge gap). This gap is likely to close only under specific conditions. Secondly, with time, the elite section of society will be saturated with knowledge (the ceiling effect), and the other public strata will have an opportunity to catch up. Thirdly, public controversy on the issue will fuel an information flow and reduce the knowledge gap, *ceteris paribus*. Finally, the more plural the public sphere, the less rapidly this gap will close even during a controversy. In other words, pre-existing pluralism in the public sphere neutralises the effect of controversy on the distribution of knowledge.

The production of knowledge in modern society has assumed historically unprecedented proportions. The emergent knowledge–ignorance paradox highlights the fact that, as the supply of new knowledge increases, levels of relative ignorance grow in parallel. Given the finite capacities of the individual mind, an individual's relative share in the stock of available public knowledge will inevitably decrease. Any one person becomes less and less well informed on any specific topic. In view of this, a premium is placed upon making intelligent selections of the available knowledge and achieving a meaningful integration of subjects and perspectives. However, in the context of societies in which ignorance is unavoidable, it is legitimate to ask: Why should one be at all concerned by the unequal distribution of knowledge relating to biotechnology? After all, considering that knowledge is now very far from static, an equality of knowledge would seem to be a far-fetched aspiration. Yet we believe that this is simply not the case. In evincing this claim, we will advance two arguments, one in the context of mass media studies, the other in the context of research into the public understanding of science. Both arguments support our contention that a well-informed public is vital to carry political decisions, and therefore imperative for the effective functioning of modern democracies.

Do the mass media influence or inform the public?

The concept of the knowledge gap is situated within the tradition of media effect studies. Media effects research was dominated, for a long time, by a focus upon purposeful short-term persuasive effects that were measured by studying how an individual's opinions, attitudes and stereotypes were influenced by the media. The key question was: how can modern mass media be used effectively to influence political or social behaviour, in general, and the acceptance of new technologies such as biotechnology, in particular? This focus upon media persuasion did not change until the early 1970s, when longer term cognitive phenomena, for instance, agenda-setting, knowledge gain and the cultivation of images of the social world, became the new focuses of media effects research (McComb, 1994: 3).

One important reason for this paradigm shift was a recognition that, before the media can begin to influence specific attitudes, the topic (or attitude object) has first to be generated by personal experience, informal conversations or formalised mass media coverage. As a result, the academic focus shifted from the question of 'How do people feel about' an issue, to 'How many people think about it at all?' This new question assumes that the public will learn about a topic, such as biotechnology, from the news media. In this respect, the most important effect of the mass media is their ability to confer salience on a topic in the minds of their audience. Thus, journalists employ news value and gate-keeping practices to select and to frame information in particular ways. People may be in a position to develop opinions and attitudes only on the basis of information that has been mediated in this way.

For this reason, the study of learning by mass media is of special interest in the domain of new and complex technologies. The technicalities of genetic engineering and modern biotechnology are complex, abstract and far beyond the majority of people's personal experience. Thus, by necessity, knowledge concerning science and technology is to a great extent cultivated and conveyed by way of media representations. Indeed, several studies of the media have demonstrated their importance as sources of information related to technology (Wade and Schramm, 1969; Robinson, 1972; Griffin, 1990), and especially where danger may be implicated (Mazur, 1981; Dunwoody and Peters, 1992; Coleman, 1993).

In 1970, Tichenor, Donohue and Olien, researchers at Minnesota University, formulated their knowledge gap hypothesis (KGH) in a programmatic paper entitled 'Mass Media and Differential Growth in Knowledge'. In the initial version, the hypothesis stated: 'As the infusion of mass media information into a social system increases, segments of the population with higher socio-economic status tend to acquire this information at a faster rate than the lower status segments, so that the gap in knowledge between these segments tends to increase rather than decrease' (Tichenor et al., 1970: 159–60). Since then, more than twenty-five years of research and more than a hundred empirical studies have dealt with knowledge gap phenomena, and this has stimulated further theoretical elaboration and refinement of the model (Bonfadelli, 1994; Gaziano and

Gaziano, 1996; Viswanath and Finnegan, 1996). The knowledge gap perspective questions normative assumptions about the role of the mass media in democratic societies by adducing empirical evidence for the existence of a chronically uninformed public. This model allows one to connect the timing of information diffusion in society to social structures. The underlying assumption is that a mere increase in information will not automatically result in a better and more equally informed public. Instead, the well educated segments of society are able to use the media more efficiently than the less well educated, and are better placed to take advantage of the abundance of mass media information. As a result, the knowledge gaps between the different social segments will increase rather than decrease. Furthermore, information dissemination conforms to the 'Matthew principle': those who have it will be given more.

Observations of a widening of knowledge gaps come from a variety of empirical studies conducted in the USA and Europe. Numerous cross-sectional studies have reported strong correlations between education and social status and between education and knowledge levels relating to such issues as politics, but especially science, technology and health. Furthermore, several studies have shown that the well educated tend to be reached more effectively by mass media information campaigns, and that mass media news items travel more rapidly in the higher social strata. But there are inconsistencies in this general picture. Several panel studies have reported both a stabilisation in the knowledge gaps and, in some cases, a narrowing.

These deviations from the original hypothesis have been explained in recent years in terms of contingent factors mediating between mass media coverage on the one hand, and audience reception on the other (Bonfadelli, 1994). In 1975, Donohue, Tichenor and Olien refined their KGH at the macro level by studying two mediating factors. They argued that gaps tend to be higher, and to increase, in more pluralistic public spheres and in the absence of controversy; whereas gaps are smaller, and tend to decrease, in more homogeneous settings and when controversy and conflict are involved. Additional factors operate at a micro level (Ettema and Kline, 1977). Audience factors that condition the knowledge gap include interest in a topic and the related motivation to seek information, communication skills, habitualised access to information-rich print media and frequent informal conversations on a topic. Factors related to media content and format that make a difference to the knowledge gap are the knowledge topic (whether it is practical or abstract), the type of knowledge (agenda, factual or structural knowledge), media channels (newspaper versus television) and the duration and intensity of media coverage.

Public controversy as a lever of public understanding of science

The KGH contributes to the discussion of controversies over new technologies in two key respects. First, in its elaborated version KGH emphasises the functionality of controversy for the development of a new technology. It identifies social controversy as a mediating factor that militates against the tendency for normal media to amplify the knowledge gap between the educated and the less educated sections of the public.

Secondly, controversy and conflict play a role in stimulating and maintaining the participation of citizens in the polity. The greatest enemies of democracy are not conflict and controversy, but apathy and indifference towards public affairs. Controversies, including those concerning new technologies, revitalise old values, redefine them through their applications to new situations, strengthen the cohesion of those engaged in conflict on all sides, and accelerate the circulation of information and knowledge. Newspapers, radio, television, magazines and other mass media play a vital role in such processes. These contributions of conflict to the sustainable development of a new technology are easily forgotten by short-sighted actors who envisage a quiet world of expert control in which time and money are saved through exclusive decision making and the consequent avoidance of public controversy. By way of example, the Swiss referendum on gene technology of June 1998 required a concerted effort by all parties involved, over a period of several years, first to launch the referendum and then to persuade the public of the rightness of their positions. Myopic, technocratic voices abound who consider these efforts to be a waste of time and money.

Controversies involving new technologies, such as biotechnology, complement the normal educational channels for the diffusion of knowledge. They provoke informal conversations on the topic in the home, at work or at leisure, and direct attention to media coverage that might otherwise be overlooked, in turn prompting further media coverage of the topic. This intensifies the controversy and mobilises an ever wider range of people to participate in the representation of the issue in the public sphere. Clearly, controversy fuels a virtuous circle of attention, awareness and knowledge. And although awareness and knowledge of a new technology cannot guarantee its future in modern industrial society, we can assume them to be necessary conditions.

In a democracy, public familiarity with an issue is a precondition for realistic and sustainable decision-taking pertaining to it. Democracy feeds on the culture of checks and control that high levels of public knowledge permit. Decisions that are unintelligible to the public are likely to raise suspicions and quickly become unpopular. An informed public is, therefore, necessary for efficient democratic government in the face of complex, non-local issues; and in a deregulated economy only informed consumers make for efficient market operations. Furthermore, a new technology requires a prominent public profile if it is to attract venture capital and to recruit enthusiastic young people for its activities.

The Anglo-Saxon debate on the public understanding of science has, to date, considered knowledge mainly as an independent variable. From the educationalist perspective, knowledge is mainly of interest as a predictor of attitudes (Miller, 1983), a common-sense expectation inspiring funding agencies and policy-makers ('to know it is to love it'). Many people therefore expect that further education on modern biotechnology will produce a public that is more supportive of its projects and products. This is also, from a polemical point of view, a convenient claim. By logical reversion it discredits the sceptics: if knowledge breeds love, ignorance breeds contempt, therefore people who challenge science and technology must be ignorant. Yet this claim is at variance with the empirical evidence.

The hopeful expectation of finding a link between knowledge of and support for science and technology (justifying, among other things, increased educational endeavours) has encouraged almost two decades of survey research into the public's understanding of science. However, the empirical evidence for a positive correlation between these factors is slim. Instead, models need to be highly context sensitive. Thus, some of us have developed a model of the public understanding of science in which this very relationship is variable and shifts as societies move from an industrial to a more post-industrial mode of production (Bauer, 1995; Durant et al., 2000).

The debate on the function of knowledge in public understanding of science research has meanwhile diversified into three areas. First, researchers are seeking to define the form of 'knowledge' that requires explication. Most of the time we are measuring the text-book enculturation of 'facts-out-of-context' rather than the local appropriation of knowledge for practical purposes. Unfortunately, in the past, conceptual confusion on this issue has led to unproductive polemics on methodology (Irwin and Wynne, 1996). Secondly, researchers analyse the context of different knowledge–attitudes relationships (Bauer et al., 1994; Durant et al., 2000). Thirdly, researchers investigate knowledge and its antithesis, ignorance, as dependent variables, and model their relations with various individual or structural determinants (Bauer, 1996). Our present concern with the KGH extends and elaborates the third concern by focusing on the distribution of knowledge in relation to structural variables, such as the intensity of media coverage, the level of public controversy and the pluralism of the public sphere.

The distribution of knowledge and its constraints: a two-level model

Based on these considerations, several research hypotheses can be formulated. The diffusion of information concerning biotechnology is constrained by the knowledge level of the single individual, and by structural factors at the level of the public sphere. In our case, the boundaries of the public spheres correspond to the political boundaries of European countries. The Eurobarometer survey of 1996 gathered data on knowledge levels, education and social status, and was coordinated and conducted in seventeen European countries. Our individual-level model stipulates a chain of influence as schematized in model I:

Model I: education > relevance > media discrimination/conversation > knowledge

General education provides the citizen with a sense that biotechnology is a relevant issue, even if remote from immediate local concerns. This sense of relevance will lead to their selecting articles in the press or programmes on television or radio for immediate attention, or it may prompt informal conversations with friends or colleagues, or in the family, related to biotechnology. Citizens with higher levels of education, who have a sense of the relevance of biotechnology, who are inclined to pay attention to the media coverage and who discuss it, will eventually arrive at an awareness and knowledge of the specific issues involved. In other words, *differences* in terms of education, estimations of the relevance of biotechnology, the selection of media topics and the frequency of conversations on this subject will, to a large extent, explain the distribution of biotechnological knowledge in a society.

We have formulated three basic hypotheses on the distribution of knowledge about biotechnology in European countries.

1. Better-educated groups will be more knowledgeable about biotechnology and will be more interested in the topic. Since the level of formal education differs in the various countries of Europe, we also expect that the level of knowledge about biotechnology will vary according to the country's overall educational attainment.
2. Knowledge levels will be related to message discrimination: the more people in a country who attend to biotechnology news items, the higher the level of knowledge in that country.

3. The more people who converse about biotechnology in a country, the higher the level of knowledge it will exhibit.

At the national level, model II stipulates structural factors that mediate the flow of information:

Model II: K-gap = *f* (media coverage/level of controversy/ pluralism of public sphere)

Testing this model suggests three further hypotheses:

4. Knowledge gaps will increase as media coverage intensifies.
5. Knowledge gaps will decrease as media coverage intensifies *if* there is public controversy over biotechnology, as this will lead to an expansion of the information flow. Controversy may encourage media channels to cover the topic more frequently, and channels that would otherwise have neglected it, may begin coverage and thus bring the topic to a wider audience.
6. Knowledge gaps will increase as media coverage intensifies as a function of the plurality of the public sphere in the country. The mass media in a more pluralistic public sphere are likely to cater for a variety of news values and concerns. In a pluralistic context a non-local topic, such as biotechnology, may be 'ghettoised' to a few channels with specific concerns.

Testing model I requires survey data; model II requires, in addition, measures of the intensity of media coverage, of the level of controversy and of the pluralism of the public spheres. The basic fact we aim to establish concerns the existence of knowledge gaps relating to biotechnology in various European countries. This constitutes the explanandum of our models.

Inequalities of knowledge of biotechnology

The Eurobarometer study of 1996 contained ten knowledge questions concerning biotechnology (see table 5A.1 in the appendix to this chapter). The percentage of correct answers varies, from item to item, between 21 per cent and 84 per cent. Correct answers were summated to provide an indicator of biotechnological knowledge for each individual and an aggregate for each country, with an overall mean score of 5 out of 10. Better-educated people gave more correct answers on every knowledge item. Table 5.1 shows that the knowledge gaps, measured as the differences in knowledge between respondents with low and high education, vary from country to country (from 1.5 in Austria, Denmark and

Table 5.1 *Knowledge levels and knowledge gaps in different countries*

Country	OECD education[a]	Knowledge level[b]					
			Individual education level[c]			Knowledge gap	
		Knowledge-10	Low	Medium	High	Diff.[d]	Corr.[e]
Germany	1	4.7	4.0	4.6	5.7	+1.7	.28
Switzerland	2	5.9	4.7	5.5	6.6	+1.9	.31
Norway	3	5.0	3.7	4.6	5.7	+2.0	.35
UK	4	5.7	4.9	5.6	6.8	+1.9	.31
Sweden	5	5.9	5.0	5.9	6.6	+1.6	.32
Austria	6	4.0	3.2	4.3	4.7	+1.5	.27
France	7	4.9	3.8	4.8	6.0	+2.2	.39
Finland	8	5.5	4.1	5.2	6.3	+2.2	.40
Denmark	9	6.0	4.8	5.7	6.3	+1.5	.28
Netherlands	10	6.3	5.1	6.0	7.1	+2.0	.37
Belgium	11	4.6	3.7	4.4	5.6	+1.9	.32
Ireland	12	4.2	3.7	4.1	5.4	+1.7	.24
Greece	13	3.8	3.0	4.0	4.8	+1.8	.36
Italy	14	5.1	4.2	5.2	6.0	+1.8	.34
Luxembourg	15	4.9	4.2	4.6	5.7	+1.5	.26
Spain	16	4.2	3.3	4.6	5.5	+2.2	.38
Portugal	17	3.8	3.1	4.9	5.5	+2.4	.44
Total		5.0	3.8	4.9	6.1	+2.3	.38

[a] Ranking according to OECD 1995 secondary-level educational attainment of 25–64 year olds.
[b] Score on ten-item Eurobarometer 1996 questionnaire.
[c] Based on school-leaving age; $N = 16{,}500$.
[d] Difference between low and high individual education.
[e] Correlation of knowledge-10 and individual level of education.

Luxembourg to 2.4 in Portugal). This picture of a direct relationship between education and knowledge holds at the country level too: the higher the country's educational attainment, the higher is the average level of knowledge of biotechnology ($r = .57$, $n = 17$).

The KGH postulates an unequal distribution of biotechnological knowledge between the most and least educated segments in the different countries of Europe. The knowledge gap, measured as the correlation between the individual's level of education and knowledge score (knowledge-10), is $r = .38$ ($n = 16{,}500$) and is discernible in every country, varying between $r = .24$ for Ireland and $r = .44$ for Portugal. Comparisons among the different countries also demonstrate knowledge gaps between countries: knowledge levels are particularly high in countries with high levels of education measured by secondary school attainment, such as the Netherlands, Denmark and Switzerland, whereas knowledge levels are low in countries with educational deficits, such as Greece, Spain and Portugal. However, the knowledge gaps in the different countries do not vary significantly with the level of educational attainment of these countries ($r = .26$, $n = 17$).

The personal relevance of biotechnology

Education functions as a motivational factor for interest in topics that are abstract and not of existential personal relevance. Therefore, one would expect that the higher the level of education individuals have received, the more they will perceive the personal relevance of biotechnology; and, at the macro level, issues related to modern biotechnology are deemed more important in countries with higher levels of education.

Clearly, knowledge is one thing, but the perceived relevance of biotechnology is quite another. People may know about it but consider it to be remote from their daily concerns. A person acquires further knowledge only on those issues that are perceived to be relevant – topics that trigger a pre-existing interest or express a particular concern. Perceived relevance can therefore be seen as the first in a chain of factors that incline a citizen to look out for more information on this novel issue.

Personal relevance was measured with question 15 in Eurobarometer 46.1 (see Durant, Bauer and Gaskell, 1998): 'We've been discussing several issues to do with modern biotechnology. Some people think these issues are very important whilst others do not. How important are these issues to you personally?' The response was given on a scale from 1, not at all important, to 10, very important.

There is only a weak correlation between self-assessments of the personal relevance of modern biotechnology and individuals' levels of education ($r = .16$, $n = 16{,}500$). Individual education does not seem to affect the interest that European citizens display in modern biotechnology. The average level of interest in the low education group registers 5.9 points on a ten-point scale and increases to 7.0 points in the high education group. This pattern at the individual level is mediated by factors at the country level, as shown in table 5.2. Switzerland, Austria, Finland and Sweden are countries with high levels of education in which biotechnology is perceived to be of high personal relevance; and in Ireland, Belgium, Spain and Portugal both factors are at low levels. However, Norway, Germany and the UK are countries with high levels of education but in which biotechnology registers low assessments of perceived personal relevance (note that this study was performed at

Table 5.2 *Relevance of biotechnology and level of education by country*

Constellation		Country	Education Per cent[a]	Education Rank	Relevance Per cent[b]	Relevance Rank
Education	High	Switzerland	82	2	43	6
Relevance	High	Austria	69	6	47	4
		Finland	65	8	43	6
		Sweden	75	5	48	1
Education	High	Norway	81	3	28	17
Relevance	Low	UK	76	4	36	11
		Germany	84	1	33	13
						10
Education	Lower	Denmark	62	9	48	1
Relevance	High	Netherlands	61	10	42	7
		Greece	43	13	48	1
						10
Education	Low	Ireland	47	12	30	15
Relevance	Low	Belgium	53	11	31	14
		Italy	35	14	37	10
		Spain	28	16	29	16
		Portugal	20	17	34	13

[a] OECD 1995 secondary-level educational attainment of 25–64 year olds; median = 63 per cent.
[b] Eurobarometer 1996 percentages of 'high' perceived relevance (8–10 points); median = 37 per cent; $N = 16{,}500$.

the end of 1996, before the controversies over GM food and Dolly the sheep). In contrast, Denmark, the Netherlands, Greece, Luxembourg and Italy have lower levels of secondary educational attainment but interest in biotechnology is relatively high.

Attention to biotechnology: message discrimination in different media

People may buy a newspaper or magazine, listen to the radio or watch television news. Yet this does not mean that they will pay attention to what they see or hear. For this reason, our survey asked whether respondents remembered having encountered anything related to biotechnology in the previous three months in newspapers or magazines or on the radio or television. This item was used as a measure of message discrimination, i.e. as an index of the amount of interest the public pays to biotechnology.

Table 5.3 *Message discrimination by level of education and relevance (%)*

		Education				Relevance			
Media channels[a]	Total	Low	Medium	High	Corr.[b] K_{tau}	Low	Medium	High	Corr.[c] K_{tau}
Television	35	24	36	44	+.20	27	35	46	.17
Newspapers	25	12	24	37	+.25	18	23	34	.15
Magazines	12	6	12	19	+.13	8	12	19	.13
Radio	11	5	10	16	+.11	7	9	15	.10
Print and AV channels	20	9	20	29		14	18	28	
Print only	11	6	11	16		9	11	14	
AV only	18	17	18	19	+.25	16	19	21	.20
Channel not known	6	6	6	7		6	6	7	
No discrimination	45	62	45	29		56	45	31	
Number of channels:									
2+	23	12	23	34		16	21	32	
1	32	26	32	38	+.24	27	34	37	.21
0	45	62	45	29		57	45	31	

Source: Eurobarometer 46.1, 1996.
Notes: $N = 16{,}500$.
[a] AV = audio-visual (radio or television); print = newspaper or magazines.
[b] Correlation between education and message discrimination.
[c] Correlation between relevance and message discrimination.

Education is a prerequisite for habitual media use and message discrimination. As such, the habitual use of media as sources of information about biotechnology is influenced by an individual's educational level. This is apparent from the fact that more people with higher education have heard about biotechnology than have those with lower education. Education correlates with the number of channels on which individuals discriminate biotechnological messages. The relationship between education and media sources is stronger for print media than for television and radio. Access to media sources is influenced not only by education but also by personal relevance of biotechnology.

Table 5.3 shows the use of different mass media channels, their combination and their number in relation to levels of education and perceptions of the personal relevance of biotechnology, expressed as percentages. Education clearly correlates with discrimination of biotechnology messages in the mass media: $K_{tau} = .25$. Whereas 62 per cent in the low education group could not remember having heard any mention of biotechnology in the mass media over the previous three months, fewer than 30 per cent of the higher educated respondents fell into this category. Furthermore, education correlates with the number of different forms of media in which respondents could recall reference to biotechnology having being made. In addition, the more highly educated consulted the print media significantly more frequently than people with a low level of education. The correlation between education and reading of newspapers is stronger than that between education and reliance upon the audio-visual media. These communication gaps are also apparent at the national level. Altogether, our data support the notion that attention to and discrimination of biotechnology media messages are unequal phenomena.

However, biotechnology message discrimination is not only influenced by education and perceived personal relevance. The combination of education and motivation shows that both factors interact in an additive way. Only 31 per cent of people with lower levels of education and a low interest in biotechnology had access to information about biotechnology in the mass media. In contrast, 71 per cent of the well-educated and highly interested group had been informed by the mass media.

As can be seen from table 5.3, 35 per cent of respondents had heard, over the preceding three months, about issues concerning modern biotechnology through television, 25 per cent from newspapers, 12 per cent from magazines and 11 per cent by way of radio. Television seems to be the most important single media channel for disseminating information about biotechnology (this is far from surprising, considering that the same was true for all non-local news items in most countries by the 1990s). Taken together, 55 per cent acquired their information on biotechnology via mass media channels. In comparison, 52 per cent derived their understanding from discussion of the topic with other people on at least one occasion. Indeed, since biotechnology is perceived, by many, to be of real importance, informal conversations have achieved a significance not far short of that of the mass media channels.

Biotechnology as a topic of conversation

The mass media are the most important sources of information in modern societies. This is especially true with respect to questions of science and technology. But interpersonal forms of communication can also be important, particularly where a topic is of high personal relevance. Furthermore, television displaced newspapers as the main information source, as many media studies in the USA and in Europe show. Our hypotheses in this section are that, first, television, rather than the print media or interpersonal communication, is the most widely utilised information source both for general information and for data concerning biotechnology; and, secondly, the higher the perceived personal relevance of biotechnology, the more people will use the mass media and interpersonal channels of communication.

Our data show that only a minority of people relied exclusively upon either the media (16 per cent) or conversation (11 per cent) in procuring information on biotechnology. However, 39 per cent used both channels and 34 per cent used neither media nor interpersonal channels. Furthermore, there is a significant correlation between the perceived relevance of biotechnology and use of media or interpersonal channels. So, the higher the personal importance of biotechnology, the more likely it is that the mass media, as well as interpersonal channels, will be recognised as important sources of information about modern biotechnology.

Slightly more than half of respondents had talked about biotechnology with other people, and the intensity of interpersonal communication varies strongly between countries. The countries associated with high-intensity discussions on biotechnology are Switzerland, Denmark, Austria, Finland, Germany and Norway, in which between 75 and almost 90 per cent have engaged in such discussions. Conversely, in Greece, Portugal, Spain and Ireland biotechnology does not seem to be a salient topic of day-to-day discussion among citizens by the end of 1996, as shown in figure 5.1.

Communication gaps exist not only in the use of mass media but also at the level of interpersonal communication. Thus, people with higher education talk about biotechnology with the greatest frequency; as a corollary of this, the amount of interpersonal communication about biotechnology is higher in those countries with higher levels of formal education. Clearly, there are significant communication gaps at the level of interpersonal communication about biotechnology similar to the gaps in media access. Table 5.4 shows that about half of the respondents with a high level of education talk occasionally – or even frequently – with other people about biotechnology. In contrast, almost three-quarters of the less well educated have not discussed biotechnology. These communication gaps also exist among countries. In those with a high educational attainment – most notably Switzerland, Denmark and Norway – informal discussions on biotechnology are frequent, with more than 70 per cent of respondents having participated in them. In countries with lower standards of education – such as Greece, Spain, Portugal and Italy – discussions of biotechnology among citizens are rare, with only about 30 per cent ever having conversed on the subject.

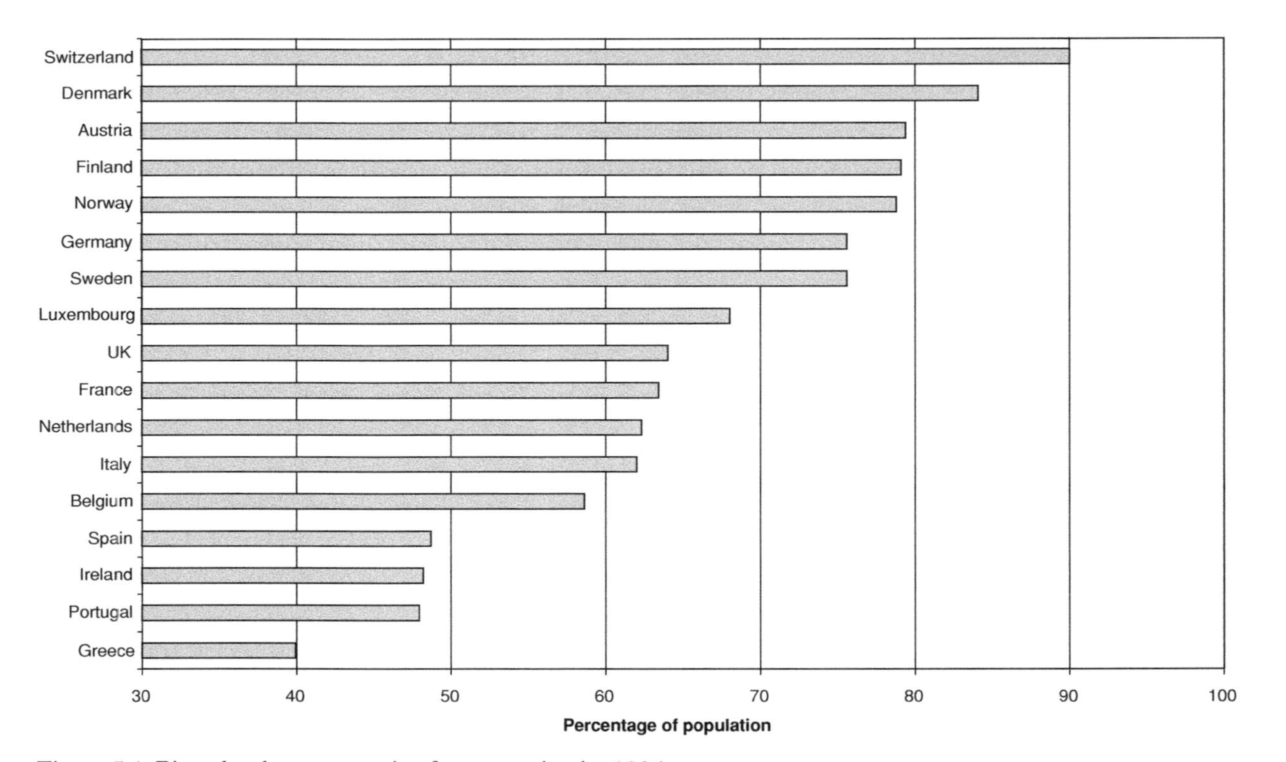

Figure 5.1 Biotechnology as a topic of conversation by 1996
Note: Percentage of yes responses to either of these questions: 'Over the last three months, have you heard anything about issues involving modern biotechnology?' or 'Before today, had you ever talked about modern biotechnology to anyone?'

Table 5.4 *Interpersonal communication by level of education and relevance (%)*

Amount of interpersonal communication	Total	Education			Relevance		
		Low	Medium	High	Low	Medium	High
Frequent	7	3	6	14	4	5	12
Occasional	27	14	26	41	18	27	39
Only once or twice	16	11	18	18	14	18	16
Never	50	73	50	30	64	50	33
Total	100	100	100	100	100	100	100
Correlation (K_{tau})			+.30			+.25	

Source: Eurobarometer 46.1, 1996.
Note: $N = 16{,}500$.

Knowledge, education and attention paid to biotechnology

Levels of message discrimination from different media provide an indication of an individual's knowledge of a given subject. In this case, people who remember having heard something about biotechnology in the media are likely to know more than people who have no recollection of having access to biotechnological information via the media. Media discrimination and level of education seem to have an additive effect upon individual knowledge. Moreover, table 5.5 shows that the mass media serve to narrow the knowledge gap somewhat, since levels of knowledge acquisition from access to newspaper or to television news are slightly more pronounced among the less well-educated segments of the data set. This notwithstanding, overall the mass media have less impact upon knowledge levels than does educational background, and therefore had little effect on the distribution of knowledge about biotechnology at the end of 1996.

Controversy and the polarisation of the public sphere

An important effect of social controversy is to direct the flow of information into regions that it would not conventionally reach. As such, in public controversies, the knowledge gap between the elite and lower strata is reduced faster than one might otherwise expect. This theoretical claim can be tested empirically using our data.

Table 5.5 *Knowledge levels by education and message discrimination*

	Knowledge level[a]					
		Education			Knowledge gap	
	Total	Low	Medium	High	Diff.[b]	Corr.[c]
Overall mean score	5.0	3.8	4.9	6.1	2.3	.38
Newspaper						
Yes	5.9	4.8	5.6	6.5	1.7	.29
No	4.7	3.7	4.7	5.8	2.1	.37
Media effect gap	+1.2	+1.1	+0.8	+0.7		
Television						
Yes	5.6	4.6	5.5	6.4	1.8	.32
No	4.6	3.6	4.6	5.8	2.2	.38
Media effect gap	+1.0	+1.0	+0.9	+0.6		
Channels						
More	6.0	4.8	5.7	6.6	1.8	.30
Print only	5.6	4.6	5.3	6.2	1.6	.29
Television/radio	5.3	4.4	5.2	6.1	1.7	.31
Source not remembered	4.8	3.8	4.6	5.6	1.8	.34
No media	4.2	3.4	4.6	5.5	2.1	.35
Media effect gap	+1.8	+1.4	+1.1	+1.1		
Combination of channels						
Media + interpersonal	5.9	4.8	5.6	6.5	1.7	.29
Interpersonal only	5.4	4.3	5.2	6.2	1.9	.33
Media only	4.8	4.1	4.8	5.7	1.6	.29
No communication	3.9	3.3	4.2	5.0	1.7	.28
Communication effect gap	+2.0	+1.5	+1.4	+1.5		

Source: Eurobarometer 46.1, 1996.
Notes: $N = 16{,}500$.
[a] Mean score on ten-item Eurobarometer 1996 questionnaire.
[b] Difference between low and high education.
[c] Correlation of knowledge score and level of education.

The reference point for a functional analysis is the national 'polity' in which citizens make daily decisions. We shop around for food, clothing and entertainment, take decisions on medication, invest trust in authorities and experts, and engage – periodically – in political activity by signing petitions, joining protests or casting votes on substantive issues. It is a basic feature of modern democracy that higher levels of knowledge on controversial issues stimulate greater political participation and improve the quality of decision-making in terms of both substance and legitimacy.

Knowledge allows one to operate with distinctions that would otherwise not make sense.

Conflicts and controversy will foster greater information flow with two measurable effects: knowledge gaps will be lowered in countries with high conflict and accentuated in countries that lack public debate on biotechnology; knowledge gaps will be higher in countries that have a homogeneous public sphere, and will be lower in countries with a segmented or pluralistic public sphere. For the purposes of this study, we have used three indicators of the level of controversy involving biotechnology in European countries: the scale of activities of the polity, the intensity of media coverage and the percentage of controversial media coverage focusing on the controversial aspects of biotechnology. We have also employed two indicators of pluralism: left–right political polarisation and religious polarisation (see the appendix for details).

Figure 5.2 shows the relationship between polity activity in any one country between 1985 and 1998 and the knowledge gap in 1996. The correlation is negative: $r = -.75$ ($n = 12$). As predicted by the hypothesis, public controversy measured in terms of polity activity is associated

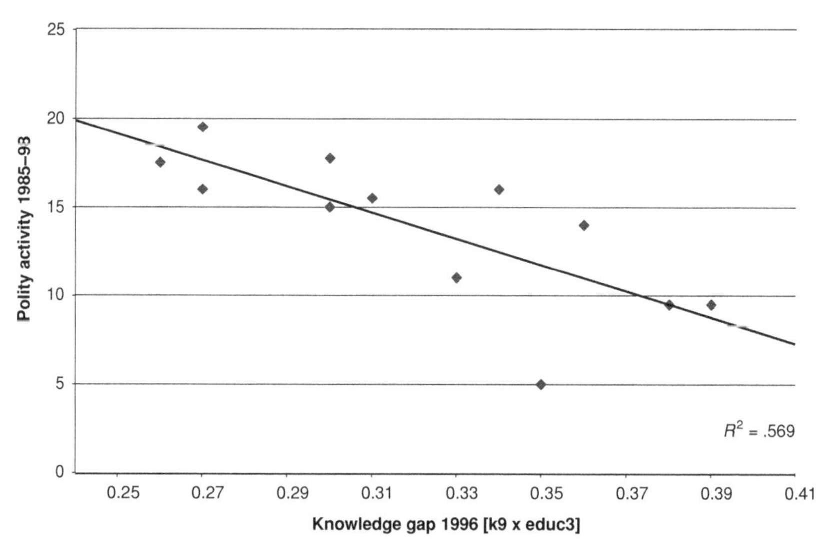

Figure 5.2 Knowledge gap, 1996, and level of polity activity, 1985–1998
Note: The knowledge gap is measured as the correlation between knowledge (k9 = nine knowledge items) and level of education (educ3 = three levels of education). For the measurement of 'political activity' see the appendix.

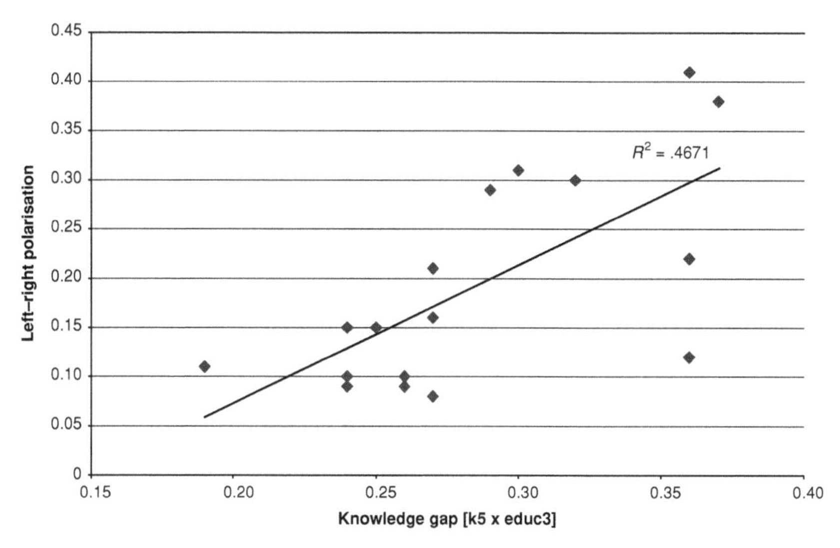

Figure 5.3 Political polarisation and knowledge gap, 1996
Note: The knowledge gap is measured as the correlation between knowledge (k9 = nine knowledge items) and level of education (educ3 = three levels of education). For the measurement of 'political polarisation' see the appendix.

with smaller knowledge gaps across twelve European countries. In countries where biotechnology and genetic engineering have been the subjects of policy-making, parliamentary (or other) debates, public hearings, demonstrations, petitions or even referenda, a higher proportion of the public will be aware of, and know about, this new technology. In this sense, public controversy surrounding biotechnology is an important factor in educating the public on this topic.

Figure 5.3 shows the relationship between the polarisation of the public sphere in seventeen European countries and the knowledge gap on biotechnology (both relating to 1996). As predicted, this correlation is positive: $r = .68$ ($n = 17$). The more plural the public sphere in terms of left–right political polarisation, the larger were the 1996 knowledge differentials. More pluralistic publics are likely to have more varied mass media, each with a different agenda and therefore promoting a more unequal distribution of knowledge on such non-local issues as the emergence of new technologies. Additionally, ethical concerns over biotechnologies would suggest that religious feelings and orientations play a part in stimulating public controversy. However, the degree of religious pluralism in a country does not affect the distribution of knowledge ($r = .16$).

Table 5.6 *Knowledge gaps, changes in knowledge gaps, intensity of media coverage, and trends in coverage*

Country	Knowledge gap[a]	Change in k-gap, 1993 to 1996[b]	Media coverage, 1995 + 1996[c]	Trend in coverage[d]
UK	.31	1.50	1,567	44
France	.39	1.43	541	46
Germany	.28	1.47	285	52
Austria	.27	–	440	57
Netherlands	.37	0.93	378	47
Switzerland	.31	–	357	47
Sweden	.32	–	130	48
Italy	.34	1.07	113	24
Finland	.40	–	82	43
Denmark	.28	1.20	28	43
Greece	.36	1.03	10	26
Spain	.38	1.16	–	–
Belgium	.32	1.13	–	–
Portugal	.44	1.44	–	–
Luxembourg	.26	1.00	–	–
Ireland	.24	0.70	–	–
Norway	.35	0.96	–	–
Median	.33	1.10	370	46.5

Source: Durant et al. (1998).

[a] The correlation between knowledge measured on nine items and the level of individual educational attainment.

[b] Ratio of k-gap for 1996 over k-gap for 1993 measured by the correlation of five-item knowledge and educational attainment. A figure >1 indicates increasing gap, a figure <1 decreasing gap.

[c] Estimates of the total number of articles in the opinion-leading newspaper for 1995 and 1996.

[d] The number of articles in 1995–6 divided by the number of articles in 1991–6. A figure >33 indicates increasing coverage, and a figure <33 decreasing coverage, by the mid-1990s.

Knowledge gaps may be either high or low in both homogeneous or more plural religious climates.

All previous results are based on cross-sectional comparisons within and among countries. Table 5.6 summarises the dynamics of the knowledge gaps within several countries. Thanks to an earlier comparable survey (Eurobarometer 37.1 of 1993), we are in a position to compare these knowledge gaps longitudinally between 1993 and 1996. This reveals that in six countries – Ireland, the Netherlands, Norway, Luxembourg, Greece and Italy – knowledge gaps remained stable; whereas in Belgium,

Spain, Denmark, France, Germany, Portugal and the United Kingdom they increased. The greatest expansions in the knowledge gap – close to 50 per cent – were recorded for the UK, France, Germany and Portugal.

For the Netherlands, Greece, Italy, Denmark, France, Germany and the United Kingdom, we are able to relate changes in the size of the knowledge gap to the level of media attention, and thereby assess the effects of media coverage and controversy on the gap. The intensity of the media's coverage of biotechnology in 1995 and 1996 relates to a widening of knowledge gaps ($r = .60$, $n = 7$). But, though intensified media coverage of biotechnology in the year before the survey accentuated the knowledge gap, it may do so with an effect that diminishes with the level of previous coverage. The knowledge gap effect in relation to already established levels of coverage is clearly a problem that requires further investigation.

In addition to examining cross-country variations in the intensity of media coverage, we were able to quantify within-country shifts in the media coverage. Countries that witnessed large increases in the media debate about biotechnology during the mid-1990s – such as Germany, France and the United Kingdom – experienced an enlarging of the knowledge gap; whereas Greece and Italy, countries with declining press coverage by the mid-1990s, saw the emergence of more stable knowledge gaps. The Netherlands presents a more ambiguous picture: this is a country with the highest average public knowledge about biotechnology, a large knowledge gap, relatively low levels of polity activity, and relatively intense (and increasing) media coverage, yet it displays a stable – or even narrowing – knowledge gap.

Conclusion: KGH is underspecified as a dynamic model

We have explored the representation of biotechnology in seventeen European countries during the mid-1990s. Proceeding from the 'knowledge gap' perspective, we compared the different representations of biotechnology: individual cognition, informal communication and mass media coverage. The hypotheses we generated stipulated the existence of knowledge gaps in the public's understanding of biotechnology, and the mediation of these gaps by various structural factors, in particular the intensity of media coverage and the degree of controversy over biotechnology in, and the pluralism of, the national public sphere.

In summary, our results present a rather ambiguous picture. Cross-sectional data clearly support the KGH. Knowledge gaps concerning biotechnology are present in all European countries, albeit to differing degrees. Individuals with a more general education are likely to know more about biotechnology than those who are less well educated. In all countries analysed, controversies on this subject in the polity, between 1985 and 1998, had the effect of decreasing these knowledge gaps, while greater left–right political polarisation worked in the opposite direction. However, longitudinal data for seven European countries show that knowledge gaps involving biotechnology increased between 1993 and 1996 as a function of both the high intensity of media coverage in the mid-1990s and an upward trend in media coverage by 1996. The Netherlands is the exception to these general rules: its already large knowledge gap remains stable despite intensified media coverage and relatively low levels of public controversy.

An assumption underlying our models concerns the nature of systemic relationships. Our claims imply linear relationships among intensity of media coverage, the knowledge gap, level of controversy and pluralism in the public sphere. However, most systemic relationships are linear only within circumscribed limits, and beyond this range it would be more realistic to assume non-linearity. Thus, it is conceivable that, once coverage reaches a certain level, the knowledge gap ceases to increase and stabilises, as in the Italian and Greek cases, or actually decreases, as in the Netherlands where there was reduced public controversy in the years up to the survey. Indeed, if controversy may, *ceteris paribus*, decrease the knowledge gap, sharp intensification of the public debate may result in the gap stabilising, or even widening. The expected effect of conflict may be the case only within a limited range; beyond this range conflict may freeze knowledge gaps in the debate, preventing rather than enhancing circulation of information in the system. It may therefore be possible to distinguish stages, or bands, of knowledge gap increase and decrease in the continuum of levels of conflict. The data to test dynamic relationships over time are currently not available. Nonetheless, by using cross-sectional data from two time-points (1993 to 1996), a rudimentary and limited dynamic comparison was performed for at least some of the countries.

Few studies looking at knowledge gaps have examined longitudinal dynamics, partly because longitudinal data of a comparative nature are hard to come by. However, longitudinal data may show that the KGH is an underspecified dynamic model. As it presently stands, the three main propositions of the model stipulate a dynamic system of interacting influences: first, increasing information flows on new technology lead to increased knowledge gaps – the educated have easier access to this information than the less educated; secondly, controversy reduces these knowledge gaps by widening the information flows; and, thirdly, pluralism in the public sphere increases knowledge gaps through a variety of media with different focuses of attention. This raises several fundamental questions, however. How do the opposing influences of public controversy and media pluralism interact over time? Is there a specific level of controversy that will necessarily reduce the size of the knowledge gaps in differentially pluralistic media structures? Is it the case that the more pluralistic the public, the larger the controversies will need to be in order to close the knowledge gaps? Is there an upper threshold of public controversy, a 'crisis' point, beyond which nobody can absorb further information? Is this crisis point different for different degrees of public plurality? These are questions that require further theoretical development and longitudinal data to allow empirical conclusions to be drawn. We leave these questions open for future research.

Fortunately, the robust longitudinal data needed for these questions to be investigated may be at our disposal in the coming years, as we continue to monitor the status of biotechnology among the European public, in terms of levels of public knowledge, media coverage and polity activity. The controversies over Monsanto's 'Roundup Ready' soybeans, GM foods in general since November 1996 and Dolly the sheep after February 1997 have ushered in a different phase in the controversy over biotechnology and genetic engineering in many countries. These new features need to be carefully compared with those of the previous phases outlined in this chapter. In this sense, we have provided a baseline against which we can compare and contrast future developments in the salience of biotechnology representations in the public spheres of Europe.

Appendix

Methodological considerations

The KGH is principally a dynamic model that makes statements about the development of variables over time: if coverage of biotechnology increases, *ceteris paribus* the knowledge gap will widen; if coverage of biotechnology increases and a public controversy arises, *ceteris paribus* the knowledge gap will decrease. Ideally, these relationships would be studied in one country over a longer period of time. In the absence of significant longitudinal data, these dynamic relationships are modelled in cross-section by making assumptions about a homogeneous trajectory of all units of comparison.

Another possibility is to compare individuals with high and low media use or countries with high and low media input on biotechnology at a single point in time. For example, we assume that different levels of media coverage represent the 'same system' at different stages of development.

We ignore the fact that different systems may have different 'natural' levels of coverage of biotechnology owing, for example, to the particular popular science culture. We also overlook the fact that such cross-sectional comparisons ignore the different contexts in which early developers and latecomers find themselves. As time progresses, the contexts are not the same for particular stages in the development of a technological controversy.

Measures

Knowledge gap Table 5A.1 lists the knowledge items used in the Eurobarometer 1996 survey. The items require either true (y) or false (f) as a correct answer. These measures are indicators of enculturation into the biotechnology system rather than a measure of any practical knowledge. The knowledge gap is measured by the correlation between knowledge and level of education. We use a ten-, nine- or five-item knowledge scale (k10, k9 or k5) and a three-point scale for education level (educ3) based on school-leaving age. The change in the knowledge gap is based on the correlations between k5 items and educ3 for both 1993 and 1996. The coefficient of 1996 is divided by that of 1993. A resulting index of >1 indicates an increase in the knowledge gap, an index of <1 a decrease (see figure 5.6).

Media intensity is the estimated number of articles in the elite press of the country for 1995–6. The changes in press coverage are measured by the number of articles in the elite press in 1995–6 as a percentage of the total number of articles during the period 1991–6. A figure of more than 33 per cent indicates an upsurge in coverage in the two years.

Controversy Measurement of controversy is based on measures of polity activity, media intensity and media controversy in each country. *Polity activity* is based on a Guttman index with nine dichotomous criteria, such as parliamentary debate, public hearings, protest, new legislation and petitions, for three periods: 1985–1991, 1992–4 and 1995–8. The judgement for each criterion is expert based and was collected mid-1998. The judgements for the three periods add up to an overall country score. *Media controversy* is measured by the percentage of articles reporting a controversy on biotechnology for 1991–6.

The level of polity activity and the intensity of media coverage are related: the more media coverage, the more polity activity; however, the required increase in media coverage is exponential ($r = .54$, $n = 12$). Polity activity and the level of media controversy are not related ($r = .15$):

Table 5A.1 *Knowledge indicators by level of education*

	Correct answers (%)				
		Education[c]			
Knowledge items[a]	Total[b]	Low	Medium	High	K-gap corr.[d]
1. There are bacteria which live from waste water (y)	84	75	85	91	.16
2. Possible to find out Down's syndrome during pregnancy (y)	80	71	80	87	.16
3. Yeast for brewing beer consists of living organisms (y)	69	55	70	80	.25
4. More than half of human genes identical to those of chimpanzees (y)	50	44	50	56	.12
5. Eating gene-modified fruits could modify one's genes (f)	49	33	49	63	.30
6. Cloning produces exactly identical offspring (y)	46	31	46	58	.27
7. Ordinary tomatoes do not contain genes (f)	36	22	35	51	.29
8. Genetically modified animals are always bigger (f)	36	20	34	54	.34
9. It is possible to transfer animal genes into plants (f)	31	26	31	36	.10
10. Viruses can be contaminated by bacteria (f)	21	11	19	34	.23
Knowledge score (0–10 points)	5.0	3.8	5.0	6.0	.38

Notes:

[a] (y) items need to be recognised to be true/yes to score a correct answer; (f) items need to be recognised to be false to score a correct answer. Item 4 is sometimes excluded from the total score.

[b] 'Total' indicates the percentage of correct answers for each item.

[c] 'Education' is the percentage of people that give a correct item for each of three levels of education measured by school-leaving age.

[d] The 'knowledge gap' for each item, and overall, is indicated by the bi-serial correlation between 'education' and giving a correct or incorrect response.

some countries have low-level polity activity and high levels of media controversy, and vice versa. Polity activity during and after 1995 and the trend in press coverage in 1995–6 are related ($r = .58$, $n = 12$); polity activities led to an increase in press coverage, or vice versa. The causality is not clear.

Pluralism We developed two indices for *pluralism* of the public sphere. Pluralism 1 reflects the left–right polarisation in a country – the number of left or right positions as a percentage of middle positions (left or right is based on a self-classification). Pluralism 2 is based on religious polarisation: the number of very religious plus non-religious as a percentage of some religiosity.

For further details on Eurobarometer survey 46.1 of 1996 and the media content analysis for 1973–96, see appendices 1–4 in Durant, Bauer and Gaskell (1998).

REFERENCES

Bauer, M. (1995) 'Industrial and Post-Industrial Public Understanding of Science', paper delivered to the Chinese Association for Science and Technology, Beijing, 15–19 October.

(1996) 'Socio-economic Correlates of dk-Responses in Knowledge Surveys', *Social Science Information* 35(1): 39–68.

Bauer, M., J. Durant and G. Evans (1994) 'European Public Perceptions of Science', *International Journal of Public Opinion Research* 6(2): 163–86.

Bauer, M. and G. Gaskell (1999) 'Towards a Paradigm for Research on Social Representations', *Journal for the Theory of Social Behaviour* 29(2): 163–86.

Bonfadelli, H. (1994) *Die Wissenskluft-Perspektive. Massenmedien und gesellschaftliche Information*. Konstanz: UVK.

Coleman, C.-L. (1993) 'The Influence of Mass Media and Interpersonal Communication on Societal and Personal Risk Judgements', *Communication Research* 20(4): 611–28.

Donohue, G.A., P.J. Tichenor and C.N. Olien (1975) 'Mass Media and the Knowledge Gap: a Hypothesis Reconsidered', *Communication Research* 2: 3–23.

Dunwoody, S. and H.P. Peters (1992) 'Mass Media Coverage of Technological and Environmental Risks: a Survey of Research in the United States and Germany', *Public Understanding of Science* 1: 199–230.

Durant, J., M.W. Bauer and G. Gaskell (eds.) (1998) *Biotechnology in the Public Sphere: a European Sourcebook*, London: Science Museum.

Durant, J., M. Bauer, G. Gaskell, C. Midden, M. Liakopoulos and E. Scholten (2000) 'Two Cultures of Public Understanding of Science' and Technology in Europe, in M. Dierkes and C. von Grote (eds.), *Between Understanding and Trust: The Public, Science and Technology*, Amsterdam: Harwood Academic Publisher, pp. 131–56.

Ettema, James S. and Gerald F. Kline (1977) 'Deficits, Differences, and Ceilings. Contingent Conditions for Understanding the Knowledge Gap', *Communication Research* 2: 179–202.

Gaziano, C. and E. Gaziano (1996) 'The Knowledge Gap: Theories and Methods in Knowledge Gap Research Since 1970', in Michael B. Salwen and Don W. Stacks (eds.), *An Integrated Approach to Communication Theory and Research* Mahwah, NJ: Erlbaum, pp. 127–143.

Griffin, R.J. (1990) 'Energy in the Eighties: Education, Communication, and the Knowledge Gap', *Journalism Quarterly* 67(3): 554–66.

McCombs, Maxwell (1994) 'News Influence on Our Pictures of the World', in Jennings Bryant and Dolf Zillmann (eds.), *Media Effects. Advances in Theory and Research*, Hillsdale, NJ: Erlbaum, pp. 1–16.

Mazur, A. (1981) 'Media Coverage and Public Opinion on Scientific Controversies', *Journal of Communication* 31(2): 106–15.

Miller, J. (1983) 'Scientific Literacy: a Conceptual and Empirical Review', *Daedalus* 112: 29–48.

Robinson, J.P. (1972) 'Mass Communication and Information Diffusion', in Gerald F. Kline and P.J. Tichenor (eds.), *Current Perspectives in Mass Communication Research*, Beverly Hills, CA: Sage, pp. 71–93.

Tichenor, P.J., G.A. Donohue and C.N. Olien (1970) 'Mass Media and Differential Growth in Knowledge', *Public Opinion Quarterly* 34: 158–70.

Viswanath, K. and J.R. Finnegan (1996) 'The Knowledge Gap Hypothesis: Twenty-Five Years Later', in Brant R. Burleson (ed.), *Communication Yearbook*, vol. 19, New Brunswick, NJ: International Communication Association, pp. 187–227.

Wade, S. and W. Schramm (1969) 'The Mass Media as Sources of Public Affairs, Science, and Health Knowledge', *Public Opinion Quarterly* 33: 197–209.

Part II

Public representations in 1996: structures and functions

6 Traditional blue and modern green resistance

Torben Hviid Nielsen, Erling Jelsøe and Susanna Öhman

In Europe, scepticism towards modern biotechnology is widespread and persistent. However, although there are considerable differences among European countries in terms of the expression and prevalence of scepticism, the data from each country suggest some common themes. In particular, nowhere do sceptics fall into distinctive groupings. And, generally speaking, neither traditional demographic categories nor political/cultural orientations seem at first glance to be very important. In addition, on the political stage, resistance against new biotechnologies is articulated by parties and political groups that represent widely differing positions on the political spectrum. In view of this, there is a need to establish whether a more fine-grained analysis of the data will reveal whether sceptics of biotechnology segregate according to alternative criteria.

It has long been known that within the related, albeit much broader, field of environmentalism there is a division between a modernist group, politically to the left, and a conservative group, oriented towards the preservation of nature and its resources. With regard to resistance to biotechnology, however, few attempts have been made to explore whether analogous, or alternative, commitments might explain patterns of scepticism. Yet such a study promises to shed much-needed light on the nature of resistance to modern biotechnology. Recently, T. Hviid Nielsen (1997a, 1997b) has shown that opponents to modern biotechnology in Norway can be divided into two camps, a traditionalist or 'blue' segment and a modernist or 'green' segment. This analysis was made on the basis of survey data from the Norwegian version of the 1993 Eurobarometer survey on attitudes to biotechnology. Further evidence was derived from documenting the arguments used by various political parties in the Norwegian debate about modern biotechnology. When the data from the 1996 Eurobarometer survey became available, Hviid Nielsen repeated his analyses, using the new data, and once again he found a division in accordance with the traditionalist/modernist dichotomy. By means of 'argumentation analysis' it was also possible to show that the 'blue' and 'green' arguments are based on fundamentally different values and concerns (Nielsen, 1997b[1]).

Moreover, there are strong arguments in favour of the hypothesis that the traditionalist/modernist pattern of segmentation is a general phenomenon. The 1996 Eurobarometer survey on biotechnology, which provides an extensive database covering seventeen European countries (fifteen EU countries, Norway and Switzerland), offers an opportunity for testing this hypothesis. The purpose of the present chapter has been to report on all of the European countries in order to explore the prevalence of the traditionalist/modernist division in the case of biotechnology. In addition, to study the features of the various nationally specific patterns of scepticism, comparisons will be made among countries, as well as, if possible, among groups within the countries.

The gap in expectations

In recent years a large, and apparently widening, gap seems to have developed between the industry's and the general public's expectations of the status of modern biotechnology.

In some scenarios, biotechnology is depicted as the new mega-technology, destined to assume an economic importance analogous to that achieved by the electronic computer, and before that by the oil and petrochemical industries (Freeman, 1995). The European Association for Bioindustries predicts that this expanding sector will be worth US$285 billion by the year 2005 (*Time*, 1998). Ernst & Young, backed by the European Commission, expect 'Europe's entrepreneurial bioscience sector . . . to continue growing at about 20 per cent a year' (Ernst & Young, 1997). And even committed critics such as Jeremy Rifkin anticipate the twenty-first century becoming 'The Biotech Century'. 'Many of the scientific breakthroughs we predicted more than twenty years ago are now moving out of the laboratory and into widespread commercial use', he argues. 'The genetic revolution and the computer revolution are just now coming together to form a scientific, technological, and commercial phalanx' (Rifkin, 1998: xv). However, these hopes and promises have clearly not yet been fulfilled.

On the other side of the expectation gap, ethical concerns and scepticism about potential risks are accompanied by scientific scepticism as to the technological feasibility of the industry meeting its objectives. Biotechnology might, it is argued, be a 'bubble' of overinvestment and

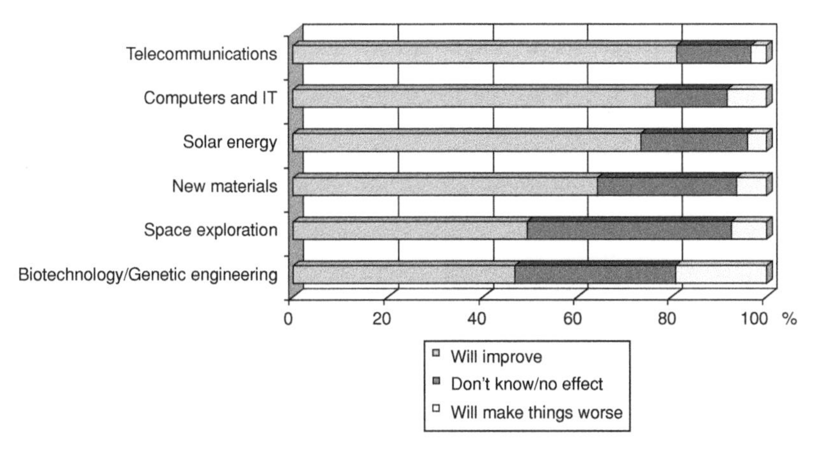

Figure 6.1 Anticipated effects of new technologies in the next twenty years: EU 15, 1996
Source: Eurobarometer 46.1 (1996).

failed investments – a dangerous hit-or-miss technology, based upon 'a crude, outmoded, reductionist view of organisms' (Mae-Wan et al., 1998). Indeed, the 1996 Eurobarometer survey (INRA, 1997) documents high and even slightly growing scepticism towards biotechnology among the European public.

In terms of the perceived ability of biotechnology to generate positive benefits, Europeans still rank modern biotechnology lowest of six new technologies: telecommunication, data and information technology, solar energy, new materials and space exploration (see figure 6.1). The largest proportion of respondents, although still less than half (47 per cent), expect biotechnology to improve our way of life in the next twenty years; but 20 per cent expect it to make things worse, 9 per cent anticipate 'no effect', and as many as 25 per cent answer 'don't know'.

Furthermore, the number of 'optimists' compared with 'pessimists' declined markedly over the five years 1991 to 1996 (see figure 6.2). The 1991 Eurobarometer survey (INRA, 1991) recorded 41 per cent more optimists (51 per cent) than pessimists (10 per cent). By 1993 (INRA, 1993) the difference had decreased to 37 per cent (48 per cent optimists and 11 per cent pessimists), and in 1996 there was a further reduction to 28 per cent (47 per cent optimists and 19 per cent pessimists).

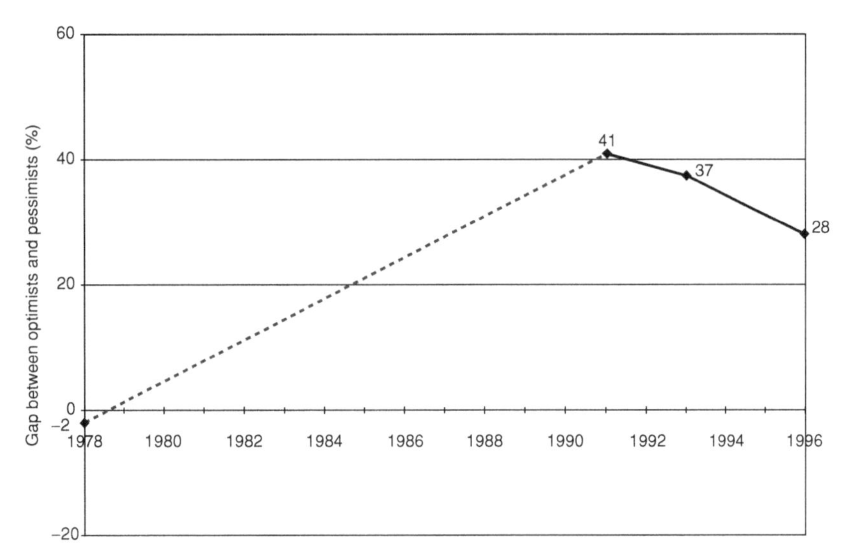

Figure 6.2 General expectations about biotechnology in the next twenty years: the balance of opinion in the EU, 1978–1996
Sources: Eurobarometer surveys 10A (1978), 35.1 (1991), 39.1 (1993), 46.1 (1996).
Note: The gap between optimists and pessimists is measured as the difference between the percentage expecting that biotechnology 'will improve way of life' and the percentage expecting that it 'will make things worse' ('worthwhile' vs. 'unacceptable risk' in 1978).

'Blue' and 'green' resistance

Expectations of biotechnology differ widely across, and within, European nations. Spain, Portugal and Italy have the highest proportion of optimists, and (apart from Greece, which has a strikingly high number of 'don't knows') Germany, Norway and Austria have the smallest proportion (see figure 6.3). This pattern clearly does not fit into the stereotype of a 'restrictive', Catholic south distinct from a 'liberal', Protestant north. Rather, expectations are generally highest in the nations where the implementation and application of the technology have been slow and on a modest scale. Conversely, expectations are lowest where the introduction of biotechnology is further advanced. An 'intuitive' public understanding of 'declining marginal utility' would seem, therefore, to be a better predictor of scepticism towards biotechnology than general religious background.

Moreover, as a group, the 'optimists' share some of the characteristics that technological-innovation theory generally expects to find among

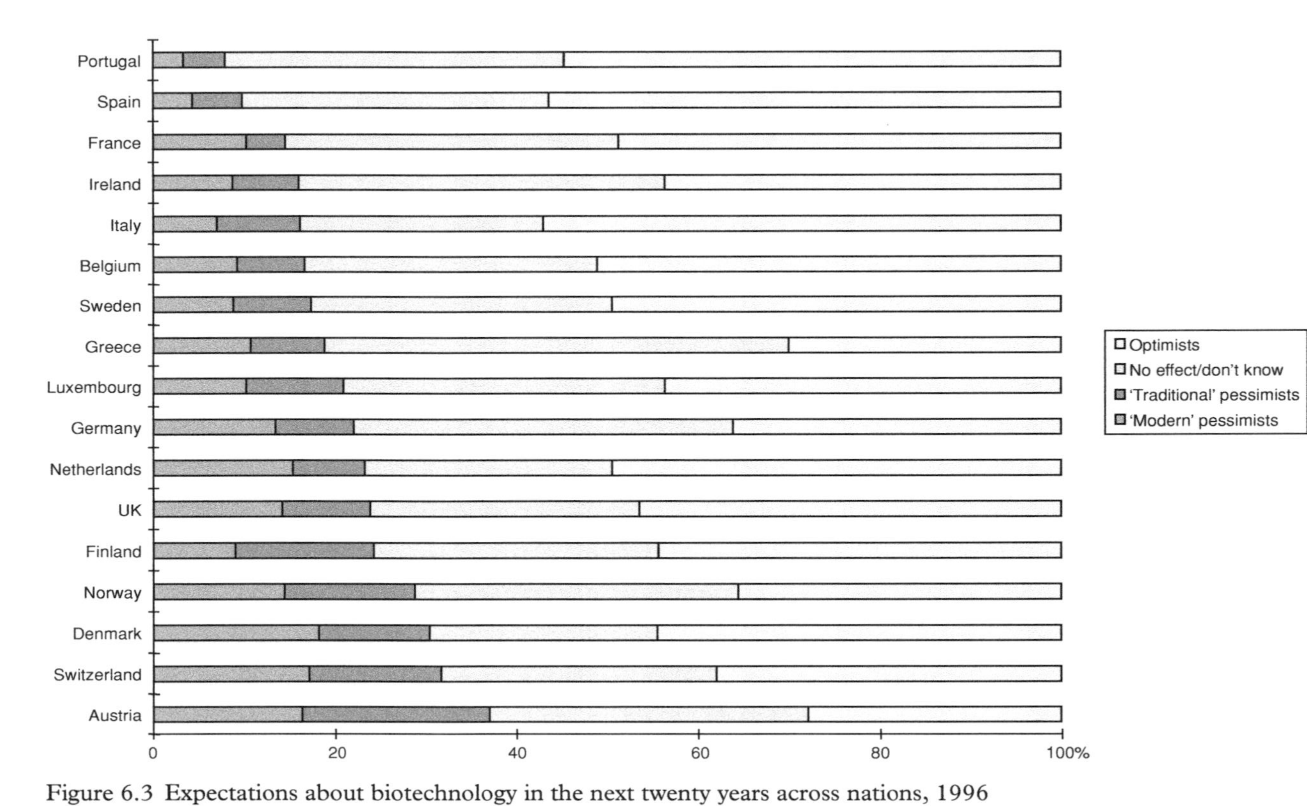

Figure 6.3 Expectations about biotechnology in the next twenty years across nations, 1996
Source: calculations based on Eurobarometer 46.1 (1996).
Note: Optimists think biotechnology will 'improve way of life'; pessimists think it will 'make things worse'.

entrepreneurs. The typical 'optimist' is a young, or younger, man with an extended education and urban residence. He combines relatively good knowledge of biotechnology with a perception of low attendant risks. And he considers himself to the right, or around the centre, of the traditional political spectrum.

The high number of 'undecided' (i.e. 'no effect' and 'don't know') respondents throughout Europe might suggest the existence of a realistic public appreciation of the fact that the future of biotechnology is contingent, and will be dependent upon the nature of regulations and the nature of its applications. The number of respondents in the 'undecided' camp is hardly affected by the ratio between optimists and pessimists, though Denmark and the Netherlands show that the number of 'don't know' optimists may be reduced in nations that cultivate high levels of public knowledge in the context of open and intense public debate. In such circumstances, a higher proportion are prepared to adopt an unambiguous point of view. It is also noticeable, however, that former 'don't knows' distribute themselves fairly evenly between the optimist and pessimist groups.

Cluster analyses on each of the national data sets for the seventeen European countries (EU countries, Norway and Switzerland) reveal that those respondents who think that modern biotechnology will make things worse can, in almost every European country, be divided into two highly distinct groups: the 'traditional' and the 'modern'. These groups can be found in fifteen of the seventeen European countries, even if the group characteristics differ somewhat among the countries (see the appendix to this chapter for details). It is only in Finland and Austria that these distinct divisions in public perception are not found.

The 'traditional' and the 'modern' groups of pessimists differ from the optimists in containing a high proportion of women, as well as high levels of rural residence and a perception of high risk concerning modern biotechnology. Aside from these three common characteristics, however, the two types of pessimists may be systematically separated. The typical representative of the 'traditionalist' group is older and will have completed his/her education after primary school rather than attending university like the typical 'modern'. The traditional group also has an inferior knowledge of biotechnology. And, where the 'traditional' inclines towards the centre and right of the political spectrum, the 'modern' is oriented towards the left. Further, the traditional tends to be strongly religious whereas the modern is inclined to be a strong non-believer. Finally, the traditional group may be described as materialists, and the modern as post-materialists.

The pattern of 'blue' and 'green' resistance in Europe

A general pattern is observable in which the number of opponents is greater in northern than in southern Europe and the number of 'green' opponents is relatively greater in the north, while 'blue' opponents are relatively more numerous in the south.

The following observations are consistent with these findings and help in explaining this pattern. Catholic nations have relatively more 'blue' and Protestant nations relatively more 'green' opponents. A greater exposure to biotechnology and/or a perception of declining marginal utility might explain why expectations are lowest in the nations where biotechnology is already established and opposition is low and expectations high in the peripheral nations (mostly in the south but also including Finland), where biotechnology is not widespread but is still seen as an integral aspect of progress and development.

Among European countries there is marked variation in the size and characteristics of the traditional and modern groups. On this basis, with the exceptions of Finland and Austria, the different states may be divided into five groups. The profiles for five representative countries – Germany, Norway, the UK, Spain and Italy – are presented in table 6.1.

In Germany, the modern 'green' group of opponents is considerably larger than the traditional 'blue' group – 14 per cent compared with 9 per cent. The traditional group is characterised by higher female than male representation; a mean age of 59 years; relatively low education; religiousness; politically right-wing sympathies; low knowledge of, and a high risk perception concerning, modern biotechnology. In contrast, the modern 'green' group is characterised by a mean age of 37 years; higher education; politically left-wing sympathies, post-materialistic values; a high level of knowledge; and a high risk perception.

In Norway, the two groups of opponents are equally large – about 14 per cent in each category. The traditional group is characterised by a mean age of 52 years; relatively low levels of education; religiousness; politically right-wing sympathies; materialistic values; and a high risk perception. Conversely, the modern 'green' group is characterised by a mean age of 35 years; higher education; a low level of religiousness; political left-wing sympathies; post-materialistic values; high levels of knowledge; and a high risk perception concerning modern biotechnology.[2]

In the UK, the modern 'green' group of opponents is also larger than the traditional 'blue' group – 14 per cent compared with 10 per cent. The traditional group is characterised by a mean age of 60 years; relatively low education; religiousness; politically right-wing sympathies; materialistic

Table 6.1 *Profiles of 'blue' and 'green' opponents in five European countries, 1996*

Country	'Blue' opponents	'Green' opponents
Germany	Size of cluster (8.6%)	Size of cluster (13.5%)
	Female (11%)	Age 15–39 yrs (22%)
	Age 55+ yrs (37%)	Higher education (18%)
	Education <15 yrs (34%)	High knowledge (10%)
	Low knowledge (10%)	Political left (14%)
	Very religious (8%)	Post-materialistic (13%)
	Political right (7%)	High risk perception (10%)
	High risk perception (11%)	
Norway	Size of cluster (14.3%)	Size of cluster (14.5%)
	Age 55+ yrs (6%)	Age 25–39 yrs (17%)
	Education <15 yrs (6%)	Higher education (14%)
	Very religious (15%)	High knowledge (9%)
	Political right (12%)	Not religious (17%)
	Materialistic (25%)	Political left (25%)
	High risk perception (6%)	Post-materialistic (17%)
		High risk perception (17%)
UK	Size of cluster (9.7%)	Size of cluster (14.2%)
	Age 55+ yrs (34%)	Female (7%)
	Education <15 yrs (26%)	Age 25–39 yrs (11%)
	Medium knowledge (15%)	Higher education (18%)
	Very religious (8%)	High knowledge (12%)
	Political right (22%)	Not religious (16%)
	Materialistic (17%)	Political left (12%)
	High risk perception (9%)	Post-materialistic (12%)
		High risk perception (11%)
Spain	Size of cluster (5.5%)	Size of cluster (4.4%)
	Female (15%)	Male (7%)
	Age 55+ yrs (14%)	Age 25–39 yrs (20%)
	Education <15 yrs (25%)	Higher education (20%)
	Low knowledge (14%)	High knowledge (23%)
	Very religious (12%)	Not religious (23%)
	Political centre (11%)	Political left (30%)
		Post-materialistic (17%)
		High risk perception (36%)
Italy	Size of cluster (9.1%)	Size of cluster (7.1%)
	Age 40–55+ yrs (11%)	Age 15–39 yrs (26%)
	Education <15 yrs (14%)	Higher education (24%)
	Low knowledge (9%)	High knowledge (20%)
	Very religious (15%)	Not religious (27%)
	Materialistic (11%)	Political left (14%)
		Post-materialistic (17%)
		High risk perception (16%)

Note: The figures represent percentage overrepresentation compared with the total population, including the 'undecided'.

values; and a high risk perception concerning modern biotechnology. The modern group is characterised by greater female than male representation; a mean age of 36 years; higher education; a low level of religiousness; politically left-wing sympathies; post-materialistic values; high levels of knowledge; and a high risk perception.

If we turn to Spain, the two groups of opponents are almost equally small – about 5 per cent in the traditional group and 4 per cent in the modernist group. The former is characterised by a higher proportion of females than males; a mean age of 48 years; relatively low levels of education; religiousness; centrist political sympathies; and low relative knowledge about modern biotechnology. The modern group is characterised by a higher proportion of men; a mean age of 34 years; higher levels of education; a low level of religiousness; politically left-wing sympathies; post-materialistic values; a high level of knowledge; and a high risk perception concerning modern biotechnology.

Finally, in Italy the traditional segment comprises 9 per cent and the modern segment 7 per cent. The traditional group is characterised by a mean age of 48 years; relatively low levels of education; religiousness; materialistic values; and a low level of knowledge about modern biotechnology. The modern group is characterised by a mean age of 33 years; higher education; low levels of religiousness; left-wing political sympathies; post-materialistic values; a high level of knowledge; and a high risk perception.

These data strongly suggest that the 'traditional' and the 'modern' categories of sceptic are well defined and distinct. Though there is some national variation, owing to differing cultural contexts, this is clearly overshadowed by the striking commonalties in the nature of scepticism across Europe.

In fact, of the seventeen countries studied, Finland and Austria are the only nations for which the data do not fit this model. To some extent surprisingly, these countries are extreme cases. Finland does not cluster with the other sceptic Scandinavian nations; instead it contains an unexpected number of optimists, analogously to the pro-development southern periphery of Europe. In stark contrast, the Austrians are not only sceptical, but extremely so.

The particular national backgrounds of these two countries might explain their deviation from the general European pattern. Thus, Finland has undergone rapid urbanisation as well as having experienced a very fast transformation from an agrarian society to a post-industrial information society. Biotechnology came late on the scene, but symbolises a positive approach to new technologies in a country that generally equates new technologies with rapid improvements in the standard of living. This optimistic attitude to technological innovation, one that tends to efface notions of the potentially negative consequences of biotechnology, is likely to have blurred the division of attitudes between optimism and pessimism.[3] The Austrian data are the more surprising because in most respects Austria conforms very closely to the Swiss pattern. Austria's high level of scepticism may be the result of the fact that the public imbibed from neighbouring countries the association between scepticism and political correctness.

Jon Miller has kindly tested this model against US data collected after the genetically modified (GM) soya and Dolly the sheep events. The data are not quite compatible: for example, the US data lack any equivalent of the materialist–post-materialist index. But his analysis does indicate that the 'blue'/'green' distinction cannot be found in America, where opposition generally is small, where religious fundamentalism is more prevalent, and where a widespread pragmatic scientific discourse seems to suppress 'green' views.

Regulation and public consultation – some differences between 'blue' and 'green'

If we turn to other characteristics of the 'blue' and 'green' opponents to modern biotechnology, the pattern is consistently stable. By means of cross-tabulations with other variables in the Eurobarometer survey, we find that the 'green' modernist group trust non-governmental organisations (NGOs) much more than the 'blue' traditionalist group do, especially environmental organisations, but also consumer organisations. The traditionalists, on the other hand, have a higher proportion of respondents who express themselves uncertain about who is trustworthy. They also place a higher degree of trust in the medical profession; and in some Catholic countries they are inclined to invest confidence in religious organisations. The modernist group display a much higher level of participation in the modern biotechnology discourse. They have both heard more about modern biotechnology than the traditionalists, and discussed it with greater regularity.

The 'green' opponents have much the strongest opinions about the regulation of modern biotechnology. They are considerably less inclined to believe that current regulations are sufficient to protect people from the potential risks linked to modern biotechnology, and they tend to reject the idea that the regulation of modern biotechnology should be left mainly to industry. The modernists also believe that GM food should be labelled and that public consultations about modern biotechnology are worthwhile exercises. In contrast, the 'blue' traditionalist opponents are more likely to argue that the public ought to accept some degree of risk attendant upon the use of modern biotechnology, so long as it enhances the economic competitiveness of Europe. Further, in most Catholic countries, the 'blue' traditional group are the more willing to defend the idea that religious organisations should have a say in how modern biotechnology is regulated.

Moreover, the 'green' modernists are more sceptical than the traditionalists of the ability of biotechnology to fulfil the expectations of rapid economic growth outlined above. Nor do they believe that modern biotechnology will usher in the scale of change over the next twenty years anticipated by the traditionalist group. The modernists are, in general, more sceptical about the levels of risk *and* benefit commonly associated with biotechnology. If they are rather less convinced that biotechnology will lead to environmental pollution, the emergence of dangerous new diseases and the production of 'designer babies', they are also sceptical of the capacity of biotechnology substantially to reduce world hunger. Indeed, the modernists are especially doubtful of the validity of the more positive expectations. For their part, the traditionalists are more likely to answer 'don't know' when questioned about the expected benefits and risks of modern biotechnology.

The conclusion must be that the 'green' opponents are more in favour of regulating modern biotechnology and of giving the public a choice, through labelling but also through more direct means of public consultation. This is probably linked to their higher degree of knowledge about biotechnology and their higher level of participation in the biotechnology debate. They do not think it a waste of time, or the issues too complicated, for the public's views to be taken into consideration.

Two arguments

'Traditional' and 'modern' sceptics share the assumption that modern biotechnology will reduce our quality of life. But the general values they articulate are substantiated and underpinned by different arguments and conclusions (Toulmin, 1958; Toulmin et al., 1979). These are represented in figure 6.4.

The 'blue' argument (figure 6.4(a)) has no external references. It is predicated entirely on the conviction that technological intervention in nature is a priori unacceptable. The position can be undermined only by challenging this underlying value judgement.

The 'green' argument (figure 6.4(b)) points to the (perceived) uncertainties and risks related to biotechnology as its principal justification. Proponents of this view focus on stressing the unpredictability of biotechnology and emphasising the level of risk compared with potential benefits.

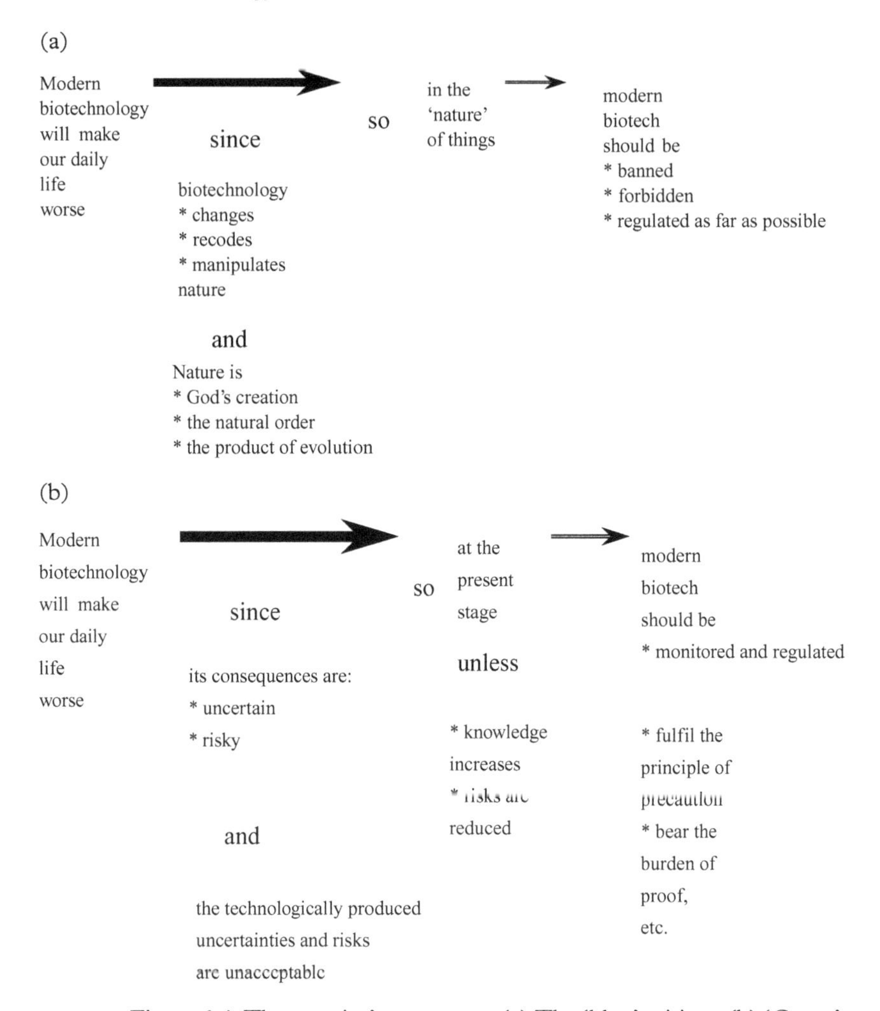

Figure 6.4 The sceptics' arguments: (a) The 'blue' critique (b) 'Green' scepticism

Its modality centres on 'the present stage' of technological development and knowledge concerning it. In this case, however, an increased knowledge of the scale of risk involved might force an alternative conclusion, one more congenial to the optimists.

Mephistopheles and Frankenstein

The underlying concerns of the 'green' and 'blue' arguments are as distinct as the different social groups that tend to express them. The 'blue' argument is supported by moral (or religious) values, the 'green' by notions of uncertainty and risk. In this sense, the 'blue' critique is 'Faustian' – the whole enterprise of biotechnology may be conceived as a modern covenant with Mephistopheles. Even though biotechnology represents technological success, insight into, and the manipulation of, Nature is considered problematic. In contrast, the 'green' scepticism is more 'Frankensteinian'. The problematic they identify concerns the insufficient knowledge of potential consequences, but intervention in nature is not held to be reprehensible per se. The danger is seen to lie in the creation of a 'monster' capable of developing in unforeseen and uncontrollable ways, and conceivably coming to usurp its creators.

Around the turn of the year 1997, two major events had the effect of mobilising scepticism, and arguments based on both ethics and risk were invoked. Shortly before Christmas 1996, the first ships carrying GM soya from fields in the USA docked in European ports. Then, in February 1997, news emerged of the cloning of the sheep 'Dolly', using a cell from a six-year-old sheep and an egg emptied of any genetic information, at the Roslin Institute in Scotland.

The soya ships on their way across the Atlantic generated fear and uncertainty about the *risks* of biotechnology. These concerns are associated with large-scale production and the consumption of GM foods. The GM soybeans were engineered to be resistant to a herbicide called 'Round-up'. The product had been approved and produced in the USA, where the soybeans had subsequently been mixed with conventional, unmodified, beans. But these shipments were to be introduced into a European market that had so far approved only field trials for, and the marketing of, a few modified products. As yet, Europe had no general rules for the regulation and labelling of GM foodstuffs.

The soya ships therefore excited fears of biotechnology as capable of unleashing a Frankensteinian monster. Could the GM soybeans spread, the public asked, and transfer their genes to other plants in the natural environment? Do we know enough about the long-term impacts of the 'Round-up' herbicide used? Does the soybean oil have the same nutritional properties? What might happen to those individuals who consume bread, chocolate, ice-cream and numerous other processed foods containing the oil? And would it be possible to reverse any unforeseen and unintended consequences?

Initially, the soybean ships were faced with direct action and symbolic protests from environmental and consumer organisations, whose campaigns were closely followed in some European countries. They subsequently managed to increase the political sensitivity of the issue at both the national and the EU level. Consumer demands for more restrictive regulation in this field could not be ignored.

The cloned sheep, Dolly, which just a few months later was pictured on the front pages of newspapers around the world (even before her creation was announced in *Nature*) rendered topical questions about the *ethical* implications of otherwise successful scientific achievements. Even though 277 trials had been performed before the researchers succeeded in their attempts to unite the young cell with the empty egg, the fact that the cloning had taken place was considered to be a scientific breakthrough. But the press soon shifted its focus away from a fascination with this narrowly scientific dimension to concerns about whether science's success portended a moral failure.

Thus, Dolly happened to mobilise the second of the classical tropes in criticising technology: the Faustian notion of an unholy bargain. This incident raised many morally pertinent questions: how old ought Dolly to be considered? who were her parents? how would a human being created in the same way form and understand its own identity? what might come next, now that humans had achieved something that a decade earlier had been considered totally unfeasible? are there now any limits? will humans now begin to create people from the genetic material of geniuses or dictators, or in the form of mannequins?

Of course, Dolly herself was only an innocent and ignorant sheep in a laboratory stable, created by techniques developed in the production of medicines for humans. But she soon acquired a symbolic importance at the highest levels of secular and religious authority. From the President of the USA to the Pope, the technology was met with condemnation, and bans were placed on the use of the same technology involving humans.

It is no coincidence that both of the classical tropes related to the dangers of new technologies – unforeseen risks and going beyond ethical limits – were conjured up as expectations of further scientific and commercial breakthroughs increased. The steps from laboratory to market, and from production to consumption, in themselves generate heightened expectations, increase levels of scepticism and serve to widen the gap between the two.

As a technological 'use of life' (Bud, 1993) and 'a cultural icon' (Nelkin and Lindee, 1995), modern biotechnology is commonly confronted by both a 'pre'-industrial critique of intervention in 'nature's order', as well as a 'post'-industrial critique of the potential risks involved with the new technology. The classic diffusion model of technological innovation is thus contradicted, even falsified, by the fact that these critical discourses have achieved such a wide and lasting appeal. And the mobilisation model demands modification in view of the two types of opposition we have elucidated: one in favour of older traditions, the other directed against the new risks ushered in by biotechnology.

The image of modern biotechnology

Do these two different segments of the public – the traditionalists and the modernists – invoke 'blue' and 'green' arguments and values, respectively, when articulating their scepticism towards modern biotechnology? In addressing this question we have analysed the Swedish data from the Eurobarometer study. In terms of people's mental associations with biotechnology and genetic engineering (in response to the open-ended question 'What comes to your mind when you think about modern biotechnology in a broad sense, that is including genetic engineering?') there are systematic differences between the two different segments of opposition. However, because the groups are not particularly large, the number of answers used in the analysis was small. In Sweden, the 'blue' group comprises 8.5 per cent and the 'green' group 8.9 per cent of respondents.

In qualitative terms, typical arguments used by the traditional group were:

'Nature should be left alone.'
'Mankind has no right to play God.'
'It is wrong, it is not moral or ethical.'
'I'm afraid that there is too much tampering with nature.'

These arguments very much mirror those encountered in the 'argumentation analysis', in which the 'blue' segment tended to raise moral and ethical issues, and used these as a basis for arguing against biotechnology.

Turning to the modernist, 'green', opponents, the typical arguments are based more on dangers and risk.

'There is a danger, the risks are not clear, not for humans, animals or plants.'
'We will have new plants and the genes will spread to the weed and then they will take over.'
'The animals suffer, their bodies are gigantic.'

These answers have also been categorised and the main differences between the two segments in the Swedish data clearly support our general claim that different arguments are employed by the different opposition groups. Among the 'green' segment, fear, risks and danger are the common themes. In fact, 42 per cent of people belonging to this group state that they are afraid of the risks and dangers involved in modern biotechnology, whereas only 34 per cent of the 'blue' group claim to be. On the

Table 6.2 *Valuation of biotechnology by the 'blue' and 'green' segments in Sweden (%)*

Valuation	'Blue' opponents	'Green' opponents	All
Negative	69	60	45
Ambivalent	17	19	22
Neutral	13	20	25
Moral undertones	35	28	23
Fear	34	42	25

Note: The question was 'What comes to your mind when you think about modern biotechnology in a broad sense?'

other hand, the 'blue' group more frequently express associations with ethical issues: 35 per cent cite moral implications in their arguments, compared with 28 per cent for 'green' opponents. The 'blue' group are also more negative than the 'green' group (69 per cent versus 60 per cent), and the 'green' opponents are rather more likely to be neutral in their description of biotechnology (20 per cent for the 'green' group against 13 per cent for the 'blue' group). Table 6.2 summarises the results for Sweden.

One conclusion to be drawn from this analysis is that, simply on the basis of asking people about their cognitive associations with biotechnology, the results reinforce those obtained from the 'argumentation analysis'. Moreover, in general, people seem to use the same lines of argument against modern biotechnology as are employed by politicians, and/or vice versa.

The traditionalist–modernist segmentation as a general scheme for the resistance against new technology?

The pattern of resistance expressed in the 'blue'/'green' segmentation seems, at first glance, to be specific to biotechnology. This might be expected considering the strong moral commitment of the 'blue' critique – revealed through 'argumentation' analysis – prompted by fears that biotechnology will enable us to manipulate the nature of life itself. One would not expect to find such a reaction to other areas of technological development. Furthermore, the debates about the various technologies have developed differently over time, and in response to different perceptions of risk. Nevertheless, technological change in general is associated with alterations in the conditions of life and always affects various groups of the population in different ways. Indeed, the kind of modernist/traditionalist pattern of reactions to new technology identified above has been an element in most debates about new technology in both the past and the present. It would, therefore, be of interest to explore whether a similar pattern of reaction exists in relation to other areas of technology.

The Eurobarometer questionnaire included a battery of questions related to six major areas in which new technologies are currently developing. One of these is biotechnology/genetic engineering. On the basis of whether respondents believed this technology would improve or worsen the quality of life over the next twenty years, they were categorised as either optimistic or sceptical (see the appendix). Opponents to the other technologies may be defined according to the same criterion. However, the questionnaire does not permit analysis of the resistance to technologies other than modern biotechnology; and the questions asked in each case do not admit of inter-comparison.

Despite these difficulties, we attempted to produce cluster analyses on the basis of the responses obtained from opponents to two of the other technologies included in the questionnaire. We excluded those variables that are specific to biotechnology and, as a control, we performed the same analysis (i.e. using the variables non-specific to biotechnology) on modern biotechnology. The other technologies selected were 'computers and information technology' and 'space exploration'. Even though the number of opponents to both of these areas was significantly smaller than for biotechnology, the numbers were still large enough to make cluster analyses feasible (in contrast to, for instance, solar energy, to which there were very few opponents).

The results of the cluster analyses for biotechnology, in which the variables specific to biotechnology were excluded, turned out to be similar to the results for cluster analyses made when all variables were included in the analysis. This was true for all nine countries for which analyses were performed.[4] Not surprisingly, the variance in the two variables specific to biotechnology, and not included in these analyses, was somewhat smaller; but that was the most significant difference between the cluster analyses made with and without the variables specific to biotechnology. We regard this finding as confirmation of the general validity of the 'blue'/'green' segmentation. At the same time, it strongly suggests that we can make meaningful comparisons among the clusterings of opponents to different areas of technology using general variables and excluding criteria specific to any one type of technology.

Looking at the results of the cluster analyses for the two other areas of technology mentioned above, the general conclusion is clear: there is in both cases a segmentation of the opponents to these types of technology that is similar to the 'blue'/'green' segmentation for biotechnology. The only exceptions to this rule were Sweden and the UK vis-à-vis 'computers and information technology', and Sweden and Norway with respect to 'space exploration'. For seven out of the nine countries the segmentation was consistently in accordance with the modernist/traditionalist pattern within both areas of technology. Yet there are some interesting contrasts in the detailed segmentation pattern between attitudes towards these other technologies and those relating to biotechnology.

Regarding 'computers and information technology', in some countries religion loses much of the importance it has in influencing the segmentation of opponents to biotechnology. This is particularly true in the case of Italy. Conversely, political views and education become more important as criteria for segmentation. With respect to 'space exploration', the differences regarding religion are generally just as important as for biotechnology. The relative sizes of the modernist and the traditionalist groups tend to follow the same pattern as for biotechnology; thus the 'blue' group is largest in the southern European Catholic countries and the 'green' group is larger elsewhere.

Given the deficits of the model used in the analyses, the significance of the more specific segmentation patterns for technologies other than biotechnology should not be overemphasised. However, our findings do suggest that there are similar segmentation patterns for both biotechnology and the other areas of technology. This would be an interesting subject for further research.

Traditional and modern 'optimists'

It is obviously relevant to ask whether a similar segmentation to the one we have found for opposition to biotechnology has a counterpart among the *proponents* of biotechnology, i.e. the 'optimists'. To answer this question we have performed cluster analyses on the responses of the proponents of biotechnology for eleven of the fifteen countries for which a 'blue'/'green' segmentation has been found among sceptics. The results show that, almost invariably, a modernist/traditionalist segmentation also exists among the proponents. However, there are certain distinctive differences between the two groups. Among proponents, the male group is larger than the female. Further, whereas this difference holds across all countries for the modernist proponents, the picture is more complex vis-à-vis the traditionalists. Here the female group is the largest in several of the countries studied. Similarly, there is no consistent picture regarding risk perception. In almost half of the countries, risk perception is slightly lower among modernists than traditionalists.

Yet certain other observed differences between opponents and proponents do not seem to have any significance in relation to the segmentation pattern. Thus, proponents are, on average, a few years younger than opponents, and politically they are also more to the right. In any case, the modernists are consistently more to the left than the traditionalists and the difference in age between modernists and traditionalists is generally just as big among proponents as it is among opponents.

Only in the case of Norway did it prove impossible to show a meaningful segmentation for the proponents, even though there is a clear 'blue'/'green' segmentation among the opponents. We will not enter into a detailed discussion or further analysis of the segmentation of the proponents found by means of the cluster analysis. But it is hardly surprising to find such a segmentation among the proponents. Yet it is clear that traditionalists are likely to experience a conflict among different considerations in formulating their attitude towards biotechnology. On the one hand, they may tend to oppose it for the reasons we have already discussed above. On the other hand, the imperative of economic growth and technological innovation, which is closely connected with the market economy, may also influence their attitudes, since they tend to be politically to the right. Consequently, a large group of traditionalists are likely to be in favour of biotechnology even though they may have some reservations concerning the manipulation of nature, etc. In terms of the modernist group, their political orientation, which is on average slightly to the left of centre, might be considered surprising since, in other respects, it seems to display entrepreneurial characteristics. But the modernist group of proponents, which is quite large in most countries, probably also contains groups of well-educated social-liberals who consider the prospect of biotechnology important, and who believe that any environmental and other potential problems caused will be ameliorated.

Conclusion and outlook: ethics and risk

The findings presented in this chapter are based on a 'secondary use' of data insofar as the Eurobarometer survey has not been designed for the kind of analysis that we have performed. Against this background it is interesting, and significant, that a 'blue'/'green' segmentation is found in fifteen out of the seventeen European countries included in the survey. The assumptions characterising the two segments have been found to be consistent across the analysis of answers to the open-ended question ('What comes to your mind when you think about modern biotechnology in a broad sense?'), as well as in cross-tabulations with other variables in the survey. This adds credibility to our general findings regarding segmentation. It is also noteworthy that the 'blue'/'green' pattern is found across Europe in countries with large differences in terms of both the strength of resistance to modern biotechnology and the history of biotechnology debates. This again points to the importance of the traditionalist/modernist segmentation in relation to attitudes to modern biotechnology.

The prevalence of the 'blue'/'green' segmentation is also unaffected by the considerable socio-cultural differences between northern and southern Europe. There are, however, differences regarding the specific features of the segmentation between the Protestant north and the Catholic south which reflect these socio-cultural differences, most notably the greater importance of the 'blue' segment in Catholic countries, and the significance religion has for the clustering pattern.

The implications of the 'blue'/'green' segmentation for political debate are clearly important. The political discourse on the question of biotechnology has concentrated on 'value-free', scientific, risk arguments, and this discourse has underpinned decisions vis-à-vis the regulation of modern biotechnology in its agricultural and industrial applications. Yet this risk discourse has, at most, an appeal to the 'green' sceptics. As a potential basis for a 'social contract' of biotechnological regulation it obviously overlooks the 'blue' opposition. This blindness to an important element in the resistance against new biotechnologies points to a major weakness in the current regulation of biotechnology.

It would probably be misleading to suggest that the scientifically oriented risk discourse is generally attractive to the 'green' sceptics. Attempts to demonstrate a stronger emphasis on moral acceptance among the 'blue' sceptics by means of statistical analysis have been unsuccessful; for 'green' sceptics, moral acceptance is also a strong predictor of their willingness to endorse applications of modern biotechnology. But there is no doubt that 'green' sceptics generally understand the far-reaching consequences of the interventions in nature that may follow from the application of modern biotechnology and, therefore, they are likely to reflect on the necessity of setting limits to these interventions. They tend to assume this position not as a sacred principle, as is the case with the 'blue' sceptics, but rather because of rational and pragmatic considerations concerning our responsibility to our descendants. The growing orientation towards ethical arguments, which is not a special characteristic of the 'blue' opponents, may also in a broader sense be linked to the emergence of reflexive modernisation, or, in the words of Giddens, 'The ethical issues which confront us today with the dissolution of nature have their origin in modernity's repression of existential questions. Such questions now return in full force and it is *these* we have to decide about in the context of manufactured uncertainty' (Giddens, 1994: 217).

Thus, an orientation towards ethical questions in relation to modern biotechnology may be a characteristic of both modernists and traditionalists, even though the specific nature of that orientation is obviously qualitatively different for the two groups. Furthermore, risk in itself has an ethical dimension that is not reflected in the scientific, expert-oriented, risk discourse, but may be apparent to laypeople when considering the far-reaching and partly unknown risks associated with the application of a new technology.

Appendix

The segmentation of the opponents to biotechnology and genetic engineering was established by means of a cluster analysis (SPSS K-means cluster analysis[5]). Through an optimisation procedure, relatively homogeneous groups of cases are identified on the basis of selected characteristics. The number of clusters must be specified. The cluster analysis begins by using the value of the first *k* cases in the data file as temporary estimates of the cluster means, *k* being the number of clusters specified by the user. The final cluster centres are found through an iterative process.

The variables used to characterise the clusters were the following:

- *Gender*
- *Age*
- *Education*: respondents are divided into three categories based on the age when they left full-time education
- *Urban/rural*: a country-specific variable indicating either degree of urbanisation or size of town
- *Religiousness*: is self-reported. In this cluster analysis the answers were divided into four categories ranging from 'very religious' to 'not religious'
- *Political orientation*: is self-reported on a ten-point left–right scale
- *Materialism/post-materialism*: as an alternative to the Inglehart index, which could not be constructed on the basis of the Eurobarometer questionnaire, *materialists* are identified by using the political importance attached by the respondent to 'criminality' and 'poverty' as an indicator, while *post-materialists* are identified by using the political importance attached by the respondent to 'environment' and 'racism' as an indicator. An index for materialism/post-materialism is then obtained by subtracting the indicators for materialism and post-materialism from each other

- *Knowledge*: an additive index of nine items designed to measure knowledge of relevant basic biology
- *Risk perception*: the sum for the perceived risk of six different applications of modern biotechnology measured on a four-point scale.

These variables represent a combination of social/demographic variables and variables based on relevant attitude measurements. It is important to note that the Eurobarometer questionnaire has not been designed for this type of cluster analysis and the choice of variables is thus a compromise between what was desirable and what was possible. The use of a self-constructed index for materialism/post-materialism as a substitute for the Inglehart index is mentioned above. Despite the ad hoc character of this variable it is a good discriminator for most countries, and the cluster means are consistently in accordance with the predicted difference between 'blue' and 'green' opponents. It would have been desirable also to include a variable for view of nature, but that has not been possible.

The scales of the variables that have been used are different. Therefore, the variables were standardised to *z*-scores before running the cluster analysis. This is the procedure recommended in the SPSS Applications Guide. Some statisticians warn against standardisation of variables for cluster analyses, because it may 'dilute' the differences between groups on the variables that are the best discriminators.[6] Alternatively, all variables could be transformed to variables with the same number of categories. This will inevitably lead to a loss of useful information. For this reason standardisation was used as the main procedure. As a control, the cluster analysis was also carried out for four of the countries with all variables converted to three-category variables (except of course the gender variable). The results of this approach showed good qualitative agreement with the results when using standardised variables.

For two countries, Austria and Finland, it turned out to be impossible to identify clusters with the 'blue' and 'green' characteristics. Furthermore, it is difficult to obtain stable solutions for these two countries. Small variations in input, for instance minor changes in the variable constructions, may lead to considerable changes in the clustering, with different variables being the most important discriminators. This seems to indicate that the data sets for these two countries are unstable with respect to the cluster analysis. It is quite likely that 'blue'/'green' segmentations of opponents to modern biotechnology exist in Finland and Austria, too, but that we cannot identify them. The cluster analysis was also carried out on the Austrian data set using the three-category variables as mentioned above, but still with negative results.

As a criterion for defining the group of opponents to biotechnology we used the answers to the question 'Do you think it will improve our way of life in the next 20 years?' Respondents who answered 'it will make things worse' to this question are defined as opponents. This group of respondents on average make up 20.2 per cent of the respondents for the EU as a whole, but ranges from 8.4 per cent in Portugal to 36.5 per cent in Austria. Generally, the group is only about one-fifth of the population, even though this result may be influenced by the 'split ballot' questioning. An alternative criterion might be based on the answers to questions regarding the willingness to endorse six different applications of modern biotechnology. An aggregated measure of 'willingness to encourage' the six applications leads to a different characterisation of the opponents, probably because they tend to answer differently when confronted with specific applications of biotechnology. Furthermore, the question about whether biotechnology or genetic engineering in general is likely to improve our way of life is part of a battery of similar questions about other major technologies and the context for this question is therefore different.

For all countries, the group of opponents to modern biotechnology defined by willingness to endorse the six applications is also larger than the group of opponents who expect biotechnology or genetic engineering to make things worse in the next twenty years. For this reason, too, it was considered appropriate to perform cluster analyses using 'willingness to encourage' the six applications as the criterion for defining opponents. This was done for four countries, and in every case a 'blue'/'green' pattern of clustering was obtained. Although it was quantitatively different from the solutions found by using the other criterion for defining the opponents, which is not surprising since the groups are different, this result indicates that the 'blue'/'green' segmentation is detectable despite such variations in the definition of the group of opponents. But these different possibilities of defining the group of opponents and the quantitative differences in the resulting clusters are also warnings against an overly detailed interpretation of the results.

NOTES

1. Some sections of this article have been reused in the present chapter.
2. Corresponding analyses of Norwegian data have been reported in two articles by Torben Hviid Nielsen (1997a, 1997b). The small differences compared with the results reported in this chapter are due to the use of 1993 data in the first article and the use of a different indicator for materialism/post-materialism in the second article. Hviid Nielsen used the Inglehart index for materialism/post-materialism on the Norwegian data. Since this indicator was

not available in the Eurobarometer survey for the EU countries, a substitute indicator had to be constructed, as explained in the appendix.

3. We are indebted to Timo Rusanen for comments on the Finnish situation. See also Rusanen et al. (1998).
4. The nine countries are Denmark, Norway, Sweden, Germany, the Netherlands, the UK, Italy, Spain and Greece.
5. See SPSS Base 8.0, Applications Guide, 1998.
6. See, for instance, Everitt (1993).

REFERENCES

Bud, Robert (1993) *The Uses of Life. A History of Biotechnology*, Cambridge: Cambridge University Press.

Ernst & Young (1997) 'European Biotech 97: "A New Economy"', *The Fourth Annual Ernst & Young Report on the European Biotechnology Industry* (April).

Everitt, Brian S. (1993) *Cluster Analysis*, 3rd edn, London: Edward Arnold.

Freeman, Christopher (1995) 'Technological Revolutions: Historical Analogies', in Martin Fransman, Gerd Junne and Annemieke Roobeck (eds.), *The Biotechnology Revolution?* Oxford: Basil Blackwell, pp. 7–24.

Giddens, A. (1994) *Beyond Left and Right: the Future of Radical Politics*, Cambridge: Polity Press.

INRA (1991) Eurobarometer 35.1, 'Opinions of Europeans on Biotechnology in 1991'.

(1993) Eurobarometer 39.1, 'Biotechnology and Genetic Engineering: What Europeans Think about It in 1993', October.

(1997) Eurobarometer 46.1, 'Les Européens et la Biotechnologie Moderne', Draft, February.

Mae-Wan Ho, Hartmut Meyer and Joe Cummins (1998) in *The Ecologist* 28(3): 146–53.

Nelkin, Dorothy and M. Susan Lindee (1995) *The DNA Mystique. The Gene as a Cultural Icon*, New York: W.H. Freeman.

Nielsen, Torben Hviid (1997a) 'Modern Biotechnology – Sustainability and Integrity', in Susanne Lundin and Malin Ideland (eds.), *Gene Technology and the Public. An Interdisciplinary Perspective*, Lund: Nordic Academic Press.

(1997b) 'Behind the Color Code of "No"', *Nature Biotechnology* 15 December: 1320–1.

Rifkin, Jeremy (1998) *The Bioetch Century*, New York: Jeremy P. Tarcher/Putman.

Rusanen, Timo et al. (1998) 'Finland', in John Durant, Martin Bauer and George Gaskell (eds.), *Biotechnology in the Public Sphere. A European Sourcebook*, London: Science Museum, pp. 43–50.

Time (1998) 'Alien Seed?' 24 August.

Toulmin, Stephen (1958) *The Uses of Arguments*, Cambridge: Cambridge University Press.

Toulmin, Stephen, R. Reicke and A. Janik (1979) *An Introduction to Reasoning*, London: Macmillan.

7 The structure of public perceptions

Cees Midden, Daniel Boy, Edna Einsiedel, Björn Fjæstad, Miltos Liakopoulos, Jon D. Miller, Susanna Öhman and Wolfgang Wagner

Social surveys are usually conducted under the assumption that people have attitudes and that the survey taps into something called 'public opinion'. The percentages of respondents agreeing or disagreeing with a particular statement then acquire the status of a social fact. However, too much weight may be ascribed to such facts, particularly when the issue is either unfamiliar to respondents or of low salience. Since biotechnology was novel and unfamiliar to most people in 1996, it is reasonable to question the validity of survey data on this issue. Thus, in this chapter, we employ survey responses in asking the question: 'Can we talk of a public opinion on biotechnology that is based on well-formed attitudes and drawn from existing knowledge and values?'

Our analysis is based on the Eurobarometer survey of 1996, which was carried out in all European Union countries, as well as in Norway and Switzerland. The main purpose of this chapter is to improve our understanding of the attitudes expressed by examining their nature and strength and by analysing underlying factors that might explain the observed attitudinal differences. The chapter proceeds from a description of the general and specific judgements on biotechnology elicited by the survey, identifying the general level of acceptance and differences in judgements between applications and between nations. After this descriptive analysis, we will evaluate various aspects of the attitudinal data: the consistency of expectations among the different countries involved in the survey; the relations between differing attitudes and the particular applications of biotechnology; relations between such factors as 'informedness' and perceptions of this technology; the effect of more general attitudes towards technology per se; the range of attitudes towards biotechnology; and, finally, the roles of knowledge and values in determining public attitudes.

Attitudes in Europe

Biotechnology was invented and developed in the context of a Western culture that is, in general, favourably disposed towards inventions and

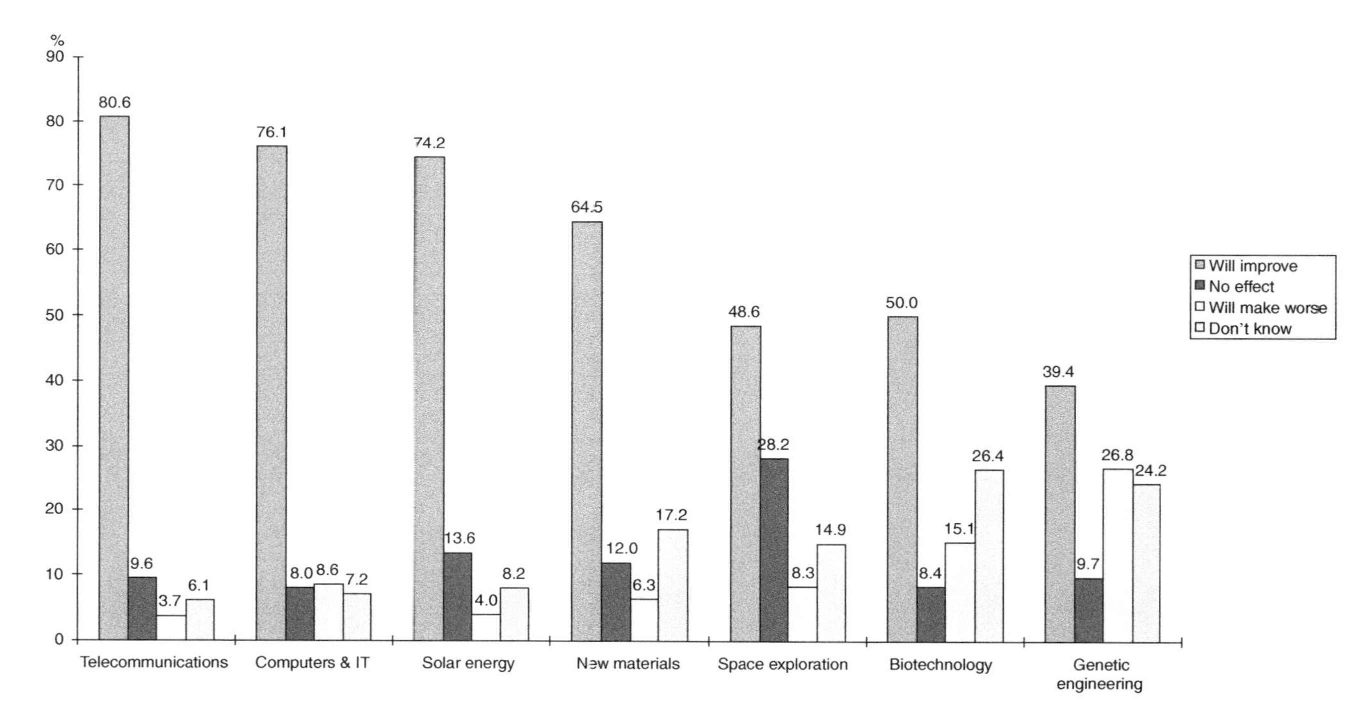

Figure 7.1 General attitudes to six different technologies

new technologies. In order to locate the position of this new technology relative to other technologies, we asked the respondents to rate six different technologies according to whether they thought they would improve our way of life in the next twenty years, have no effect, or make things worse.

As can be seen in figure 7.1, two-thirds or more of the European respondents are optimistic (in declining order) about telecommunications, computers and information technology, solar energy, and the development of new materials and substances. A mere 3 to 8 per cent are pessimistic. Half of the respondents are also optimistic about space exploration and, again, very few are pessimistic. Questions relating to our target technology, modern biotechnology, were formulated in two different ways. Half the sample were asked about biotechnology, the other half were asked about genetic engineering. The latter received the lowest score in terms of optimism (39 per cent) and the highest score in terms of pessimism (27 per cent) of the entire range of technologies cited. For both wordings, about a fifth of EU citizens expect modern biotechnology to make things worse; as many as a quarter don't know. Italy, Spain, Portugal and Belgium contain the highest percentages of optimists (all over 50 per cent), and the highest proportions of pessimists were found in Austria, Denmark, Norway, the Netherlands and the United Kingdom (all over 25 per cent). The tendency to answer 'don't know' is relatively high in some countries, especially Greece (47 per cent) but also Ireland and Portugal (33 and 31 per cent, respectively), indicating that in these countries unfamiliarity with biotechnology and genetic engineering is especially prevalent. The lowest proportions of 'don't knows' were reported from the Netherlands and Denmark (both 16 per cent).

Applications of genetic engineering

Genetic engineering comprises a number of biotechnological procedures used in a wide variety of applications. Thus, in order to refine our attitudinal measurements, we identified six different applications and asked respondents to judge them according to four criteria with relevance to the European debate: the usefulness to society of each application; how risky they are; their moral acceptability; and the extent to which their further development and use should be encouraged.

The mean results are reported in figure 7.2 on a five-point scale running from +2 to −2. The value +2 indicates maximum usefulness, minimum perception of risk, maximum moral acceptability, and maximum encouragement. Overall, EU citizens are clearly in favour of the two medical applications: the development of vaccine and drug-producing bacteria and

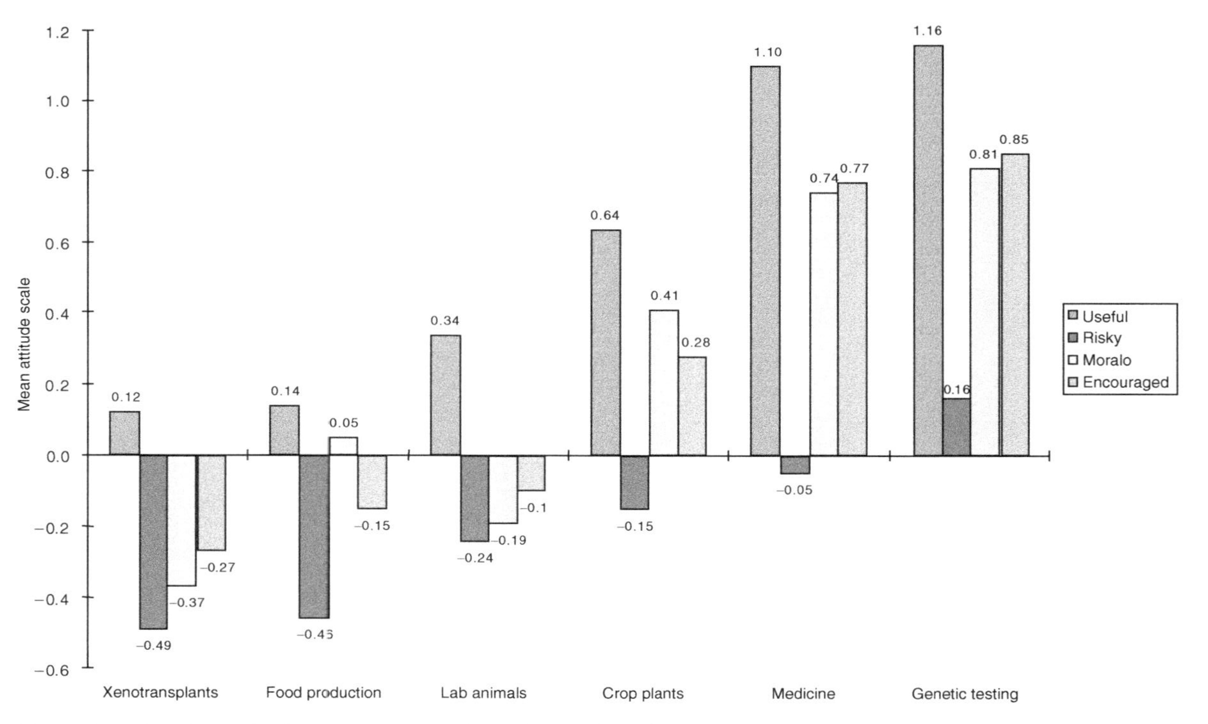

Figure 7.2 Specific attitudes to six applications of genetic technology
Note: Attitudes are measured on four different criteria on a scale from −2 to +2.

the testing for hereditary diseases. To a lesser degree, Europeans would also encourage introducing foreign genes into crop plants to protect them against pests. The remaining three applications (xenotransplantation, genetically modified (GM) food and laboratory animals) all registered close to, or below, the zero line. Least approval is accorded to the breeding of GM animals for xenotransplantation.

All six applications are judged as being useful, but to varying degrees. Again, the two medical applications and GM crops are considered to have the highest utility. These three applications are also evaluated as being the most morally acceptable. Least morally acceptable are xenotransplants, which are also seen as being most risky, together with the use of modern biotechnology in food production.

To permit comparison of the attitudes in the different EU countries, we have calculated the mean 'encouragement' scores for the six applications for each country (see figure 7.3). Most inclined to encourage the use of genetic technology is Portugal, followed by the other southern countries of the EU and the one furthest to the north, Finland. The common characteristic of these countries would seem to be their peripheral location to central Europe. The middle group comprises the western–central parts of the EU; and below them, in terms of the level of encouragement expressed, are a group of countries in the northern half of the EU. Finally, Austria registers the lowest level of encouragement. Thus, with a few notable exceptions, there is a clear south–north dimension to attitudes towards biotechnology and genetic engineering.

Expectations of genetic engineering in general

Modern biotechnology has been hailed by some as a panacea for a wide array of world problems, and by others as a threat to both humanity and nature. In order to gauge the variety of these expectations, ten possible consequences of genetic engineering were presented to the respondents, who were then asked to judge whether they were likely or unlikely to materialise within the next twenty years. The results are presented in table 7.1.

The perceived likelihood of genetic engineering bringing about both good and bad effects is striking. A majority of EU citizens believe that this technology will be able to cure most hereditary diseases, while, at the same time, two out of three hold that it will probably create dangerous new diseases. Many, but not a majority, are also optimistic about impacts on the environment and on world hunger, and are sceptical of the notion that babies will soon be made to order.

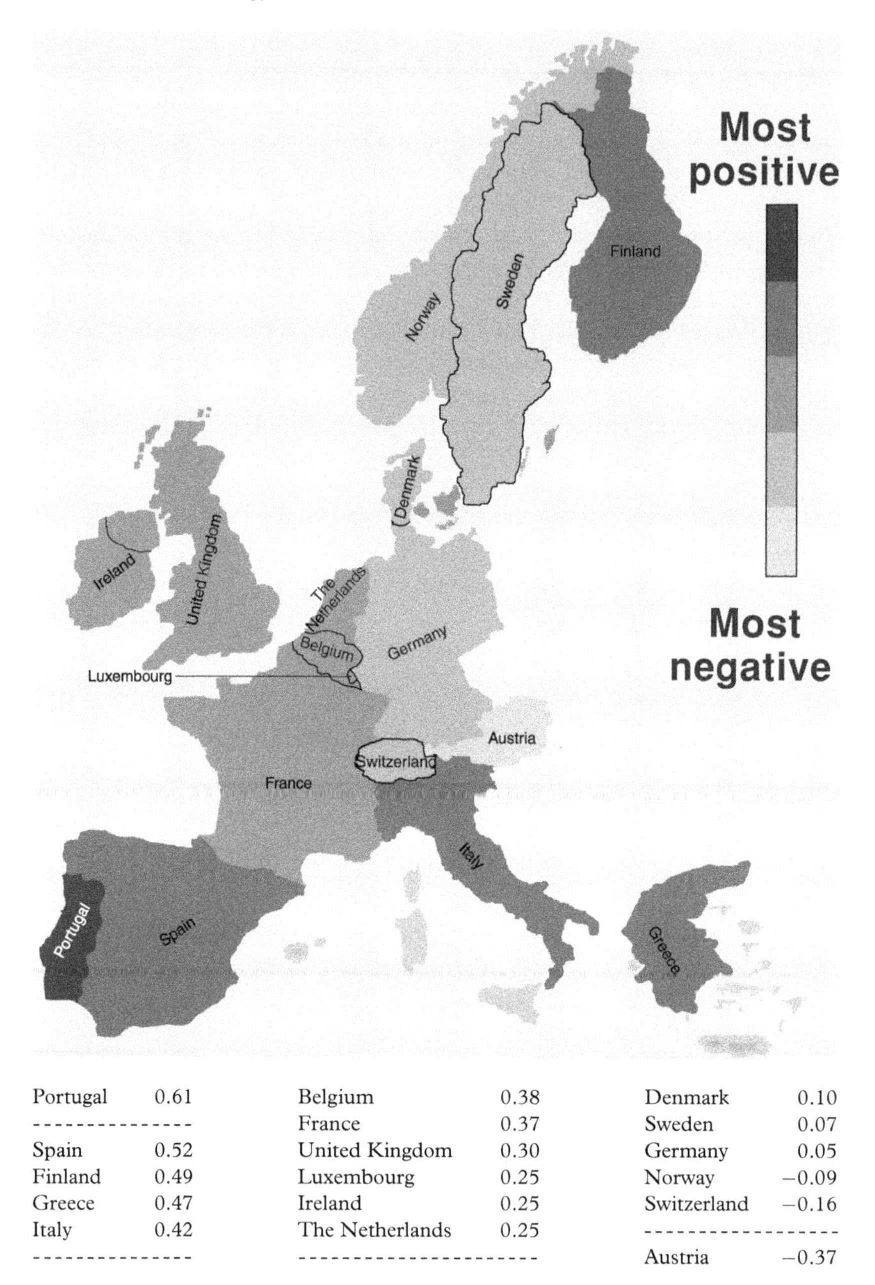

Portugal	0.61	Belgium	0.38	Denmark	0.10
----------		France	0.37	Sweden	0.07
Spain	0.52	United Kingdom	0.30	Germany	0.05
Finland	0.49	Luxembourg	0.25	Norway	−0.09
Greece	0.47	Ireland	0.25	Switzerland	−0.16
Italy	0.42	The Netherlands	0.25	----------	
----------		----------		Austria	−0.37

Figure 7.3 Mean aggregate attitudes across Europe

Table 7.1 *Positive and negative expectations about biotechnology*

	Likely	Unlikely
Positive consequences		
Solving more crimes through genetic fingerprinting	71	19
Curing most genetic diseases	56	33
Substantially reducing environmental pollution	47	46
Substantially reducing world hunger	37	56
Negative consequences		
Creating dangerous new diseases	68	21
Allowing insurance companies to ask for genetic tests	43	42
Producing designer babies	40	50
Reducing the range of fruit and vegetables we can get	28	60
Other consequences		
Getting more out of natural resources in the Third World	54	32
Replacing most existing food products with new varieties	45	43

Table 7.2 *The positions of European countries in terms of evaluative consistency*

Evaluative consistency	High negative expectations	Low negative expectations
High positive expectations	SP, P, GR	
Low positive expectations	A	A, NL, S, D

Note: A = Austria; D = Germany; GR = Greece; NL = the Netherlands; P = Portugal; S = Sweden; SP = Spain.

It would be reasonable to assume that respondents who consider positive outcomes likely would also tend to see negative predictions as unrealistic, and vice versa. And to some extent this is borne out by our data. A quarter of Europeans belong to the positive group, and 14 per cent to the negative group – close to 40 per cent in total. Table 7.2 shows the national differences.

Congruent with the attitude results reported above, the lowest proportion of positive respondents (see table 7.2) was found in Austria (15 per cent). Yet, surprisingly, the country with the lowest share of negative respondents was also Austria (10 per cent). This apparent paradox is largely resolved if one recognises that the highest proportion of respondents with few expectations of biotechnology is again to be found in Austria, where close to 70 per cent belong to this group. This group of people with 'non-expectations' is by far the largest in the EU total. Other countries with a high proportion of respondents with low expectations are the Netherlands (64 per cent), Sweden (60 per cent), and Germany (58 per cent). In general, countries in the north of Europe tend to have lower expectations than those in the south, with west–central Europe located between the extremes. Ambivalence, i.e. the tendency to attribute both good and bad effects to biotechnology, is also correlated with the south–north axis, with the highest proportion of ambivalence found in Portugal, Spain and Greece (20–21 per cent), and the lowest found in Finland, the Netherlands and Sweden (all 5 per cent).

Attitude structures related to specific applications

As described above, respondents were asked detailed questions about six specific applications of biotechnology. They were asked to assess these six applications according to the criteria of utility, risk, morality and the proper level of encouragement. In this section we assess the coherence of the attitudes articulated. Where attitudes towards the six applications are more or less related, one can impute the presence of a meaningful general concept of biotechnology. A weak relationship, however, would suggest that specific applications tend to be judged more individually. In addition to this analysis of the structural characteristics of attitudes, we will seek to identify factors and predilections that underpin the level of support expressed for particular applications.

A confirmatory factor analysis (Lisrel8 procedure) was conducted in which twenty-four judgements were brought together (four characteristics per application). The analysis shows that the twenty-four judgements can be condensed into *two* main factors (Chi-square = 157.9, d.f. = 123; RMSEA = .017; 90 per cent CI = .007; .024). Factor 1 combined all the judgements on use, moral acceptability and encouragement. Factor 2 combined the six risk judgements alone. These findings carry two implications. The first is that the six applications are, indeed, judged in a highly consistent manner. The second is that there are two independent components to the attitudes expressed: the first comprises judgements on the proper level of encouragement integrated with judgements on moral acceptability and utility; the second concerns solely judgements as to perceived risk.

Overall, these results demonstrate that respondents employ the four criteria (use, risk, moral acceptability and encouragement) as a two-dimensional structure, which is applied across the range of biotechnological applications. Thus, when thinking about risks related to medical applications, the same basis for judgement is involved as when the respondent is judging agricultural and food applications.

One consequence of this analysis is that we are able to create an aggregate measure over the six applications. A further outcome is that we can construct two attitude indicators: a combined summary score for the judgements relating to use, moral acceptability and encouragement, and one indicator for perceived risk, all based on the respondents' evaluations of the six applications.

The core attitude structure

The correlations found among perceived risk, moral acceptability, perceived utility and desired encouragement suggest that the general evaluative reaction towards the six applications reflects a subjective assessment including each of these criteria, with perceived risk forming a separate dimension. Yet this does not necessarily imply that these factors are integrated, by respondents, in a thoughtful manner, since intuitive responses might just as well comprise these criteria. The unrelatedness of risk to the core attitude structure is a surprising finding considering the centrality of risk and safety issues in the regulation of biotechnology. A possible explanation is that the risk debate, which usually consists of technical and scientific arguments about the effects of processes and products on health and the environment, is not well understood by the average member of the public; debate on these complex and contested issues requires specialised knowledge beyond the experience of the majority of people. However, this is not to say that the public believes biotechnology to be without risk. As we have previously observed, the overall perception of risk is high with respect to most applications of biotechnology. The present analysis, however, shows that risk perception is not related to the overall evaluation of the technology.

The core attitude – composed of judgements as to utility, moral acceptability and encouragement – also requires explication. The strong correlation among these criteria shows the appeal arguments of these types have to the wider public. Moral acceptability will tend to comprise an assessment of whether an application is either 'good' or 'bad'. Combined with the utility judgement, one has a general indication of the public's sense of the rightness of biotechnological applications and their value for society at large. The high correlations suggest that this general attitudinal expression should be interpreted not as the result of a trade-off among various judgements, but as the outcome of a fundamental, irreducible impression.

This conclusion forms the entrée into the next stage of our analysis, in which we try to explain this core attitude component, as well as the perceived risk component. The objective is to develop a structural model that describes the determining factors of the attitudes expressed. This model will take into account the interconnections among the criteria identified, and their relative importance in affecting respondents' judgements as to the overall acceptability of biotechnological applications.

Modelling attitude-influencing factors

It seems unlikely that public evaluations of the six applications were usually based on a rational process of judgement based on an elaboration and integration of relevant beliefs. In view of this, we hypothesised that more general attitudes play a role in the formation of attitudes towards specific applications. In particular, we anticipated that general expectations about biotechnology would, in general, be employed by respondents where their knowledge of particular applications was ambiguous and/or incomplete. Beyond this, even more general perceptions of technology per se might be activated and applied to more specific technologies and applications.

This expectation is consistent with the notion of schematic processing (see, for example, Smith, 1998). A schema can be understood as a knowledge structure represented in the form of memory that is related to a certain class of objects. Schemas are based on earlier experiences and help a person to understand the world by linking new information to existing knowledge. In this way a person can make use of earlier experience and does not have to conceptualise new objects and situations *de novo*. Thus, schematic processing refers to a process of attitude formation in which objects are judged not according to object-specific observable characteristics, but with reference to characteristics that are connected to a general category of objects. This may occur if the specific object at hand is categorised as being connected to an existing category. Schematic processing is, therefore, an efficient means of apprehending a new concept and of obviating the need for time-consuming deliberation. Indeed, it is all but essential when a person lacks the capacity to process information thoroughly because of, for example, time constraints, distractions or a lack of required basic knowledge. Schematic processing is likely to be the customary way of forming judgements. Object-specific processing, on the other hand, requires not only cognitive capacity but also the motivation to process, and this latter factor will be contingent on the degree of personal relevance the individual ascribes to the object (see, for examples, Eagly and Chaiken, 1993). Consistent with this, we would expect the level of available knowledge and information on the topic of biotechnology to influence judgements relating to its application. Knowledge can be seen as a precondition for thoughtful object-specific judgement.

Finally, we may expect that the process of attitude formation will be different for people from different backgrounds. More specifically, we expect education to be a major determining factor, especially insofar as it may relate to levels of prior information and basic knowledge. Education might also increase an individual's sense of the importance and relevance of new technologies. Other factors are not so readily identifiable – studies in different countries report divergent results. However, the effects of age and gender will be explored.

A structural model of antecedent factors of attitudes to biotechnology

In order to test the expectations advanced in the previous section, a structural model was developed and tested applying LISREL8 (Joereskog and Soerbom, 1994). Figure 7.4 depicts a fitted LISREL solution that describes the relations among the attitude factors and the following explanatory factors: positive expectations of biotechnology in general, negative expectations of biotechnology in general, optimism or pessimism towards technology in general, level of 'informedness', and the group characteristics of education, age and gender. The model shows both direct and indirect effects upon attitudes towards biotechnology. Before discussing the relationships we must first explain the factors involved in the model.

Attitudes. The attitudes are represented, first, by the core attitude component (composed of perceived utility, moral acceptability and desired level of encouragement) and, secondly, by the perceived risk component. Both components represent the aggregation of specific attitudes toward the six applications, as

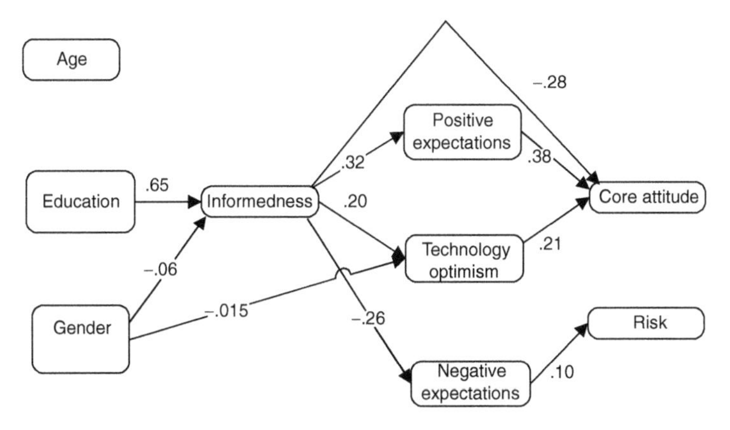

Figure 7.4 A structural model representing antecedent factors of attitudes toward biotechnology
Notes: Results of the Lisrel analysis: RMSEA = 0.26; 90 per cent CI = .014; .038; Chi-square = 38.8; d.f. 19; $R^2_{\text{core attitude}} = .32$; $R^2_{\text{risk}} = .06$.

indicated by the Principal Component Analysis analysis described above.

Positive and negative expectations of biotechnology in general. As discussed above, a set of questions asked respondents to assess the likelihood of certain positive and negative outcomes of biotechnology, in general, over the next twenty years. From these items two scales were created: one for positive and one for negative attitudes (Cronbach's alpha for scale 1 = 0.68; for scale 2 = 0.66).

Optimism or pessimism about technology in general. As discussed above, respondents were asked if they were optimistic or pessimistic about six main areas of technology. These judgements (with the exclusion of biotechnology) were aggregated into an indicator representing optimism or pessimism about the future of technology in general (Cronbach's alpha = 0.66).

'Informedness'. This factor represents the level of knowledge (based on the knowledge scale as described in the section 'Knowledge and attitudes' below, and the level of awareness of biotechnology as measured by questions on having previously talked, or heard, about biotechnology). Combining these factors gave a three-level indicator of 'informedness'.

Age, education and **gender** were also included, using direct measures.

The model shows that the core attitude, reflecting perceptions of the general utility of biotechnology for society, is directly explained by three factors: the extent of positive expectations of biotechnology in general; levels of optimism or pessimism concerning the use of technology in general; and levels of 'informedness'. The pattern and direction of these effects is interesting. 'Informedness' has a direct negative effect upon general attitudes, suggesting that the more information the public has about biotechnology, the more critical its attitudes will tend to be. At the same time, however, we see that 'informedness' also exerts an influence in the opposite direction. Higher levels of knowledge and comprehension seem to foster technological optimism, and thereby this factor generates a positive effect on the core attitude. A parallel finding is that 'informedness' has a positive effect on the level of positive expectations, which also implies an indirect positive effect on the core attitude. Finally, 'informedness' is negatively correlated with public expectations of biotechnology, and thus has an indirectly negative influence on perceived levels of risk. In sum, the core attitude is influenced by positive expectations about biotechnology and by a more general technological optimism/pessimism;

'informedness' does not exert a simple effect upon attitudes to biotechnology. Its influence is mixed in the sense that knowledge of biotechnology directly promotes negative attitudes, but *in*directly it can favour a more positive core attitude and risk perception. The risk component of the attitude is directly influenced in the model by negative expectations of biotechnology.

The group characteristics (age, gender and education) have, at most, indirect effects. There is no evidence that age has any effect on attitudes towards biotechnology. Gender is weakly and negatively related to 'informedness' and to optimism or pessimism about technology in general. Education is the most relevant characteristic. It is, obviously, strongly connected with 'informedness'. Yet, the total explained variance for the risk component is very low ($R^2 = .06$). For the core attitude component the results are meaningful, albeit moderate ($R^2 = .32$).

Nonetheless, it can be concluded, first, that general ideas and beliefs about both biotechnology *and* technology were able to affect the public's attitude towards specific biotechnological applications. This suggests that respondents applied general schemas relating to biotechnology and technology to the more focused applications about which they were questioned. Secondly, we have found that 'informedness' has an important effect on attitudes. The highly informed expressed more negative views on the utility of the specific applications of biotechnology. At the same time, the model indicates that high 'informedness' is related to higher levels of optimism concerning technology *in general*, which contributed indirectly, in a positive sense, to public attitudes towards biotechnology. The highly informed also tend to have fewer negative expectations of biotechnology in general, contributing to a lower level of perceived risk.

In general, this analysis indicates that more highly informed groups are critical of specific applications. However, this critical attitude appears to be softened by positive attitudes towards technology and biotechnology in general. These findings would suggest that initially positive opinions of biotechnology, as a new technology, will recede once the public becomes better informed on the specific attributes and implications of this technology. When such information becomes available, the effect of general schemata will be reduced.

Differentiation and the strength of attitudes

The foregoing analyses have shown that attitudes among European citizens to biotechnology are unstable. This finding may be explained if we look in more detail at the strength of the attitudes expressed. New technologies are often not well known and difficult to grasp for many people; and it takes time for the public to become sufficiently well informed to make a judgement. The meaning of attitudes differs according to the strength of an individual's convictions. Superficial attitudes are unstable and susceptible to change in response to new information. They are also sensitive to peripheral cues such as suggestive headlines and the content of public relations campaigns. Furthermore, superficial attitudes are poor predictors of people's actions. Therefore, attitudes towards new technologies can be properly understood only if attention is paid to the strength of the attitudes articulated.

In social psychology, numerous indicators for attitude strength have been employed (see, for an overview, Krosnick et al., 1993), but in this study two indicators for attitude strength have been constructed: attitude 'extremity' and attitude 'differentiation'. Extremity indicates the extent to which respondents give answers at the extremes of the scale, on the assumption that extreme answers reflect more convinced and certain attitudes than answers that appear in the middle of the scale. This assumption is supported by many attitude studies in social psychology (see, for example, Eagly and Chaiken, 1993). The extremity scale is based on the six scores on levels of encouragement deemed appropriate for the specific biotechnological applications. A neutral answer was given a score of 0, a score tending towards agreement or disagreement received a score of 1, and strong agreement or disagreement was assigned a score of 2. Attitude differentiation reflects the difference in the scores for the six applications for each respondent. It is assumed that persons whose responses to the six applications are differentiated have more carefully thought out attitudes. The scale is constructed by calculating the variance of the six encouragement questions per respondent.

The frequency distribution shows that the majority of the respondents have a moderate extremity score between 6 and 9 (range 1–12), 20 per cent have a high extremity score and 10 per cent a low score. The differentiation index shows that about 15 per cent exhibit no variation and 66 per cent express a variance of less than 1 (range 0.0–4.5). We can conclude that there is a large group of respondents who have a low differentiation in their attitudes. That people find it hard to differentiate among the applications is also suggested by the high interrelations between the specific attitudes.

Figure 7.5 shows country differences in average attitude extremity. The findings suggest that the countries in Western Europe whose populations seem to express opinions on biotechnology with the greatest conviction also display the most extreme attitudes. This is especially true for Denmark, the Netherlands and Sweden. However, their scores do not differ strongly from those of other nations. Moreover, this analysis reveals some unexpected results; in particular, Greece, with a rather low

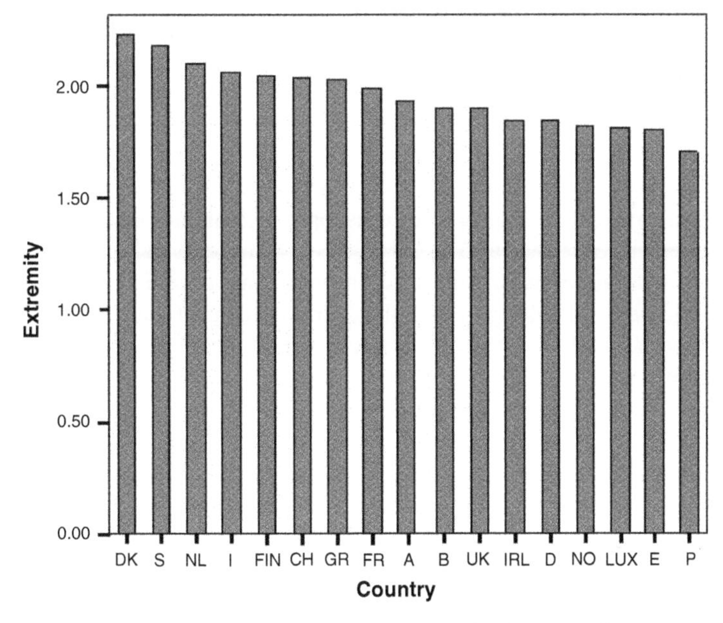

Figure 7.5 Country differences in average attitude extremity

level of public knowledge on biotechnology, also exhibits rather extreme scores. Obviously there are other factors influencing extremity. In the next section the relations between attitude extremity and knowledge will be analysed at the individual level.

Knowledge and attitudes

School education and academic knowledge are traditionally considered to be prime predictors of attitudes towards new technologies. Many scientists have the expectation that opposition to science and technology is grounded in a lack of knowledge and familiarity. Education, it is thought, may counter this alleged deficiency. From a social psychology point of view, however, one need not assume that knowledge and attitudes are related in so uncomplicated a manner. According to one alternative view, the relationship depends more on the evaluation of attributes connected to the particular technology. Knowledge may then influence the attributes that are included in the forming of an attitude, and will thus have an indirect effect upon this process (Hisschemoller and Midden, in press). In the present case we analysed the knowledge items in the 1996 Eurobarometer survey and their relationship to attitudes.

Knowledge indicators

The questionnaire comprised nine knowledge items, six of which were designed to measure 'objective' knowledge and three 'subjective knowledge'. Objective knowledge items were defined as questions whose correct answer can be found by studying a basic modern textbook on biology. These include the uses of microscopic organisms (e.g. cleansing wastewater via bacteria, and brewing beer with yeast cells), technical issues (e.g. cloning of organisms, and prenatal diagnosis of Down's Syndrome), and basic knowledge (e.g. that viruses cannot be 'contaminated' by bacteria, and that humans share a very high proportion of genes with chimpanzees).

Subjective knowledge items were defined as questions whose correct answer is either presupposed or implied by academic knowledge but cannot ordinarily be found by consulting a textbook index. Examples of such knowledge items are that all tomatoes – not only those that have been genetically altered – contain genes; that genetically modified organisms are always bigger than ordinary ones; and that genes from 'higher' species (e.g. animals) can be inserted into the genomes of 'lower' species (e.g. plants). The term 'subjective' refers to the fact that not knowing how to answer such questions makes one more prone to adopt inaccurate images of biotechnology, for example that of genes being injected into natural tomatoes to alter them genetically, a graphic illustration that was, in fact, used in some newspapers (see plate 9.1 in chapter 9).

To generate a knowledge score, the number of items answered correctly by each respondent was counted and the scores trichotomised for use in correspondence analysis. A correspondence analysis of a matrix cross-tabulating the fifteen EU-countries by the three 'objective' knowledge (OK) categories plus the three 'subjective' knowledge (SK) categories (low, medium high) reveals that both knowledge scores are highly correlated. The first dimension captures 90 per cent of variance, with low objective and subjective knowledge on one pole and high knowledge on the other (figure 7.6). The correlations between the two knowledge scores are also rather high ($r = .56$).

From this analysis, three clusters of countries appear quite clearly. At the high end of the knowledge pole are Denmark, the Netherlands and Sweden, in the middle are France, Germany, Italy and Luxembourg, and on the low end of the knowledge pole are Austria, Greece, Portugal and Spain. Finland, Ireland and the UK assume peculiar positions, outside of the principal clusters. The UK, for example, seems to be high on the scale of objective knowledge, but low in terms of subjective knowledge, whereas Finland is the reverse.

Table 7.3 *Relation between knowledge and general and specific attitudes*

	Perception of applications			
Knowledge	Useful	Risk	Morally acceptable	Encouragement
Knowledge total	.21	.20	.20	.18
Subjective knowledge	.17	.16	.17	.16
Objective knowledge	.19	.18	.18	.16

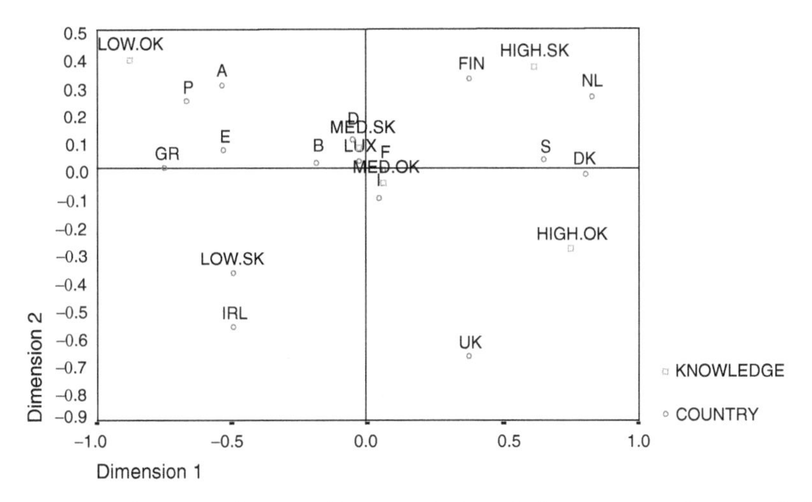

Figure 7.6 Country differences in knowledge: correspondence analysis, canonical normalisation

It should be noted that the so-called subjective knowledge items were more difficult to answer than the objective knowledge items. On average, only a third of respondents got the three subjective items correct, but the number of correct answers to the objective items ranged from 30 per cent to over 80 per cent. Neither the objective nor the subjective knowledge scores are correlated with the general attitude of the populations (optimism/pessimism) and they are only slightly (positively) correlated with perceptions of usefulness, risk, moral acceptability and appropriate encouragement. A positive correlation would indicate that the more a respondent knows, the more he or she agrees that applications are useful, risky, morally acceptable and worthy of encouragement, but this relationship never explains more than 4 per cent of the variance (table 7.3).

Knowledge and attitude extremity

It might be expected that people with high levels of knowledge would have the more definite attitudes towards biotechnology, because they will have a greater ability to process new information on this subject. Thus, we calculated the correlations between the general knowledge scale (nine items) and the extremity and differentiation indicators. Knowledge and attitude differentiation appear to be unrelated ($r = .04$). The expectation that more knowledge enhances the differentiation in judgements appears, therefore, to be invalid. Nonetheless, we found a modest, but meaningful, positive relationship between knowledge and attitude extremity ($r = .24$), which confirms that, in general, more extreme attitudes tend to be based on a higher degree of knowledge.

Values and attitudes

To what extent can attitudes to biotechnology be explained by taking the value structures of the respondents into account? Briefly put, the answer is 'not very much'. When looking at both general attitudes towards biotechnology and genetic engineering (figure 7.1 above) and a summation of the specific attitudes towards six applications (figure 7.2 above), a remarkably small percentage of the variation can be explained in terms of ordinary background variables. Entering political position (on a ten-point left–right wing scale), religiosity (on a four-point scale) and environmentalism (on a two-point scale) into a multiple regression enabled us to explain only 1 per cent and 3 per cent, respectively, of the variation in the two attitudinal measures. Some characteristics emerge as contributing significantly to this 1 per cent, mainly owing to the unusually large sample: persons with non-green values and persons to the right of the political centre tend to be more positive than those with green values and those to the left of the political centre. But, when looking at additional specific value indicators, it emerged that neither views on risk-taking nor views on the regulation of biotechnology contribute significantly to explanatory power. On the other hand, 'breeding conservatism' does. This was measured by two questions relating to the wisdom and the efficiency of relying upon traditional breeding methods. Introducing this extra variable into the regression analysis produced beta coefficients of .14 and .21 respectively, more than doubling the explained variance. However, it might be argued that this measure of 'breeding conservatism' in itself could be construed as an indicator of attitudes to modern biotechnology and genetic engineering.

Conclusions

What have we learned from our exploration of these attitudinal data? The most important and striking conclusion is that the public at large did not appear to have, as yet, well-formulated or crystallised attitudes towards biotechnology. It seems likely that the attitudes of many respondents were, at most, superficial. This chapter has underscored the crucial point that, although an issue may appear to be intensely contested, this does not necessarily mean that the public *at large* is heavily involved. Debates on political issues can be restricted (at a certain point in time) to what may be called political elites or to what have been identified as 'attentive publics' (Miller, 1983). Indeed, it would seem that only recently have clear and resilient attitudes begun to emerge, largely in response to the news coverage of new biotechnological applications and GM products entering the market.

Rather than expressing pre-existing judgements, evaluations and perceptions, many respondents will have rapidly formed attitudes when confronted with the necessity of cooperating in the survey. Although responses elicited in such a context can still be meaningful, they should be interpreted with the utmost caution. People's answers may have been influenced more strongly than is usual by circumstantial factors and by information that became available as the result of the survey process. (Yet the fact that the data show large differences between European citizens and between the various countries suggests that background factors were far from unimportant.)

How do Europeans feel about biotechnology?

Half of Europeans had few expectations about biotechnology, a trend that was stronger in southern Europe than in the north. With a few notable exceptions, there was also a south–north axis of decreasingly favourable attitudes. A quarter of Europeans expressed a basically positive attitude, and about 14 per cent a negative one; 12 per cent felt ambivalent.

At the level of content structure we find that attitudes towards biotechnology broke down into two independent components. The first, and most dominant, aspect can be understood as deriving from a recognition of the potential economic and health benefits of biotechnology, which is at least tempered by a sense of the social and moral threats attendant upon these developments. The majority of people did not seem consciously to calculate trade-offs among these various factors; instead, these aspects were conjoined at an intuitive level and produced a general reaction to the survey questions. The final response may be interpreted as an indication of how 'good' or 'bad' the general effect of biotechnology on society was considered likely to be.

The second component in our analysis constitutes the perceived level of risk. Surprisingly, this factor appears to be independent of the core attitude. Yet our modelling efforts were unable to shed light on the underpinnings or consequences of risk perceptions. Instead, we were able to find only some *effects* of risk perception upon respondents' positive and negative expectations of biotechnology in general. This suggests that people's risk perceptions were poorly formulated and were, at least partly, based on more general notions related to biotechnology.

Looking at the structure of attitudes it would seem that these are, to a large extent, uncrystallised. A majority of the respondents have a moderate extremity score and a large majority have low differentiation scores. Moreover, there was no differentiation in public attitudes among the different biotechnological applications: although different applications produced different levels of acceptance/opposition, neither the decision-making process nor the attitudinal outcome tended to be affected by the nature of the particular application.

Our structural modelling analysis shows that specific attitudes are predicted by more general evaluative expectations of biotechnology and are related to an unspecified feeling of optimism or pessimism towards technology.

We have also shown that the level of 'informedness' has interesting effects. It appears that the degree of 'informedness' has, via different paths, both positive and negative effects upon attitudes to biotechnology. People who are better informed are more sceptical about the possible benefits and moral acceptability of this technology, but at the same time they are less inclined to consider biotechnology to be high risk and they exhibit greater optimism towards technology in general.

Similarly, knowledge of biotechnology does not have a unilateral effect upon attitudes. We find that people with a higher level of knowledge express more extreme convictions, suggesting that their attitudes are more refined; but knowledge does not seem to correlate with attitude differentiation. Independently of knowledge level, we find that attitudes and the attribution of risk, moral acceptability and utility to the six specific applications are highly consistent.

Our general conclusion, that attitudes were not very well crystallised, is also reflected in the fact that attitudes towards biotechnology do not seem to have been anchored to standard value systems. We did not find strong relations between perceptions of this technology and such value orientations as political preference, religiosity and environmental consciousness.

What attitudinal changes can be expected in the near future? It seems plausible that, as biotechnology further develops, clearer attitudes will emerge. It should be noted that our data were collected before the important media events of, for example, the modified soya imports and the birth of Dolly the sheep, issues that projected biotechnology to the forefront of social debate. As it becomes apparent to more people that biotechnology is a diverse field, capable of producing a variety of very different products and services, people are likely to develop more differentiated attitudes.

The public's currently rather undifferentiated view nonetheless deserves the attention of policy-makers and companies. The introduction of new products may be hampered because these products are likely to be evaluated according to more generalised perceptions of biotechnology. Moral objections to certain applications may in this way also inhibit the development of applications that are, in themselves, highly desirable.

It seems important for us to gain a better understanding of the factors that determine perceptions of utility, moral acceptability and risk. Such an enterprise promises to be an important contribution to public policy. At the same time, our study indicates the need for well-designed and targeted information campaigns through which to inform the European public, in greater detail, about the potential pros and cons of biotechnology, and to stimulate the emergence of well-elaborated attitudes towards technology and to biotechnology in particular.

REFERENCES

Eagly, A.H. and S. Chaiken (1993) *The Psychology of Attitudes*, Fort Worth, TX: Harcourt Brace Jovanovich College Publishers.

Hisschemoller, M.H. and C.J.H. Midden (in press) 'Improving the Usability of Research on the Public Perception of Science and Technology for Policy-Making', *Public Understanding of Science* 8(1): 17–34.

Joereskog, K.G. and D. Soerbom (1994) *Lisrel8: Structural Equation Modeling with the SIMPLIS Command Language*, Hillsdale, NJ: Lawrence Erlbaum.

Krosnick, J.A., D.S. Boninger, Y.C. Chuang, M.K. Berent, and C.G. Carnot (1993) 'Attitude Strength: One Construct or Many Related Constructs?' *Journal of Personality and Social Psychology* 65(6): 1132–49.

Miller, J.D. (1983) *The Role of Public Attitudes in the Policy Process*, New York: Pergamon Press.

Smith, E. (1998) 'Mental Representations and Memory', in D.T. Gilbert, S.T. Fiske and G. Lindzey (eds.), *Handbook of Social Psychology*, Boston: McGraw-Hill.

8 European regions and the knowledge deficit model

Nick Allum, Daniel Boy and Martin W. Bauer

Prototype cultures of public understanding of science

Previous chapters have explored the individual or national variation in knowledge and attitudes to modern biotechnology. In this chapter we will go a step further and introduce European regions as a level of comparison. Regional analysis is a speciality of geographers who map the similarities of regions across national borders in terms of their economic and social dynamic (e.g. Rodriguez-Pose, 1998a, 1998b).[1] The globalisation of capital and other productive factors leads to the expectation that regional variation is levelled and geo-spatial factors in social explanation lose their power, hence the question: do regions matter?

There are contradictory expectations about the significance of European regions in the future. Visions of European integration are tied to a rise in the political and economic importance of regions. As a correlate to European integration, regional autonomy will be strengthened, while the nation-state weakens. The sociological imagination concerning the dynamics of industrial development provides some specification of the contexts for our exploration into attitudes towards and knowledge of modern biotechnology in the 1990s. In recent years, much has been written about the cultural, economic, social and political changes associated with the transition from industrial to what may be termed post-industrial or knowledge societies. In phrases such as 'post-industrialism', 'post-materialism', 'postmodernity' or 'the risk society', sociologically minded writers have sought to evoke different axes of this transition. Post-industrialism emphasises the dominance of knowledge-based activities and industries (Bell, 1973; Touraine, 1969); post-materialism emphasises changing societal values (Inglehart, 1990); postmodernity emphasises a demise in the certainties about the direction of history (Lyotard, 1984); and the notion of the risk society emphasises the globalisation of risks and their distribution (Beck, 1992; Boehme, 1993). The prefix 'post' may indicate a transitional period between the waning industrial age and the establishment of a new social formation.

> The use of the term 'post-industrial' is perhaps deliberately chosen to affirm the lack of a deterministic center by calling attention to the idea that this social formation stands in transition between more distinctive types of societies, namely the industrial society and a future social formation, not yet endowed with the same degree of specificity. (Stehr, 1994: 89)

In all of these arguments the claim is made that major structural transformations accompany the progress of industrial societies, and that this transition is likely to occur first in the most advanced industrial countries.

Macro-sociological theory concerning this transition observed as a function of economic development provides a specified context for interpreting our data on the public perception of science in general and of biotechnology in particular. The literature on public understanding of science has hitherto tried to support or reject the so-called deficit model (for example, Irwin and Wynne, 1996), which in our view expresses a common-sense idea among policy-makers rather than a comprehensive conceptualisation of the phenomenon of public perceptions. According to the deficit model, knowledge of science and technology encourages more positive attitudes. In other words, the more science you know, the more you like it. Negative attitudes towards science reflect an ignorant public, hence the notion of a knowledge 'deficit'. Empirical evidence and conceptual thinking are somewhat at odds with this simplistic idea (for example, Evans and Durant, 1995). We suggest a formulation of a model of public perception of science that focuses on varieties of perceptions of science in different contexts. For this purpose we need to characterise differences in public perception as well as differences in the contexts in which they are socially embedded. Our effort could therefore be labelled a 'specified' contextual model of public perception. The call for contextual models is not new (Wynne, 1993); what is new is the attempt to specify the contexts and to work with testable hypotheses for which evidence is available.

We reconsider a model of scientific cultures in Europe previously suggested by one of us (Bauer, 1995; Bauer et al., 1994): the post-industrialism hypothesis of public understanding of science. The model projects a two-dimensional matrix onto a set of analytic units. The first dimension represents the hypothetical transition from industrial to post-industrial societies, and the second dimension represents what may be termed the ordered deviation from the idealised path of this transition in response to local circumstances. In a sense, the first dimension attempts to capture a general process, and the second dimension attempts to take account of the particularities of place and time. We have previously tested this model with fairly good fit to national variations in knowledge, interest and attitudes to science in general (Durant et al., 2000). Here, we test

Table 8.1 *Public understanding of science in the transition from industrial to post-industrial society*

Industrial society	Post-industrial society
Scientific knowledge is confined to a small social elite	Scientific knowledge is widely distributed in society
Scientific knowledge is strongly socio-economically stratified	Scientific knowledge is weakly socio-economically stratified
Public interest in science is relatively high	Public interest in science is relatively low
Popular scientific knowledge is unified	Popular scientific knowledge is specialised
The relationship between knowledge and support is positive	The relationship between knowledge and support is increasingly chaotic
Univocality: TO KNOW IS TO LIKE IT	Ambiguity: TO KNOW IS TO LIKE IT; FAMILIARITY BREEDS CONTEMPT

this model on variations in attitudes and knowledge of a particular new technology and across regional variation.

The features of the hypothetical transition on the first dimension are set out in table 8.1 as a dichotomy of industrial and post-industrial society. In industrial societies, it is expected that scientific knowledge is confined to a relatively small, well-educated elite; socio-economic factors play a relatively large part in determining the distribution of scientific knowledge; interest in science (as a symbol of industrial and economic progress) is relatively high in all sections of society; a relatively unified canon of popular scientific knowledge exists; and there is a correlation between knowledge of and support for science. The image here is of a society in which science and technology, having achieved limited penetration into society, are extensively idealised as the preferred route to economic and social progress.

In post-industrial knowledge societies, by contrast, it is expected that scientific knowledge is more widely distributed; socio-demographic factors such as education, income, age or gender play a relatively small part in determining the distribution of scientific knowledge; interest in science is relatively low, because science is more generally taken for granted as part of the everyday stock of knowledge; there is a proliferation of specialist knowledges in the public domain; and the relationship between knowledge of and support for science becomes unpredictable, as different sections of the community develop different points of view on particular issues. The image is of a society in which science, having already achieved a high level of penetration into the public sphere, is no longer idealised but is critically evaluated by a public that expects to obtain continuing benefits but is also increasingly alert to the possibility of problems or disbenefits. Thus there is an apparent paradox: as the public understanding of science expands and scientific knowledge becomes more diffused, science becomes more problematic for the public.

A further refinement of the model proposed here is an extension of these hypotheses to the development of European regions. As European integration continues apace, accompanied by rising labour mobility and the development of a highly efficient communications infrastructure, it is likely that regions within nation-states will become more economically autonomous. As a corollary to this, one would expect regional political autonomy to increase, as it has done recently in the UK with the devolution of Scotland and Wales. The thesis of this chapter is that, just as we can observe differences in public understanding of science according to the stage of post-industrialisation at a national level, so we will observe the same phenomenon between European regions that are presently at different stages on this developmental continuum. This assumes that the representations of science and technology (in the present case, of biotechnology) that circulate within the public sphere are to some extent produced in relation to the local as well as the national social and economic climate. Regions that share similarities in their stage of economic development are expected to share similarities in their publics' attitudes to biotechnology, independently of the country in which they are located.

Needless to say, this model of transition of the structure of public understanding of science is a simplification. To characterise different scientific cultures adequately, more dimensions are necessary. However, radical simplification may be justified by its heuristic value of generating testable hypotheses. A second limitation of our argument will be our data. Whereas the hypotheses are longitudinal in nature, our data are cross-sectional. We are assuming a uniform transition from industrial to post-industrial mode, and any moment in time will show how countries and regions are differentially advanced in this transition.

The relationship between knowledge and attitudes

In order to measure the relationship between attitudes towards biotechnology and the level of knowledge, we use an indicator that comprises nine knowledge items, and a combined indicator across six applications of biotechnology (see Durant et al., 1998). For each application, responses regarding usefulness, moral acceptability and encouragement were coded as +2 (strong agreement) or +1 (agreement), and disagreements were coded respectively as −1 or −2; non-responses coded as 0. The assessment of risk of any of these applications was coded in the following way: perception of risk was coded −2, denial of risk was coded +2. The final index is the sum of individual responses across six applications and four criteria of judgement. Our indicator of post-industrialism is GDP per capita in purchasing power parities in 1996.

For Europe as a whole, measured at the level of the individual, the relationship between knowledge of and attitudes to biotechnology is positive but weak, although statistically significant: Pearson's r is .14 ($p < .01$). Europe-wide, knowledge explains less than 2 per cent of the variance of attitudes towards biotechnology. This would be consistent with current social psychological reasoning, according to which knowledge is not a predictor of attitudes but an indicator of their quality: attitudes based on knowledge are more elaborated and hence more resistant to change (Eagly and Chaiken, 1993).

This low correlation might be taken as an indication that, at the level of Europe as a whole, the post-industrial mode of public perception prevails; a substantial relationship between knowledge and attitudes is not present. However, we know that from north to south and east to west, European countries vary considerably, and the characteristics of a post-industrial condition are not equally distributed. How does the relationship between knowledge and attitudes vary, taking into account the different developmental contexts?

National-level analysis

Country-by-country analysis of the correlation between knowledge and attitudes shows considerable variation. How can we explain this variation? Can we verify our post-industrial hypothesis, according to which the correlation is high for countries close to an industrial mode and low for countries close to a post-industrial mode?

Our data, aggregated to mean country levels, show a negative correlation between GDP and the attitude score ($r = -.48$, $p = .05$) and a positive correlation between knowledge and GDP ($r = .43$, $p < .10$). Figure 8.1 plots the within-country correlation of knowledge and attitudes against GDP per capita. The figure clearly shows that the knowledge–attitude correlation is stronger in countries where GDP is lower, for example Greece, Spain and Portugal. This link is much weaker in the wealthier nations, for example Switzerland and Luxembourg. In between these extremes is a band of richer countries between which GDP per capita is not a good discriminator (Finland, the UK, Sweden, Italy and France).

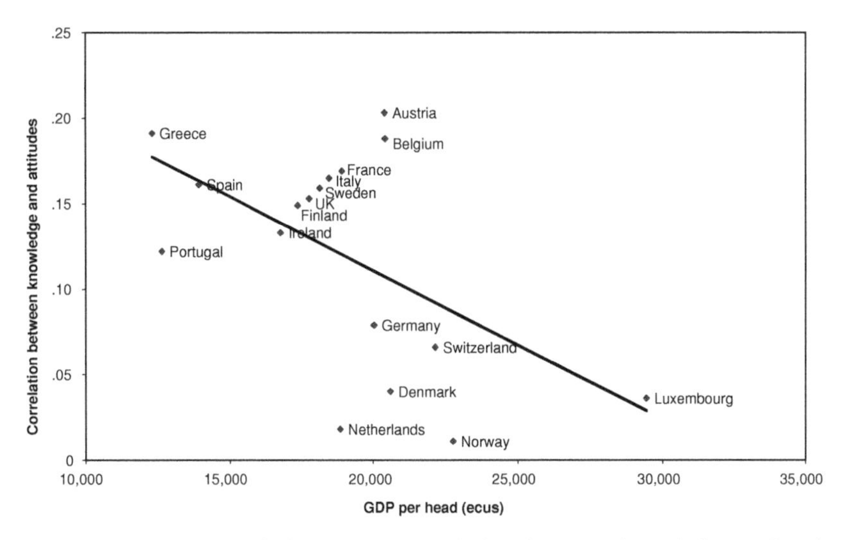

Figure 8.1 Within-country correlation between knowledge and attitudes to biotechnology in relation to GDP per head
Source: EUROSTAT 1996.

It is important, of course, that we do not make the 'ecological' error of inferring from these aggregated data that a different cognitive mechanism operates at the level of the individual in different countries (although that cannot be ruled out). However, assuming that the level of GDP (1996 in purchasing power parities) is a valid measure of the transition from an industrial to a post-industrial mode of the economy, this is exactly what our hypothesis predicts. In post-industrial mode the relationship between knowledge and attitudes becomes more complex when we make comparisons at the societal level.

Looking at figure 8.1, we observe that some countries are outliers in the linear model. Austria and Belgium have a higher correlation than their level of development might suggest, while, on the other hand, for the Netherlands, Norway and Denmark the correlation is well below expectations. In this middle range of development, the variance between countries is very large. To explain this variation we have to consider the second, as yet unspecified, dimension of our model: explaining the variations in the post-industrial model according to local circumstances.

A possible factor could be the status of the public debate on biotechnology in these countries. The Netherlands and Denmark have had intensive public controversies on biotechnology (see Jelsoe et al., 1998; Midden et al., 1998), whereas Austria and Belgium had not had significant public debates by the end of 1996 (see Wagner et al., 1998). We therefore conclude that our post-industrial model needs to include an auto-catalytic factor: public controversies. In the aftermath of public controversies over a new technology, which in themselves are likely to be a feature of post-industrial societies, the correlation between knowledge and attitudes disappears. Public controversies enhance the circulation of new knowledge in society, while not prejudicing attitudes in post-industrial mode (see chapter 5).

Multilevel regression analysis

As we have seen, there are significant differences between European countries in the relation between textbook knowledge of biology and the evaluation of a set of biotechnological applications when looking at aggregated data within the sample. It also appears that the pattern of correlations between knowledge and attitudes aggregated at the regional level is rather weak and not statistically significant in most cases. Aggregating the data tells an ambivalent story, and another form of analysis is required to disentangle the complex relationships that exist within and between different analytic levels. A suitable form of analysis for this problem is hierarchical multilevel linear regression (Goldstein, 1995; Bryk and Raudenbush, 1992). The multilevel approach treats individual cases as existing within hierarchical levels that correspond to real world situations (where individuals live in regions and those regions are situated in countries). Using this approach, no aggregation of data is necessary because the estimation procedure takes into account the hypothesised similarities between individuals within the same higher-level units and provides measures of the relative importance of factors at each level that affect the outcome variable.

A multiple regression model was set up with knowledge as the dependent variable. Attitude towards biotechnology appears alongside several new variables as predictors. The assumption here is not that this is a causal model, with attitudes driving knowledge, although this is, of course, one of the possibilities. We are viewing the set of relationships between these variables simply as associations, not causes and effects. Our primary interest is to explore the distribution of variations in these associations according to the various hierarchical levels in the analysis – individual, regional and national.

New variables

The number of people living in a household is known to correlate positively with the number of school-age children in the household. This

Table 8.2 *The variables used in the analysis of the multilevel model groups*

Level	Variables
Dependent variable (individual level)	Textbook knowledge about biotechnology
Level 1 (individual level)	Attitude towards biotechnology (negative to positive)
	Number of people in the household
	How much the respondent has talked about biotechnology in the past
Level 2 (regional level)	Long-term unemployment in the region (percentage of working population)
	Ratio of tertiary-sector employment to total employment in the region

simply reflects the fact that most households comprising more than two people are those in which there are families with children. The type of knowledge that is measured in the present survey – knowledge of a 'textbook' nature – would be more likely to circulate within a family that contains school-age children whose education will generally include basic biology (see table 8.2).

The extent to which someone has talked about biotechnology in the past can be taken as a measure of active engagement with the subject. One might consider it as a type of informal, subjective knowledge insofar as to talk about something, or simply to refer to it, involves conceptualising it in some way, anchoring it in a familiar frame of reference. Hence this variable is expected to be positively associated with the formal 'textbook' knowledge measured.

Individual, family and household relationships can be constituted in a multitude of ways. That said, there is less heterogeneity between individuals and groups of whatever kind who occupy the same social milieu. In the present case, we expect to see that the way knowledge about and attitudes towards biotechnology relate to the type of household, and the extent to which the topic is discussed, are mediated to some degree by the socio-economic climate of the region as well as the country. The general thesis of this chapter is that the relationship between attitudes to and knowledge about science, and, in this case, biotechnology, changes according to the degree of post-industrialisation. While this is usually conceptualised at the level of the country, here the aim is to see if these differences are also present according to the socio-economic characteristics of regions, independently of their national location.

Two variables are used as economic indicators for the European regions. The proportion of long-term unemployment in the region is assumed to be a salient feature, impacting as it does on the types of economic opportunities that its residents enjoy, and most likely on attitudes of optimism and pessimism about future prospects. As far as the post-industrialisation thesis is concerned, the level of long-term unemployment in any given region may signal transition towards post-industrialisation or, alternatively, economic stagnation. Thus the direction taken by any association between long-term unemployment and knowledge is an interesting empirical question. The proportion of employment in the tertiary sector of the economy (service sector) is another of the indicators of post-industrialisation. This measure is by no means perfect. There are a few regions in the sample where there is a strongly developed service sector but very little industrialisation and a heavy reliance on agriculture. These are primarily tourist areas in Greece, Portugal and Spain and do not resemble post-industrialised economies. Nevertheless, from the data available, it is the single best proxy measure for post-industrial economic development.

Method

The economic data were taken from the European Commission's statistical service Eurostat. A reduced sample of 9,228 respondents was used for the analysis owing to missing data in the Eurostat data set. This left nine countries and fifty-one regions covered.

The models presented here are 'variance componence' models with up to three levels – individual, regional and country. In a single-level analysis, at the individual level, the assumption in a standard regression model is that each individual's score on the dependent variable is independent from all other individuals' scores. However, we hypothesise here that individuals share similarities in their knowledge about biotechnology according to the area in which they live. We would expect there to be similarities between respondents from the same region and country that remain undetected in a single-level model. One way to approach this would be to add a separate term for each country and region. This is clearly inefficient in that a huge number of coefficients would need to be estimated. Another approach is to aggregate data to the regional or country level, taking the mean score on all the variables over all individuals in the region or country. Although this is often done, and with useful results, any causal explanation at the level of the individual is not strictly warranted by discarding individual-level information, which is what aggregation does. Also, it is a well-documented statistical phenomenon that correlations tend to be exaggerated and can even change their sign when aggregated data are used (the so-called 'ecological fallacy').

The variance componence models presented here assume that the regions and countries (levels 2 and 3) are a sample of a possible population of regions and countries. The variation between countries and between regions is modelled as random variance between regression intercepts.

The general model is of the form:

$$y_{ijk} = a_{jk} + b + e_{ijk}.$$

There are random disturbance terms for each level and these are conceptualised as variance at each level. As the level 2 and 3 (j and k) variance is random only for the constant, a, in the model, one can visualise the result as a set of regression lines having the same slope but different intercepts corresponding to each region and country. If no systematic differences exist between individuals in different regions and countries, we would not expect the intercepts of the regression slopes for each of these to vary significantly. The converse is to think of this as demonstrating that there are similarities between individuals within these regions and countries. If we see significant variance between regions and/or countries (undetected heterogeneity), we can then go on to add our regional (level 2) variables to the model, and to see how much, if any, of that heterogeneity can be accounted for by these economic indicators. This would be observed as a reduction in the unexplained variance at that level.

Results

Full details of each model are reported in the appendix to this chapter.

Stage 1: two-level model with individual-level predictors

All three independent variables have a significant effect on level of knowledge, all in a positive direction. Whether a respondent had talked about biotechnology to anybody prior to the survey is the single best predictor of their textbook knowledge about it. The more they had talked about it, the better is their knowledge ($r = .64$). There is a small association between the number of people in a respondent's household and their knowledge ($r = .15$). There is also a positive association between knowledge and attitude. Those with a favourable attitude towards biotechnology are likely to know more about it. The variance between regions (see table 8.3) is significant at $< .05$. Approximately 13 per cent of the variance in people's knowledge is associated with unknown factors (outside of the model) that are active within regions.

Table 8.3 *The variance explained in the multilevel models*

	Level 1 (individuals)		Level 2 (regions)		Level 3 (countries)	
Model	δ^2_e	% variance	δ^2_u	% variance	δ^2_v	% variance
Model 1	3.285	87.4	0.472	12.6	–	–
Model 2 (with level-2 variables)	3.359	92.2	0.284	7.8	–	–
Model 3	3.361	89.7	0.088	2.3	0.297	7.9
Model 4 (with level-2 variables)	3.363	91.8	0.053	1.4	0.248	6.8

Stage 2: two-level model with individual- and regional-level predictors

In this model, two variables at the regional level have been introduced. Both are significant at < .05. Higher tertiary-sector employment is associated with higher knowledge about biotechnology ($r = .22$). Higher long-term unemployment is associated with lower knowledge. The magnitude of the other predictors' coefficients is slightly reduced with the economic data added in. Overall, this model is significantly better at explaining changes in knowledge than the previous one (reduction in -2^* log likelihood: 19976; 2 d.f.).

The most interesting difference between models 1 and 2 is the difference in between-region variance. In model 1, between-region variance was approximately 13 per cent of the total variance. In model 2, it has fallen to approximately 8 per cent (.28). A significant proportion of the unexplained regional-level heterogeneity can be explained by the employment conditions in the area.

Stage 3: three-level model with individual-level predictors

To test whether these regional effects are disguised national effects, a further level was added. Model 3 shows the partitioning of variance across three levels – individual, regional and national – with only the level-1 independent variables, as in stage 1. Here we can clearly see that heterogeneity between countries accounts for most of the variance attributed to the regions in model 1. Approximately 8 per cent (.30) of the unexplained variance is between countries, while only 2 per cent (.09) is now accounted for by the regions. However, it is important to note that, even when country-level variance is modelled, regional effects are still significant.

Stage 4: three-level model with individual- and regional-level predictors

To complete the analysis, our level-2 predictors are introduced into the model. The addition of these reduces the amount of between-region and between-country variance compared with the model at stage 3. Reduction in between-country variance is approximately 16 per cent between stages 3 and 4, while about 40 per cent of between-region variance is concurrently reduced. This demonstrates that the economic conditions in regions are still correlated with inhabitants' knowledge even after controlling for country-wide similarities.[2] These results are summarised in table 8.3.

This analysis supports the hypothesis that regional economic factors are significant in predicting people's knowledge about biotechnology. It also suggests, unsurprisingly, that most of the unexplained factors that affect people's knowledge operate at the level of the individual. Which country a respondent comes from is more important than which region they live in, when considering all nine countries. We might therefore want to say that national cultures are better predictors of an individual's knowledge about biotechnology than regional cultures are. But there is still a significant variation in people's knowledge according to the region in which they live. A substantial proportion of this is associated with differences in local economic conditions.

Regional clusters of public understanding of biotechnology

We have thus far identified a statistical relationship between GDP and attitudes towards biotechnology, and between GDP and knowledge about biotechnology across European regions. We have also presented estimates for the significance of regional variation in explaining people's cognitive involvement with biotechnology. We now focus on the regional level of analysis to explore the structural features that determine people's engagement with modern biotechnology beyond the socio-political unit of the nation-state.

For this purpose we conducted a cluster analysis of European regions. The variables used were regional means on the following measures: people's general level of education measured by school-leaving age, their knowledge of, and attitude to, biotechnology, and the frequency of their previous conversations about biotechnology. The analysis yielded

Table 8.4 *Some European regions classified by economic dynamism*

Capital	Industrial declining	Intermediate	Peripheral
Berlin (D)	Wales (UK)	East/West Midlands (UK)	Sardegna (I)
Brussels (B)	Wallonie (B)	Emilia Romagna (I)	Galicia (E)
Île de France (FR)	Nord Westfalen (D)	Anatoliki (GR)	Northern Ireland (UK)
Madrid (E)	North West (I)	Mediterraneé (FR)	Ipeiros (GR)

Source: Rodriguez-Pose (1998a,b).
Note: B = Belgium; D = Germany; E = Spain; FR = France; GR = Greece; I = Italy; UK = United Kingdom.

a three-cluster solution.[3] In region type A, people have high knowledge and education, have talked only moderately about biotechnology and have moderate attitudes to biotechnology. In region type B, people have moderate levels of education and knowledge of biotechnology, have talked about it a lot and have rather negative attitudes. Finally, in region type C, people have lower levels of general education and knowledge and have not talked much about biotechnology but hold very positive attitudes towards it.

The second classification considers the socio-economic structure and dynamism of the regions as shown in table 8.4. Here we follow the regional classification developed by geographers in the context of the debate on regional convergence (Rodriguez-Pose, 1998a, 1998b, 1999). European regions are classified by level of GDP in 1996 and their rates of economic growth over the period 1980–96, standardised within each country. European regions face the secular challenge of globalisation of capital, speed of innovations and new production technologies in different ways. Rodriguez-Pose identifies four types of regional responses to these challenges. Response pattern I is characteristic of urban capital regions, world cities or megalopolises. GDP is above the European average and there is a high concentration of service industries, in particular financial sectors. Response pattern II is characteristic of declining industrial areas. Traditional heavy industries are concentrated here, with over 33 per cent of employment in the secondary sector, and major difficulties in changing the production system. Response pattern III defines intermediate areas that are neither urban centres nor old industrial or agricultural areas, but show clear economic dynamism often based on small enterprises. Finally, response pattern IV comprises the peripheral regions of Europe where the remaining agricultural employment is concentrated. Economic

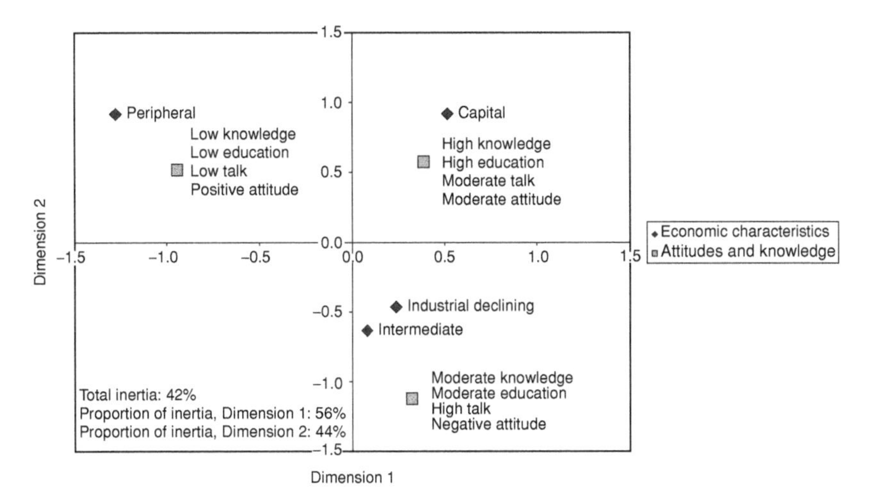

Figure 8.2 Correspondence analysis of clusters of engagement with biotechnology and patterns of regional development across Europe
Notes: The measures of engagement with biotechnology were knowledge, level of education, frequency of conversations about it and attitudes. Total inertia = 42%; proportion of inertia, dimension 1 = 56%; proportion of inertia, dimension 2 = 44%.

dynamism, if at all, arises mainly from tourism, creating a single-sector dependency.

For our problem it is now of interest to see whether these patterns of regional development over the past twenty years are systematically associated with patterns of the public's engagement with biotechnology in these regions.

Figure 8.2 shows the joint plot from a correspondence analysis (Greenacre and Blasius, 1994) of clusters of engagement with biotechnology and regional types. We can observe from this plot that the attitudinal and economic regional typologies are quite closely associated. Urban capital regions show a dominance of an engagement pattern that is characterised by high levels of general education and knowledge about biotechnology, moderate intensity of talking about it and moderate attitudes; 72 per cent of the urban population shows this response pattern. Peripheral regions are clearly associated with a pattern of engagement that is characterised by lower levels of education and knowledge, little talking about biotechnology and very positive attitudes towards biotechnology; 87 per cent of the population in peripheral regions show this response pattern. Industrial-declining and intermediate regions share a very similar pattern of engagement with biotechnology: moderate levels of education and knowledge of biotechnology and high levels of talking, but rather negative attitudes. However, the pattern is not as dominant as in other regions; only 38 per cent and 41 per cent, respectively, of the population show the characteristic pattern of response.

This is apparent from the correspondence analysis too. Looking at the first dimension of the joint plot, one can see that it is anchored by the peripheral regions at one end and the urban capital at the other. In fact the contribution of these two points to the total inertia of the dimension is 97 per cent. Intermediate and industrial-declining areas contribute to only 3 per cent of the inertia of dimension 1 and 44 per cent to dimension 2. Dimension 1 could be interpreted as an axis of post-industrialisation, but, given the small contribution of the intermediate and industrial-declining regions, it is perhaps better conceptualised more simply as marking out the contrast between peripheral and urban capital economies.

Notwithstanding the ambiguity of the intermediate and industrial-declining regions, we further hypothesised that the individual-level correlation between knowledge and attitudes would decline as one moves from the peripheral to the urban capital clusters. The evidence does not confirm such an expectation. The correlation, after pooling all respondents into the regional clusters, is approximately $r = .15$ in all four clusters. Although there is a clearly differentiated typology of attitudes and knowledge amongst the regions, this does not appear to be associated with the knowledge–attitude correlation predicted by the post-industrialisation hypothesis.

Conclusion

In this chapter a number of analyses have been presented that explore a complex of attitudes towards and knowledge about biotechnology at various levels – individual, regional and national. The post-industrialisation model of public understanding of science (PUS) receives some corroboration. The correlation of knowledge and attitudes on a country-by-country basis reveals the pattern of correlations predicted by this model. The correlation between knowledge and attitudes is lower in countries that are closer to the post-industrial ideal-type than in those where the economy is less developed. Evidence directly corroborating this particular hypothesis when extended to the level of European regions was not found. However, the multilevel analysis shows that there is a significant intra-regional correlation in the relationship between knowledge, both formal and informal, and attitudes. It has also been shown that part of this within-region homogeneity is associated with its modal type of economic production and the relative prosperity of the region. Furthermore, in using a multilevel approach we have shown that *individual-level* similarities exist between respondents in the same region that cannot be attributed to artifacts of aggregation.

The patterns of correspondence between the knowledge–attitude clusters developed here and the independently developed economic typology also lend support to the notion that regional variation in attitudes towards biotechnology across Europe is not entirely random. Although the precise basis of this regularity is as yet unclear, local social and economic conditions clearly play a part. For the extreme ends of the economic spectrum in the regions of Europe, the post-industrialisation hypothesis of PUS is also corroborated. In the peripheral regions of Europe, predominantly in the south, biotechnology is seen as progress, part of a vision of modernisation that is welcomed for its positive benefits. Attitudes are likely, though, to be rather unstable, based as they are on a low knowledge-base. The picture in the capitals of Europe is of citizens at once more knowledgeable about biotechnology but more circumspect in their expectations of what benefits it may bring.

Some implications for further research are suggested by the results of our investigations here. First, a base-line has been established that can serve as the beginning of a programme of longitudinal research on the role of the regions using future Eurobarometer surveys on biotechnology. We are already undertaking such research. Secondly, the results presented here beg a more general question: do attitudes towards science in general follow the same pattern of regional variation, or is what has been established here limited to the domain of biotechnology as a very new and unfamiliar technology?

We started this chapter with the question 'do regions matter?' The answer is 'yes, they do matter', in Europe at least, but we have only just begun to understand how, why and when. It will be the interesting task of future research to establish some firmer answers to these questions.

Appendix

Variables

know textbook knowledge about biotechnology
cons1 constant
ztotatt attitude towards biotechnology (negative to positive)

zpeople	number of people in the household
ztalked	how much the respondent has talked about biotechnology in the past
zltunemp	long-term unemployment in the region (percentage of working population)
zemp03r	ratio of tertiary-sector employment to total employment in the region

All independent variables were standardised (mean = 0, s.d. = 1) in order to facilitate comparisons of the relative effect of each. Because the regression analysis can only be carried out using listwise deletion in software package MLwiN, approximately one-third of cases from the survey are treated as missing because of the gaps in the regional economic data. This left 9,228 cases in nine countries to be analysed (Belgium, France, Germany, Italy, Luxembourg, the Netherlands, Portugal, Spain and the UK).

Model 1 (only individual-level variables)

$$
\begin{aligned}
\text{know}_{ij} &\sim \text{N}(XB, \Omega) \\
\text{know}_{ij} &= \beta_{0ij}\text{cons1} + 0.236(0.016)\text{ztotatt}_{ij} + 0.149(0.016)\text{zpeople}_{ij} \\
&\quad + 0.643(0.016)\text{ztalked}_{ij} \\
\beta_{0ij} &= 4.336(0.075) + u_{0j} + e_{0ij} \\
|u_{0j} &\sim \text{N}(0, \Omega_u) : \Omega_u = 0.472(0.075) \\
e_{0ij} &\sim \text{N}(0, \Omega_e) : \Omega_e = 3.285(0.039) \\
&\quad -2^{*}log(like) = 57481.840
\end{aligned}
$$

Model 2

$$
\begin{aligned}
\text{know}_{ij} &\sim \text{N}(XB, \Omega) \\
\text{know}_{ij} &= \beta_{0ij}\text{cons1} + 0.208(0.020)\text{ztotatt}_{ij} \\
&\quad + 0.693(0.020)\text{ztalked}_{ij} + 0.159(0.020)\text{zpeople}_{ij} \\
&\quad + -0.155(0.067)\text{zltunemp}_{j} + 0.218(0.077)\text{zemp03r}_{j} \\
\beta_{0ij} &= 4.445(0.079) + u_{0j} + e_{0ij} \\
|u_{0j} &\sim \text{N}(0, \Omega_u) : \Omega_u = 0.284(0.061) \\
e_{0ij} &\sim \text{N}(0, \Omega_e) : \Omega_e = 3.359(0.050) \\
&\quad -2^{*}log(like) = 37505.350
\end{aligned}
$$

Model 3

$$\text{know}_{ijk} \sim \text{N}(XB, \Omega)$$
$$\text{know}_{ijk} = \beta_{0ijk}\text{cons1} + 0.210(0.020)\text{ztotatt}_{ijk} + 0.697(0.020)\text{ztalked}_{ijk} + 0.156(0.020)\text{zpeople}_{ijk}$$
$$\beta_{0ijk} = 4.407(0.189) + v_{0k} + u_{0jk} + e_{0ijk}$$
$$|v_{0k} \sim \text{N}(0, \Omega_v) : \Omega_v = 0.297(0.152)$$
$$u_{0jk} \sim \text{N}(0, \Omega_u) : \Omega_u = 0.088(0.024)$$
$$e_{0ijk} \sim \text{N}(0, \Omega_e) : \Omega_e = 3.361(0.050)$$
$$-2^{*}\mathit{log(like)} = 37482.070$$

Model 4

$$\text{know}_{ijk} \sim \text{N}(XB, \Omega)$$
$$\text{know}_{ijk} = \beta_{0ijk}\text{cons1} + 0.211(0.019)\text{ztotatt}_{ijk} + 0.695(0.020)\text{ztalked}_{ijk} + 0.158(0.020)\text{zpeople}_{ijk} + 0.177(0.054)\text{zltunemp}_{jk} + 0.150(0.049)\text{zemp03r}_{jk}$$
$$\beta_{0ijk} = 4.406(0.172) + v_{0k} + u_{0jk} + e_{0ijk}$$
$$|v_{0k} \sim \text{N}(0, \Omega_v) : \Omega_v = 0.248(0.125)$$
$$u_{0jk} \sim \text{N}(0, \Omega_u) : \Omega_u = 0.053(0.017)$$
$$e_{0ijk} \sim \text{N}(0, \Omega_e) : \Omega_e = 3.363(0.050)$$
$$-2^{*}\mathit{log(like)} = 37470.510$$

NOTES

1. We thank Andres Rodriguez-Pose of the Department of Geography at the London School of Economics for access to the regional database he had compiled.
2. From a technical point of view it might be argued that, with only nine level-3 units, the normality assumption for these country-level effects is questionable. We therefore tried representing the nine countries as fixed effects in the two-level models (stages 1 and 2) by creating dummy variables for each. In both models, the unexplained level-2 variance and the coefficients of the economic variables took on similar magnitudes to those in stages 3 and 4. Both remained significant at $< .05$. Hence we conclude that the multilevel analysis discussed here is a suitable method for representing the relative contributions of individual, regional and national variations.
3. A hierarchical cluster analysis was carried out in SPSS using squared Euclidian distances to create the distance matrix and the 'between groups' linkage method

for the clustering. The resulting dendrogram indicated that a three-cluster solution was optimal. In order to assign each region unambiguously to one cluster, a *k*-means cluster analysis was carried out, requesting a three-cluster solution. In all the cluster analyses, Austrian regions were omitted owing to an extreme configuration of attitudes and knowledge, giving them the status of 'outliers' for our analysis.

REFERENCES

Bauer, M. (1995) 'Industrial and Post-Industrial Public Understanding of Science', paper presented to the meeting of the Chinese Association for Science and Technology, Beijing, October.

Bauer, M., J. Durant and G. Evans (1994) 'European Public Perceptions of Science: an Exploratory Study', *International Journal of Public Opinion Research* 6(2): 163–86.

Beck, U. (1992) *Risk Society: Towards a New Modernity*, London: Sage.

Bell, D. (1973) *The Coming of Post-Industrial Society: a Venture in Social Forecasting*, New York: Basic Books.

Boehme, G. (1993) *Am Ende des Baconschen Zeitalters*, Frankfurt am Main: Suhrkamp.

Bryk, A. and S. Raudenbush (1992) *Hierarchical Linear Model: Applications and Data Analysis Methods*, London: Sage.

Durant, J., M.W. Bauer and G. Gaskell (1998) *Biotechnology in the Public Sphere: a European Sourcebook*, London: Science Museum Publications.

Durant, J., M. Bauer, G. Gaskell, C. Midden, M. Liakopoulos and E. Scholten (2000) 'Two Cultures of Public Understanding of Science and Technology in Europe', in M. Dierkes and C. von Grote (eds.), *Between Understanding and Trust. The Public, Science and Technology*, Amsterdam: Harwood Academic Publishers, pp. 131–56.

Eagly, A.H. and S. Chaiken (1993) *The Psychology of Attitudes*, Fort Worth, TX: Harcourt Brace College Publishers.

Evans, G.A. and J.R. Durant (1995) 'The Relationship between Knowledge and Attitudes in the Public Understanding of Science', *Public Understanding of Science* 4: 57–74.

Goldstein, H. (1995) *Multilevel Statistical Models*, London: Arnold.

Greenacre, M.J. and J. Blasius (eds.) (1994) *Correspondence Analysis in the Social Sciences. Recent Developments and Applications*, London: Academic Press.

Inglehart, R. (1990) *Culture Shift in Advanced Industrial Society*, Princeton, NJ: Princeton University Press.

Irwin, A. and B. Wynne (eds.) (1996) *Misunderstanding Science? The Public Reconstruction of Science and Technology*, Cambridge: Cambridge University Press.

Jelsøe, E., J. Lassen, A.T. Mortensen, H. Frederiksen and A.W. Kamara (1998) 'Denmark', in J. Durant, M.W. Bauer and G. Gaskell (eds.), *Biotechnology in the Public Sphere: a European Sourcebook*, London: Science Museum Publications, pp. 29–42.

Lyotard, J.F. (1984) *The Post-modern Condition: a Report on Knowledge*, Manchester: Manchester University Press; Canadian French original, 1976.

Midden, C., A. Hamstra, J. Gutteling and C. Spink (1998) 'The Netherlands', in J. Durant, M.W. Bauer and G. Gaskell (eds.), *Biotechnology in the Public Sphere: a European Sourcebook*, London: Science Museum Publications, pp. 103–9.

Rodriguez-Pose, A. (1998a) 'Social Conditions and Economic Performance: the Bond between Social Structure and Regional Growth in Western Europe', *International Journal of Urban and Regional Research* 22(3): 443–59.

(1998b) *Dynamics of Regional Growth in Europe*, Oxford: Clarendon Press.

(1999) 'Convergence or Divergence? Types of Regional Responses to Socio-economic Change in Western Europe', *Tisdschrift voor Economische en Sociale Geografie* 94(4): 367—78.

Stehr, N. (1994) *Knowledge Societies*, London: Sage.

Touraine, A. (1969) *La Société post-industrielle,* Paris: Denoel.

Wagner, W., H. Torgersen, P. Grabner, F. Seifert and S. Lehner (1998) 'Austria', in J. Durant, M.W. Bauer and G. Gaskell (eds.), *Biotechnology in the Public Sphere: a European Sourcebook,* London: Science Museum Publications, pp. 15–28.

Wynne, B. (1993) 'Public Uptake of Science: a Case of Institutional Reflexivity', *Public Understanding of Science* 2: 321–37.

9 Pandora's genes – images of genes and nature

Wolfgang Wagner, Nicole Kronberger, Nick Allum, Suzanne de Cheveigné, Carmen Diego, George Gaskell, Marcus Heinßen, Cees Midden, Marianne Ødegaard, Susanna Öhman, Bianca Rizzo, Timo Rusanen and Angeliki Stathopoulou

Why do some Austrians, when asked to talk about biotechnology, conjure up images of children being carried to term by genetically modified pigs in place of human mothers? Why do Swedes associate biotechnology with the monster bull 'Belgian Blue', even though this breed of cattle is a product not of biotechnology but of traditional breeding methods? In this chapter, we explore how citizens of all different European countries express the concern that industrial biotechnology involves playing God by tampering with nature. Nature, the argument goes, will eventually strike back with unforeseen consequences. Be it monster bulls, playing God or other associations the public might have with modern biotechnology, these images are the fundamental bases upon which public rejection and acceptance are based.

When a new technology enters the marketplace of everyday life, it frequently creates ambivalent feelings among the general public. On the one hand, technologies are meant to make everyday chores easier and life more comfortable. On the other hand, many technologies, and particularly radically new ones, involve operations that are underpinned by complex scientific achievements. This was the case with satellite-related space technology in the 1950s and with nuclear energy technology in the 1960s, and it is the case with biotechnology's use of genetic manipulation to tailor the properties of living organisms in the 1990s. An in-depth understanding of the science involved, however, is not the prime interest of the majority of the general public, and rightly so.

However, if an understanding of the scientific basis of a technology is not necessary, people wishing to assess the value of a new technology will nevertheless have to struggle in comprehending what its effects will be. This is especially the case with modern biotechnology because it so profoundly impinges upon everyday life. It has, for instance, led to the introduction of crops and foodstuffs that are, or will be, offered on supermarket shelves to ordinary people who have only the vaguest understanding of industrial genetics. Moreover, people's own bodies might eventually be subjected to genetic tests and cures, and there are even experts who – for whatever reasons – talk of producing clones of humans, who might one day be an individual's closest relatives.

This struggle for understanding may occur at different levels. Some people, i.e. the more scientifically literate ones, might read the relevant columns in newspapers or buy and consult professional literature on the topic. This is the straightforward way. However, the proportion of people who have at their disposal the necessary educational resources or the time required for such study is unlikely to be very high. Hence, many people need to resort to other means of understanding, most commonly those governed by common sense.

The present chapter attempts to highlight some of the modes by which everyday people arrive at a common-sense understanding of modern biotechnology. This is rarely if ever an individual's achievement, but rather the result of collective processes beyond the individual's realm of influence; and it involves events at the level of policy-making and implementation, extended media discourse, and conversations in everyday life.

Such complexity requires a conceptual model that is capable of integrating social as well as individual processes, such as social representation theory (Moscovici, 1984; Jodelet, 1989; Bauer and Gaskell, 1999; Wagner et al., 1999). When a new phenomenon, such as modern biotechnology and its products, appears to threaten a social group's normal course of practice, it needs to be collectively coped with in both material and symbolic terms (Wagner, 1998). 'Material coping' involves technical and, to some extent, legalistic measures that are aimed at containing the potential risks implied by the novel technology. 'Symbolic coping' refers to the naming of the new phenomenon and attempts to understand its qualities and consequences. In other words, it involves assigning the phenomenon a place in the symbolic universe of everyday thinking and common sense. Symbolic coping results in a social representation: an ensemble of beliefs, images and feelings about a phenomenon that is shared by the members of a social unit.

A social representation of biotechnology, for instance, will rarely if ever be veridical in the sense of scientific correctness. Rather than serving a scientific understanding, the beliefs and images that constitute a representation are products of personal and media discourse, which unfold in the course of symbolic coping and serve the purpose of everyday communication. Representations straddle the interface between the individual and the collective because they are generated in collective discourse and their elements are shared to a large extent by the individuals of the group concerned.

Representations that comprise attitudes, beliefs and feelings are frequently structured in a pictorial way. Using images and metaphors is one way of symbolically coping and involves attempts to understand an unfamiliar object in the light of a more familiar one. In this view, images are not just a matter of illustrative words, but rather a way of understanding and experiencing one kind of thing in terms of another (Lakoff and Johnson, 1980). Images not only describe the object and its features, but are also interpretative and evaluative. Thus, a well-suited metaphor is 'good to think with' (Wagner et al., 1995). Such images are not a property of the individual mind but are frequently shared by a group or society when addressing an object. Their use is automatic and effortless and they are perceived not as illustrations of the unfamiliar but as the object itself. They appear as pictures, text or talk in personal and collective discourse.

Overview and method

Asked about what comes to mind when thinking about modern biotechnology, Austrians, Britons, Dutch, Finns, Germans, Greeks, Italians, Norwegians, Portuguese and Swedes show distinct patterns of response. These responses were categorised according to content and evaluative tone, then cross-tabulated and correspondence analysed. In this sample of European nations, the evaluative tone of respondents ranges from neutral (Germans) to negative and/or ambivalent (Austrians and Swedes) to distinctly positive (Italians and Portuguese) (figure 9.1).

People from countries with a more negative evaluation also raise moral concerns in their responses. Although agricultural, medical, economic and fertilisation issues are common to all countries, the more negative or ambivalent respondents are, the more they also mention the possibility of monsters being created by applying biotechnology (Sweden, Finland and Austria). Similarly, the more neutral the respondents, the more likely they are to think in terms of food issues (Norway and Germany). A belief in the potential for biotechnology to foster progress is expressed by the positively inclined Italian, Greek, Portuguese and, to a lesser extent, UK respondents. In general, it appears that the less people know about biotechnology ('don't know' responses), the more a country's population assumes a positive attitude towards this technology.

Figure 9.1 can give only a sketchy overview of the situation in Europe.[1] Results presented in the following sections will fill in this sketch with an in-depth analysis of discourses from diverse sources in six European countries.

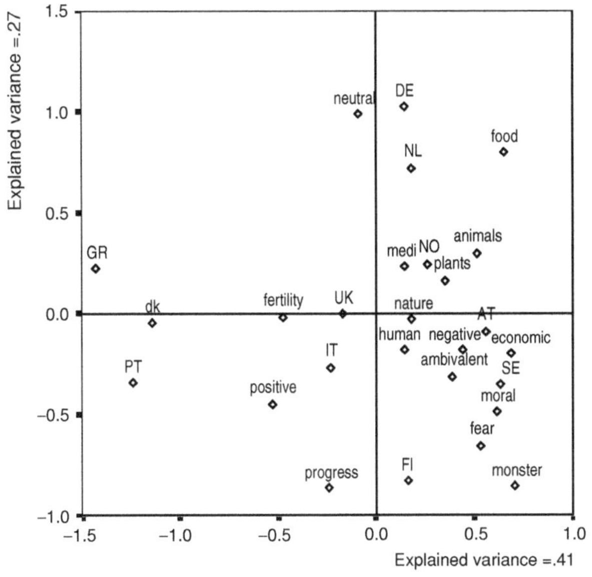

Figure 9.1 Evaluations and topics mentioned by respondents from ten European countries when asked about biotechnology: correspondence analysis, canonical normalisation
Note: AT = Austria, DE = Germany, FI = Finland, GR = Greece, IT = Italy, NL = The Netherlands, NO = Norway, PT = Portugal, SE = Sweden, UK = United Kingdom; dk = 'don't know', 'human' = human monsters, 'medi' = medical applications, 'monster' = non-human monsters.

Method

The aim of analysing different data corpora from Austria, France, Germany, the UK, Norway and Sweden is to address the following questions: what is the public perception of biotechnology? and what attitudes, images and linguistic repertoires guide thoughts on and discussions of this topic?

The qualitative data analysed in this chapter comprise a representative survey in various European countries (Eurobarometer 46.1) in which open-ended answers to the question 'What comes to your mind when you think about modern biotechnology in a broad sense, that is including genetic engineering?' were collected. Data were also gathered *via* focus groups and in-depth interviews, which provided extensive verbal responses about different aspects of biotechnology.

Besides being categorised (see figure 9.1), the survey data for each country were subjected to separate ALCESTE analyses (Reinert, 1983,

1990, 1998; Kronberger and Wagner, 2000; see the methodological appendix). Interview and focus group data were available only for Austria, the UK and Sweden. Data from the first two countries were classified using ALCESTE, while data from Sweden were analysed according to the grounded theory approach (Glaser and Strauss, 1967).

Working with different sets of qualitative data we faced the problem of depicting and comparing complex repertoires of arguments. Thus, in order to ensure an intelligible presentation of the data, we decided to focus upon similarities and differences with regard to the contents of national discourses rather than quantitative or methodological aspects. Beginning with the classification results for the open question responses, the dominant representations in different countries are described and discussed. The answers to the open question are short and include the most salient aspects of the topic. It is not surprising, of course, that many aspects mentioned in these associations also occur in the more in-depth discourses of the interviews and focus groups. Therefore the results of the latter data sources are used to complement the results of the open question.

Discursive frames

Overview

Table 9.1 gives a very general overview of the discursive classes found in the responses to the open question from Austria, France, Germany, the UK, Norway and Sweden.[2] One has to keep in mind that the presence or absence of a discourse in a category of the table does not mean that this aspect is not mentioned in a country, but rather that it is not mentioned frequently or explicitly enough to form its own class. It is also possible that one class comprises two or even more categories of the table (if this is the case, it is listed in between the two relevant categories).

Taking a look at the table, it can be seen that, in different countries, similar aspects of biotechnology are mentioned. At the same time, however, these are addressed with differing emphasis and intensity. There is only one discourse class that is clearly country specific: the Swedish association of biotechnology with the 'monster bull', Belgian Blue.

On a general level, we find that people interpret the survey question in two different ways. The two central questions – implicitly posed and explicitly answered by the respondents – produce different frames in which the topic of biotechnology is addressed. First, there is the question 'What is biotechnology?' Answers to this question produce statements that are not primarily evaluative, but rather descriptive. This is a form of 'knowledge discourse', that is, descriptions of what respondents believe

Table 9.1 *Comparison of the analysis results of the open question responses for six European countries*

	What is biotechnology? (Focus on content) General (rather neutral)		What is biotechnology? (Focus on content) Specific: domains of application (evaluation involved)			Is biotechnology good or bad? (Focus on evaluation) Positive	Ambivalent	Negative				Lacking knowledge		
Country	Research (progress)	Manipulation/ alteration	Food	Reproduction	Medicine	Good	Good but risky[a]	Risky/ dangerous[b]	Expression of fear	Interfering with nature	Country Specific	Echo[c]	Guessing[d]	Don't know
Austria		Biotechnology is a scientific activity applied to plants, animals and humans (food, reproduction, medicine) (27%)					Good but risky/ Dangerous (fear) (22%)	Unknown effects/ Dangerous (16%)		Interfering with Nature STOP! (36%)			*see* Interfering with Nature	*see* Unknown effects
France	Research (11%)		Food/ Agriculture (15%)	Reproduction (2%)	Medicine (14%)	Improve-ment (10%)	Dangerous/risky although there can be good effects (also morally dangerous) (8%)		Fear/ Against nature (18%)			Echo (3%)	Guessed (16%)	Don't Know (3%)
Germany		Manipulation of plants, animals, humans/ Agriculture (16%)	Food (also Medicine and Reproduction) (15%)		Medicine (12%)		Good but risky/ Risky/dangerous (fear) (37%)			Interfering with Nature STOP! (11%)			*see* Medicine/ Good but risky	Don't Know (10%)
Norway	Research (8%)	Alteration of plants, animals, humans (21%)	Food (8%) Food and Reproduction (16%)		Medicine (14%)		Good but frightening/ Non-specific worry (22%)			Interfering with Nature (10%)			*see* Medicine/ Good but frightening	
Sweden	Research (19%)	Manipulation of plants and animals (11%)	Food and Reproduction (7%)				Good if used the right way/Dangerous (15%)		Fear too fast (19%)	Interfering with Nature (21%)	Belgian Blue (9%)		*see* Research	
UK			Food (21%)	Reproduction (7%)	Medicine (21%)			Unspecific worry/Dangerous (fear) (16%)		Interfering with Nature (18%)			*see* Medicine	Don't know (17%)

Note: Numbers in parentheses indicate for each country the percentage of responses classified in a specific discourse.
[a] Good but risky: may have good effects but is risky and dangerous, therefore must be applied properly; demand for control.
[b] Risky and dangerous: biotechnology is unpredictable and therefore dangerous; fear of loss of control.
[c] Respondents repeat technologies mentioned in the preceding question ('telecommunication', 'solar energy', etc.)
[d] Associations evoked by the terms 'bio', 'gene' and 'technology' (mostly positive; e.g. ecologically beneficial or optimistic view of science).

biotechnology to involve. The second question focuses on an evaluation of biotechnology ('Is biotechnology good or bad?'). This question can be answered according to different concerns: a 'moral discourse' evaluates biotechnology according to ethical standards ('Is it morally acceptable?'), whereas a 'risk/danger discourse' considers the advantages and disadvantages as well as the potential benefits and dangers of applying genetic engineering ('Will it have salutary effects for us?'). These spontaneously emerging evaluative discourses show that, in the public view, biotechnology is not only an issue that must be understood, but also one that has to be integrated into an existing evaluative framework.

Of course, there is no clear distinction between descriptive or evaluative in the responses. Nonetheless, a certain emphasis on one or the other is discernible when people are talking about biotechnology. Some respondents evaluate biotechnology in general terms, while others refer to specific aspects and contents of this new technology (which can, nevertheless, be depicted in a more or less evaluative tone).

Different countries reveal differing emphases on these descriptive and evaluative aspects. In the Austrian and Swedish samples, for example, more than 50 per cent of the statements (for Austria, even more than 70 per cent) can be judged as primarily evaluative in nature. These are countries in which the public had been exposed to a high level of public discourse prior to the Eurobarometer survey (see figure 9.2).

In Austria, for example, a country in which media coverage started late in comparison with other countries, a people's initiative in early 1997 obliged the public to consider the 'pros' and 'cons' of biotechnology even though few had a profound knowledge of the subject (see figure 9.1). In most of the other countries, respondents highlighted different applications of biotechnology, such as food, medicine or reproduction, more frequently and in a more differentiated way than, for example, the Austrians did in 1996.[3]

In the following section, different kinds of discourse on biotechnology will be described. After taking a look at representations of what biotechnology constitutes, different ways of evaluating biotechnology are discussed. Then the most salient fields of application in public discourse (medicine, food, reproduction and xenotransplantation) will be presented by considering evaluative as well as metaphorical forms of addressing the topic.

What do Europeans associate biotechnology with?

When asked what comes to mind when thinking about biotechnology, a range of respondents in all countries described what they took this term

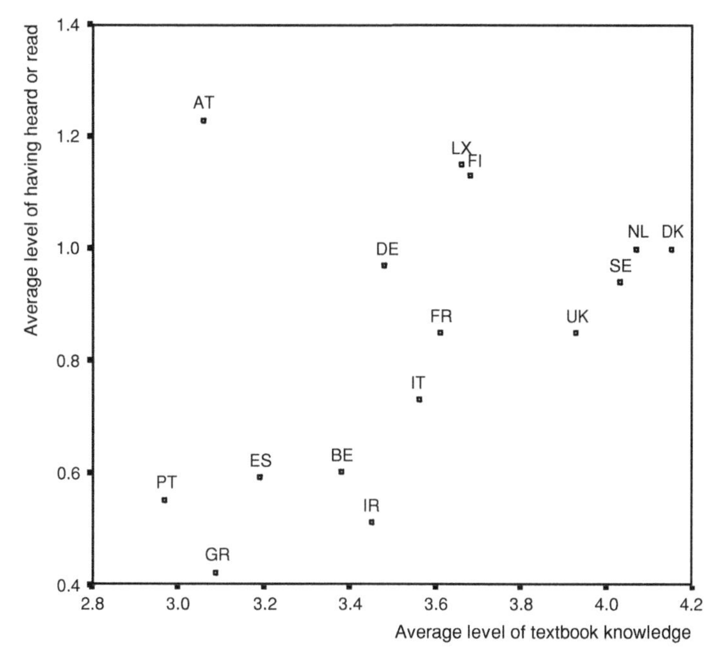

Figure 9.2 Scatterplot of European countries by average level of textbook knowledge and average level of having heard or read about biotechnology
Note: AT = Austria, BE = Belgium, DE = Germany, DK = Denmark, ES = Spain, FI = Finland, FR = France, GR = Greece, IR = Ireland, IT = Italy, LX = Luxembourg, NL = The Netherlands, PT = Portugal, SE = Sweden, UK = United Kingdom.

to mean. These statements represent a thumbnail sketch of biotechnology by referring to this topic as an action (scientific research) as well as by mentioning objects and domains of application. When biotechnology is interpreted in terms of research, statements tend to be of a general nature, related to scientific investigation and developments, and are frequently expressed in the form of optimistic views of science and new technologies. Biotechnology conceived as manipulation/alteration is taken to imply the modifying, manipulating and cloning of plants, animals and humans. The most salient objects discussed in this context are fruit and vegetables (with tomatoes being prototypical), but also animals and children. As a result of this scientific manipulation, respondents imagine the products to have become 'bigger' and 'artificial' in contrast to 'natural' products, which are free of genetic interference. The manipulation of plants, animals and humans is associated with specific domains of application, with food and medicine being the most commonly cited in all countries. The fields of reproduction and embryology are less frequent associations, but still represent salient topics in relation to biotechnology in most of the countries studied. Of course, the descriptions of biotechnological applications in the different domains also comprise evaluative (moral as well as risk-related) aspects.

On closer inspection one can see that this descriptive 'knowledge' need not correspond with scientific definition. Examples such as 'with the help of wind and solar energy, as well as with hydroelectric power plants, one can provide jobs and preserve nature' (AT, 006405[4]) or 'biotechnology also comprises winning new food stuffs out of the sea, plankton and the like' (GE, 1367) suggest that a term such as 'bio'-technology evokes certain images even in people who lack knowledge about its technical use (see the category 'Guessing' in table 9.1). To some of the respondents the word biotechnology seems to be connected with environmental responsibility and scientific progress, that is, with being in harmony with nature ('bio'); and for some the term 'bio' refers to pesticide-free agriculture. For others, biotechnology, even when poorly understood, is situated within an optimistic view of science and technology: 'things to do with the vegetable kingdom; food and cloth textiles; growing tomatoes in fluid rather than earth; trying to grow food in space or those sorts of conditions' (UK, 2219).

Another group of respondents admits to not having any idea about biotechnology at all. It is not a topic of salience or interest for them. Furthermore, a small percentage of respondents simply repeat technologies mentioned in the preceding question of the survey without referring to biotechnology in a specific sense (category 'Echo' in table 9.1).

How Europeans evaluate biotechnology

Optimism Table 9.1 shows that there is relatively little unquestioning acceptance of biotechnology in the countries covered by this analysis. Although there are optimistic evaluations concerning biotechnology in most of the countries, only in France does one find a considerable percentage of people favouring biotechnological development. This group generates its own class of statements, such as 'improvement of the life of the individual, health, better living, better food and better ageing' (FR, 542).

Ambivalence This discourse depicts biotechnology as a new technology that can, to some extent, improve our way of life, but is nonetheless risky and dangerous for mankind. 'I think that's good, there are so many people in the world that want to live and to be well-fed, but the control of biotechnology must be increased' (GE, 3240). People fear that the new possibilities offered by biotechnology could be exploited by unscrupulous and irresponsible people, and they are therefore concerned about the 'proper' application of this scientific knowledge: 'Biotechnology is positive but it should be dealt with in a responsible way. It can improve a lot but applied by the wrong hands it is dangerous' (GE, 0285). Therefore a strong demand for control, delimiting activities in relation to biotechnology by legal as well as moral standards, is expressed. 'I think man has got to conserve and preserve with technology the skills he has been given. It has to be used properly, not abused. He can just as easily destroy as create' (UK, 0227).

Rejection based on risk This discourse involves worrying about the future effects of genetic engineering in terms that suggest these to be far-reaching, incalculable and risky: 'The risk is too big because one cannot estimate the consequences' (GE, 1269); 'very dangerous subject they're getting into. It will create more problems than it will solve' (UK, 0300); or 'We will poison ourselves altogether, nobody knows what will happen, unfortunately one can't do anything about this' (AT, 016701). Although most of the time potential consequences are not explicitly identified, the consequences of biotechnology are judged to be dangerous in nature. The respondents frequently view themselves to be in the hands of science, industry and politics without having any power themselves. This discourse involves a clearly negative evaluation of biotechnology.

An image that illustrates this fear of conjuring up a scenario beyond our control is Goethe's literary figure of the Sorcerer's Apprentice. This image is most frequently invoked in France: 'It's playing the Sorcerer's Apprentice, it is necessary to elaborate some ethics before doing it practically' (FR, 260); 'Mad cows, considering animals as things, the pride of humans playing the Sorcerer's Apprentice' (FR, 937). The wish to make life easier by using only a little-known (magic or scientific) formula, it is asserted, may end up producing unexpected and uncontrollable results and, ultimately, chaos. In one scenario, this involved the 'uncontrolled spreading of artificial life' (NW, 362). According to this picture, scientists are seen as powerful on the one hand (like the apprentice, they have the necessary information at their disposal) but lacking the master's wisdom to apply their knowledge in a useful and responsible manner. On a less literary level, statements such as the following refer to the same concern: 'They just do experiments all the time and destroy earth, it would be better to spend the money in a more useful way' (AT, 006403); 'I think that there are already enough bad inventions in this world that we aren't able to control anymore like the hydro- or the atom bomb. I think this is a field in which we can't estimate the risks, we should not give these inventions, like computer-machines, too much power over us' (GE, 1100). This fear of creating far-reaching problems, without being able to handle them, is also expressed by a further reference, this time to Greek mythology's Pandora: 'It raises a whole Pandora's box of medical and ethical questions' (UK, 2048).

Ideological rejection This discussion, one that has clearly emerged in all the countries covered by this analysis, judges biotechnology against the background of general thoughts, values and assumptions about the nature of humans and their relationship with their environment. The cultural practices of science and technology are perceived as tools that help humans to shape nature according to their own ideas. Regarding biotechnology, this discourse therefore deals with the question of whether or not humans should be permitted to interfere with the 'natural' harmony of nature. Many respondents, in all countries, think that genetic engineering constitutes an inappropriate tinkering with life and, more specifically, with the meaningful order of nature in which every species has its place and its purpose and where natural boundaries should not be transgressed by unnatural means: 'I'm against genetic engineering. If nature wanted any changes it would already have made them itself. Any interference in natural developments brings about risks that can't be estimated by anyone' (GE, 2466).

Statements such as this are predicated on a view of nature as an organism with its own laws and rhythms and, therefore, one that can never be totally predictable or controlled. An important dimension here is time; nature is associated with slow continuous development and humans are posited as interfering with this 'natural' evolution: 'It is scary, the development is too fast' (SE, 890245); 'They should go more slowly and think about the consequences' (SE, 050585). This view of biotechnology, as a development that is proceeding too fast, is stressed most commonly by Swedish respondents.

The idea of nature as some kind of 'living being' is conceived in two ways: as a sagacious, benevolent being on the one hand, and as a wild force that will take revenge on the other. It is wise, many assert, to live in 'harmony with nature', or to live 'within' nature. From this point of view, biotechnology and its applications are perceived as a harmful intervention in natural processes. Therefore, one must expect nature to 'hit back' and take revenge: 'When man interferes too much in nature it always comes back like a boomerang' (GE, 1290); 'Tampering with nature may be all right in the present, but it will hit back at us in a couple of years' (NW, 716). In countries such as Austria and Germany, respondents frequently accompany their rejection with an imperative to 'stop', prompting statements such as 'One should leave everything as it is, one should leave Nature alone' (AT, 009607), or 'The manipulating of normal things should be stopped' (GE, 3009).

This 'interfering with nature' discourse also appears in the interview and focus group data. A morally challenging point that emerges here is that biotechnology allows not only the modification of plants and animals, but also the manipulation of a very special kind of 'nature': the nature of being human. One has to note the double interpretation of 'human' used in this discourse. On the one hand, 'human as scientists' are seen as actively shaping nature and, on the other, 'humans' are considered to be the object of that manipulation. The moral concern, therefore, is humans interfering with human nature. This means not only modifying specific human characteristics such as intelligence, physical appearance and ageing, but also altering the very essence of what it is to be human. Respondents anticipate this bringing about an unpredictable change in the nature of human life. For example, with the introduction of cloning, biological fathers or mothers will become superfluous. To be born as a result of a scientist's intervention impacts upon human dignity and, according to many, contravenes any moral sense of 'good and right': 'Humans originate from humans according to a development which cannot be planned. If anybody could determine what I will become, then I would have the impression that I lack something of being human' (AT, focus group). As will be seen later in this chapter, questions of how biotechnology might undermine the essence of being human also extend to such issues as the importance of a 'natural' end to life and concerns about human uniqueness.

After all, this notion of humans 'designing' humans contradicts the claim that there are things that only a 'god' can do: in this view, humans are able to do god-like things now. This is generally judged to amount to the grossest impudence, a conviction usually expressed in the form of the persuasive image of humans 'playing God'. This also appears in the open question data: 'It is this manipulation of man, it is this alteration of nature. This is an interference in Creation, man plays God' (AT, 008002). In this view, scientists and industry are consistently viewed as playing with our lives. In accordance with the metaphor of the Sorcerer's Apprentice, they are viewed as doing something that they are not entitled to do. But, whereas the apprentice's impatience to do things before he has the requisite knowledge to do so is viewed as dangerous (because of the potential for bringing about an uncontrollable situation), the wish to 'design' and to determine the world like a god is judged as morally unacceptable: 'I'm fearful about it, it's in men's hands the use to which it's put, it could be a tool of the devil to do with altering the genetic make-up in humans' (UK, 2295).

In this context, people involved in genetic research and industry are considered not to be acting responsibly, but rather to be thoughtlessly following their – mostly egoistic – desires for power and economic gain: 'We've already come to know different technologies, modern ones like nuclear energy, the atom bomb, chemical weapons, weapons in general – did these improve human life? Only for the few who make profits out of them, totally neglecting the suffering of innocent people. With biotechnology it will be the same because again there will be some powerful people taking advantage in an unscrupulous way' (GE, 3338).

Domains of application

Fighting diseases

Medical application is one of the most common references to biotechnology appearing in the public discourse of most countries. The potential for biotechnology to help cure hitherto incurable diseases is generally judged to be an important potential improvement. Unsurprisingly, biotechnological applications in relation to medicine are evaluated in a mostly positive tone: 'curing diseases' (AT, 008806); 'fighting hereditary diseases, early diagnosis of diseases' (GE, 0190); 'development of new drugs for control of cancer, arthritis and other diseases' (UK, 0061).

In this domain, the image of 'progress' clearly emerges. Diseases such as cancer and AIDS are frequently cited by respondents expressing the hope that biotechnology will provide new remedies. 'For coping with diseases like AIDS, BSE and cancer, technologies like that are really necessary and must be encouraged' (GE, 1028). Only a few negative consequences of applying biotechnology to the domain of medicine were mentioned.

'I'm for genetic engineering because my wife suffers from cancer and maybe they will find new remedies against diseases like that' (GE, 1099). It is not surprising that personal suffering and the hope for medical advances encourage the endorsement of this aspect of biotechnology. Interestingly, this positive evaluation changes to ambivalence when different potential medical applications are distinguished. An Austrian interviewee stated: 'the term "for medical purposes" is a very vague one because being able to cure cancer is different from surgery for cosmetic purposes and is different from breeding perfect humans' (AT, focus group). This more differentiated discourse takes place in interviews and focus groups and only to a small extent in the context of answering the open question. Surprisingly, however, medical applications involving reproduction and xenotransplantation are more commonly discussed as moral than as medical questions.

Unnatural food

The food discourse occurs in the context of evaluating the advantages and disadvantages of genetically manipulated (GM) food. The validity of applying genetic manipulation techniques to food is often assessed by opposing the consumer's and producer's points of view. On the producer's side it is suggested that GM food is easier, more efficient and more profitable to grow. More frequently, however, responses are framed around the supermarket consumer shopping for the best-quality produce, and making decisions about what constitutes healthy, good food. Respondents, therefore, mainly refer to characteristics of the products, invoking words such as 'better', 'cheaper', 'bigger' or more 'beautiful', and having a longer shelf life. The disadvantages of GM food mentioned focus on the effects of genetic food manipulation for humans as well as for the ecosystems. Generally speaking, the French, British, German and Norwegian discourses on GM food are expressed in a fairly neutral, sometimes ambivalent tone, whereas evaluations in Austria and Sweden are rather negative. The Swedes stress potentially negative consequences for the biological diversity of plants and animals. Respondents from Austria and Germany more frequently assume that eating GM food is unhealthy and that it will bring about new diseases and allergies, or produce resistance to antibiotics.

An important issue from the consumer's point of view is having a choice of which products to buy and to eat: 'I don't buy it any more because now I know that it is genetically modified rape. I tried it, it had a good taste, but I won't buy it any more' (AT, focus group). This is a question of labelling, and comes down to the fear of not being able to determine whether or not a product is genetically modified. The feeling of being denied the ability to choose results in statements like those expressed by the Swedish media, in which GM food products are referred to in such terms as 'leaking uncontrollably over the borders', 'gene food sneaking into the shelves' or 'mixing modified and natural beans so that we will not know'.

As mentioned above, the objects of genetic manipulation that appear most conspicuously in the public discourse are fruit and vegetables, with tomatoes as the ideal-type. These new products are imagined to be 'bigger', 'unnatural' and 'artificial', in contrast to 'natural' foodstuffs that have not been genetically manipulated: 'giant fruit, bigger potatoes, cattle with more meat, wheat yielding more flour' (GE, 0454); 'manipulated giant tomatoes, . . . artificial colours, artificial children' (GE, 0453). The paradox in this image is that products artificially produced by genetic manipulation are conceived of not in their own terms but as products whose inherent characteristics have been intensified: 'making tomatoes rounder – apples larger and greener' (UK, 0034); 'sugar-beets were genetically modified, the proportion of sugar is increased' (GE, 1194). Less frequently, people mention fruit and vegetables that have been visibly altered: 'I know they can make tomatoes of all colours, as big as pumpkins and melons' (FR, interview); 'a tomato that is a mixture of tomato and potato and that tastes like a banana' (FR, interview); 'square tomatoes' (AT, 001204).

Topics that are regularly associated with biotechnology in several countries are nuclear energy and radioactivity (the two issues seem to be perceived as similar because in both cases the technology involved is potentially dangerous and may have incalculable effects). People in different countries talk of 'irradiated' food: 'fear of BSE and irradiated vegetables' (GE, 0242); 'Nature should be left alone, otherwise we'll soon have the atom bomb in our vegetables' (AT, 015202). Understood in metaphorical terms, statements such as these express a perception of biotechnology as something dangerous and threatening. And, as was the case with GM fruit and vegetables above, irradiated food is understood as an extreme form of the original product.

In addition, biotechnology is sometimes associated with hormones. This is most typical among Norwegian respondents: 'disturbance in animals' growth hormones' (NW, 004); 'messing with the hormones in food production' (NW, 149). Hormones are most often mentioned when respondents are referring to the size of manipulated objects. This is not surprising since one of the first biotechnological applications of which the Norwegians heard was the implantation of growth hormones in salmon, an important Norwegian export product. In terms of public perception, genes and hormones are conceptualised in similar ways: both are small and invisible, and seem to provide a sound, scientific basis for understanding otherwise inscrutable aspects of organismic biology.

Although the molecular alterations are invisible to the human eye, GM food is perceived as having been changed from its natural to an unnatural and artificial state. The ubiquitous opposition of 'natural' and 'artificial' subsumes two of the themes within the discourses of health and illness as well as discussions of biotechnology: first, the association of 'artificial' products with an 'infringement of the natural state' (Herzlich, 1973: 32); and, secondly, the notion that this infringement has connotations of 'heterogeneity' or the introduction of foreign elements into nature. Moreover, if taken a little further, these heterogeneous elements may be interpreted as pollution, and this explains the use of 'unnatural' and 'unhealthy' as synonymous in some countries. The representation of the

Plate 9.1 Photo of tomato in Austrian newspaper

products of biotechnology as big and beautiful, but at the same time artificial and unnatural products (prototypically the tomato), containing heterogeneous elements, is expressed in Plate 9.1, taken from an Austrian newspaper.

Similar representations are discernible with respect to the manipulation of animals. As was the case with manipulated fruit and vegetables, GM animals are imagined to be bigger while being, at the same time, unnatural. A typical association among the Swedish public is that of biotechnology and the 'monster bull Belgian Blue',[5] a cattle breed that has no connections with genetic engineering from a scientific point of view. Nonetheless, 9 per cent of the Swedish respondents refer to this 'giant bull' when thinking of biotechnology. In the public mind, the Belgian Blue is a living embodiment of all the characteristics attributed to GM creatures: it is big and monstrous on the one hand, but weak, bizarre and unnatural on the other. This image of the Belgian Blue was also reported extensively in the Swedish mass media, attention being focused on the breed's physical characteristics, such as its giant body with large muscles, weak legs and small internal organs, together with its calving problems (see Plate 9.2). Only a few days before the survey, the Swedish newspaper *Aftonbladet* published a story with the headline 'Calf slaughtered with chain saw – inside the cow'. Similar descriptions are provided by respondents in the

Plate 9.2 Photo of 'Belgian Blue'

survey: 'I think about the Belgian Blue – monster bulls which must be delivered by sawing the cow in half' (SE, 980247).

Images such as the 'monster bull' Belgian Blue or 'mad cows' (BSE), neither of which issues is directly related to biotechnology, nevertheless occur in the public discourse and provide persuasive images informing the public's perception of the new technology. The fact that cattle are 'mad', 'degenerate' and 'bizarre' is perceived to be a result of breeders striving for better profits while paying insufficient regard to the limits set by nature. Mad cow disease is frequently quoted when respondents are referring to the unpredictable risks of applying genetic manipulation. This image recurs when respondents talk of scientists, industrialists and political authorities, which are viewed as failing in their duty to protect their country's citizens: 'I do not agree with it at all, we wouldn't have had BSE if they hadn't bloody well interfered. No, it's just a feeling of interfering, you can't alter nature without paying for it. Nature always balances' (UK, 2404). (According to the logic of the images discussed in the evaluative discourses, this could be conceived in terms of the plagues coming out of Pandora's box, as well as a first sign of Nature's revenge.)

Testing and designing humans

Upon being asked to think about biotechnology, a significant proportion of respondents mentioned aspects of reproduction and embryology. This domain is less frequently associated with biotechnology than either food or medicine, but is still of remarkable salience. Interestingly, as far as the public is concerned, this discourse is separated from, rather than connected to, the medical context. (Table 9.1 shows that in most countries the medical discourse is separated from the others; the vocabulary and associations employed in talking about reproduction are more similar to those used in discussing food.)

The most frequently mentioned reproductive associations involve 'artificial fertilisation' and 'test-tube babies'; yet neither of these is directly related to biotechnology in the current understanding. Nonetheless, these images are well suited for encapsulating a whole set of concerns relating to both technology and the creation of life. The image of a baby created in a test tube captures the 'unnatural' and 'artificial' character of biotechnology. Statements such as 'artificial children' (GE, 0453), 'the lady having a baby from her dead husband's sperm' (UK, 0213) or 'a woman who had eight babies' (UK, 0429) all depict putatively 'unnatural' aspects of human reproduction. In this way, the stress on the 'unnatural, weak or bizarre' products of the genetic manipulation of food and animals is recapitulated in the discourse surrounding the possibility of human genetic manipulation. This gives rise to such remarks as 'test-tube babies are often handicapped' (GE, 0453), 'it will lead to manipulation and the creation of degenerate species, both humans and plants' (SE, 462362) or 'it frightens me, all the horrific changes we could do with it; we could end up making little monsters' (UK, 2047).

Besides the 'test-tube babies', there are other images that contrast reproductive applications of biotechnology with 'natural births'. An Austrian image mentioned during the interviews involves a pig carrying a child (the pig is styled as a 'loan mother'). Here biotechnology is seen to offer those mothers who want to preserve their bodily youth and beauty the possibility of having children without the burden of a pregnancy. The sarcastically uttered statement 'my mother the pig' (AT, focus group) can be understood in the double sense of imagining a child being confronted with a mother-pig (What defines being a mother? Who will be the biological mother, then – pig or human?) as well as judging a mother who would do such a thing as morally unacceptable.

In the same way as depicted in the 'interfering with nature' discourse, the topic of an 'unnatural' beginning of life (whether being 'produced' in a test tube, originating from conserved sperm of an already dead person or being carried to term by a pig) is morally challenging since it questions the 'nature' of being human and therefore also the social identity of human beings. Furthermore, there is the topic of cloning, a practice viewed as a very special kind of reproduction. As mentioned earlier, cloning makes it unnecessary to have both a biological father and mother. Indeed, the possibility of asexual reproduction prompts people to wonder why there should be men and women at all: 'It must not go so far because then we wouldn't need males and females anymore, no, I'm against that. Natural reproduction, at least this should be true. I'm very much against that' (AT, focus group).

Genetic engineering in relation to reproduction is also often perceived as a matter of life and death, being seen as a tool with which to select those human beings considered worth living and to eliminate those that are not. This discourse is embedded in a general discussion about abortion and focuses on genetic engineering as enabling the early diagnosis of illnesses and handicaps: 'terminating pregnancy if foetus has illness' (UK, 2397); 'they will use their knowledge to abort babies with only small defects' (SE, 462362). In the interviews this discourse focuses on moral questions along the lines of: What makes a life worth living? What are we doing when we abort a child diagnosed as handicapped? Is the prevention of suffering a justification for overriding an organism's 'right to live'?

This moral discourse is embedded in a subjective context of personal involvement. The questions are not dealt with in a general or a neutral way; instead they tend to elicit a personal point of view: a concrete decision framed in terms of 'What would I do?' In the UK and Austria these discussions can mostly be traced back to interviews with women, and involved the issues, practical and ethical, that would inform their decisions if faced with the possibility of genetic screening and other biotechnological applications. Being a moral issue, motives and interests are of crucial importance; and because a moral conflict is often involved in deciding in favour of a single option, evaluations are generally ambivalent. In this case, the moral concern is a matter of life and death, and more specifically it is the right of humans to decide about the lives of others. Furthermore, the discussion deals with the question of what defines a 'good' life, for parents as well as for children, and which forms of interfering with the 'natural' processes of living and dying are morally legitimate.

A topic related to abortion is that of biotechnology enabling the selection of humans for certain attributes: 'makes a woman have a boy or girl if she wants it, there could be production of babies by test-tube creation, it's a lot of rubbish' (UK, 0489); 'not good: making children in tubes and sorting the children according to if they are good or bad' (NW, 174).

The images of 'perfect children', 'humans made-to-measure' or 'designer babies' suggest that parents (or scientists) can choose from a 'catalogue' of characteristics for their offspring's attributes, for example physical appearance and intelligence. This kind of discourse also links the 'production' of humans, equipped with certain features, to financial and economic factors. An English interviewee imagines a future scenario of buying children in shops: 'I mean, you just visualise, you know, the typical child, saying "where are we going to get the baby from", do you go to Marks & Spencer, eventually you're going to be able to do that' (UK, focus group). Clearly, the buying of children contradicts any sense of 'normal' birth.

Moreover, imagining some kind of industry 'producing' babies implies that people will have to pay for what they order: 'But to be able to go one step and choose your blue eyed, blonde haired, intelligent child, is going to come down to money because you're going to have to pay somebody because it's not a necessity' (UK, focus group). This impinges upon another moral issue: not everybody will be able to afford these services, and new social injustices will consequently arise.

Of course, the 'designing' of babies is not far away from creating humans for specific purposes. When discussing cloning, and emphasising the dangers of the new power relations that might emerge, science fiction scenarios are often depicted: 'This is the story of subhuman creatures, of cloning human robots, of being cloned from some physically extraordinary humans but provided with less intelligence, so they can do dangerous jobs' (AT, focus group). Huxley's 'Brave New World' is sometimes mentioned in this context. Images are commonly invoked that involve subhumans and supermen, armies bred only to fight, or living machines created to do dirty jobs. An association with eugenics is frequently cited. And it is far from surprising that in some countries the name 'Hitler' is strongly associated with cloning: 'It's frightening, in the end they will manage to do what Hitler failed to do' (UK, 2208). This raises fundamental questions: What should one be allowed to manipulate and where should the line be drawn? Who should make this sort of decision?

Monsters and matters of life and death

An application only rarely mentioned in the open question data is animal–human transplantation. This – at least in 1996 – was not a salient association in relation to biotechnology for most respondents. During the interviews, the topic was introduced by the interviewer whenever it was not mentioned by the interviewees themselves. It is interesting to note that the issue is discussed only marginally in relation to medicine. Here the motive of 'rescuing human life' is of crucial importance. But, above all, the issue evokes divergent moral questions.

The topic of xenotransplantation stimulates fantasies about combining the body parts of different species, and thereby constructing monsters.

This discourse deals with the 'natural' make-up of beings, as well as with the 'natural' borders between species: 'mice with an ear on their back, pigs with two more ribs' (GE, 0208); 'pigs with cows' heads' (GE, 0227).[6]

In the public mind, the aim of xenotransplantation to provide healthy organs is visualised in terms of 'spare parts': 'If we accept that we're going to breed a spare parts bin, do we accept that it is right to take those parts and put them into anything we so desire, be it a human being, another animal, whatever?' (UK, focus group). Animals are no longer perceived as living organisms but rather seen as sets of body parts that can be combined in different ways. But this raises the question of whether such organisms should still be considered to be living beings, or just machines: 'We produce living machines, and this must be rejected' (AT, focus group).

The idea of animals providing 'spare parts' is also extended to humans (for example, human beings without a head existing to provide organs and body parts): 'and then they will take "inferior" humans, they will say that they are a bit handicapped but they have a healthy heart, and so on, it will go on like this only for the reason that we can live eternally' (AT, focus group); 'I don't like genetic engineering, I don't want to end up with other people's bits of body' (UK, 0070). People come to think of human and animal 'monsters', and the idea of mixing different humans or combining parts of humans and animals seems to infringe 'natural' boundaries.

In the interviews, this topic is also discussed as an issue of animal welfare. Respondents question the ethics of breeding, manipulating and slaughtering animals for the purpose of curing disease and prolonging human life. Genetic engineering is here posed as a matter of life and death for humans as well as for animals. Most of the time, the motive of 'rescuing human life' is acknowledged as an important value that should be endorsed; but the problem with xenotransplantation, in the respondents' view, is that saving the life of a human costs the life of an animal. In a broader sense, the discourse also deals with the topic of organ donation in general. Respondents refer to the fact that an individual continuing to live with an implanted organ always demands the death of another (whether this is an animal or a person dying in an accident). The discourse therefore focuses on questions such as the following: Is human life worth more than that of an animal? Why not breed and slaughter animals with the purpose of obtaining organs? After all, we do so in order to produce meat. Which is more morally acceptable: waiting for a seventeen-year-old motorcyclist to die or using a pig as a 'spare parts bin'?

Xenotransplantation is viewed as a tool for prolonging human life by delivering healthy organs. A belief that emerged in the Austrian interview data is that everybody has the right to live his or her 'whole' life. That is, when the 'natural' lifespan of an individual is threatened by illness, or by other factors, then help should be available. This subsumes the motive of wanting to be immortal or 'forever young', a dream that some feel might be realised once we are capable of 'producing' young and healthy organs, and of using animals as 'spare parts bins'. However, the fulfilment of this extreme form of the wish is considered morally unacceptable. At this point, the question concerns the proper limits of xenotransplantation. Do we have the right to deny help to an ill person if this technology could be developed and employed? What is the difference between implanting an organ into a fifty-year-old person or into an eighty-year-old person? More fundamentally, when is a person old enough to die? And who should make the decision on this question?

Generally speaking, respondents' evaluations change when different motives and perspectives are considered. The motive of curing disease and enabling individuals to live 'normal' lives of 'normal' length is generally considered in a positive light. The wish to be 'forever young', however, is considered egoistic and unacceptable. Using animals to, as it were, pay for our sins (our unhealthy way of life) is also considered to be improper. But if our life is in danger, not through our own fault but because of an 'unjust' illness, then the case is different again.

Like any moral discourse, the discussion focuses upon 'limits'. What is the difference between having a new set of artificial teeth and having a new heart grown in a pig? What is the difference between receiving a donor heart from a person who died and one from a pig? What is the difference between organs bred in animals and those bred in humans? What is still acceptable and what is not?

'National' images

Figure 9.3 shows the associations between different countries and the typical images evoked by biotechnology. Generally speaking, one can see that there are few country-specific images, a major exception being the 'Belgian Blue', which is a typically Swedish association. Most of the images, though, are located around the centre of the graph and occur in most, or all, of the countries. The general (moral) images of 'interfering' or 'tampering' with nature, life and humans are central and ubiquitous. These are images that refer to biotechnology in general and do not explicitly identify specific objects of manipulation. As mentioned above, this 'interfering with nature' discourse occurs in all six countries: it depicts biotechnology as an unwise, harmful or dangerous interference in nature and life.

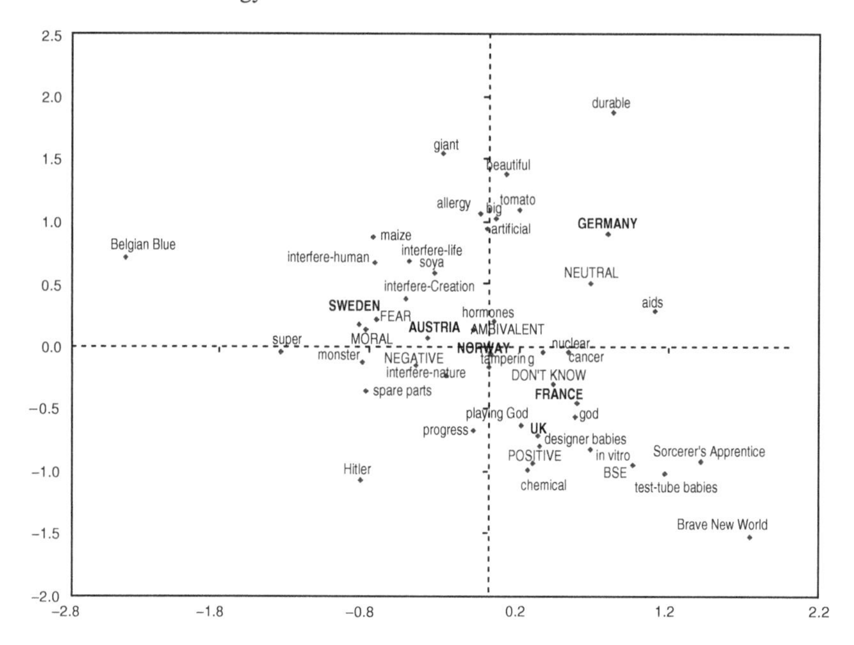

Figure 9.3 Correspondence analysis of open question for countries, evaluative tone and images.
Note: Global categorisations are in capital letters; labels in lower case letters indicate images referred to in respondents' answers to the open question.

Most of the images are mentioned in every country studied, and the national variation comprises respondents placing different emphases on the various domains of application. Images concerning food more frequently occur in Germany and Austria (tomato, big, artificial, giant, durable, etc.); Swedish respondents more often refer to animals (Belgian Blue, spare parts, monster); whereas the domain of reproduction and embryology is more frequently mentioned by British and French respondents (test-tube babies, *in vitro* fertilisation, designer babies). Comparisons to literary scenarios such as Huxley's 'Brave New World' or Goethe's 'The Sorcerer's Apprentice' are mainly drawn by French respondents. Norway, being located in the centre of the graph, does not show any special association with any of these diverse forms of application.

What do the differences and similarities in imagery among these countries mean? It must be noted that, in this context, images or metaphors not only constitute means of expressing ideas, but also function as socio-cognitive devices that help to organise the understanding of a phenomenon. Because they are metaphorically structured at a cognitive level, the manifestations of imagery in language use are also metaphorical. For example, in all six languages we find a range of respondents describing biotechnology as 'interfering', 'messing', 'tinkering' or 'tampering' with nature or life. At a more general level, all these statements express the concept that 'biotechnology is a (harmful) intervention'. This rather general concept can be understood as an image schema that structures the perception as well as the evaluation of the issues in question. The concept therefore delimits all the possible interpretations of this technology by focusing on some aspects and neglecting others. From this point of view, it is unlikely that biotechnology would be evaluated as something beneficial that will improve our quality of life. But the image schema are flexible enough to allow the expression of various evaluations: 'interfering' implies a certain degree of neutrality, whereas 'messing' and 'tampering' are clearly negative.

However, respondents in all of these countries employ very similar images of biotechnology. Therefore, it is reasonable to suppose that, because of the limited number of organising concepts, biotechnology discourses are heavily circumscribed.[7] These concepts 'establish a range of possible patterns of understanding and reasoning. They are like channels in which something can move with a certain limited, relative freedom' (Johnson, 1987: 137). Yet, although the number of interpretations of biotechnology available in the public discourse may be delimited, within the limits of any structuring concept there is a measure of freedom or variability of understanding that is heavily context dependent. The concrete words and images that are used to allude to these underlying schemas will be mainly determined by local, historical and personal contexts.

These contexts determine which images are most salient for the public in thinking about a topic such as biotechnology. Thus the images stem from different sources: first, there are images such as BSE or the Belgian Blue that reflect current discussions in political, social and everyday life in different countries; secondly, there are historical references to the past, such as to Hitler and the Nazi regime; thirdly, people may refer to mythology or literature, in this case in the form of Frankenstein, Brave New World, Pandora's box or the Sorcerer's Apprentice; and, finally, there are images that cannot be linked to any of the sources mentioned, but that grasp the essence of biotechnology with everyday words such as 'artificial food', 'interfering with nature' or 'designing babies'.

It is clear that underlying concepts and feelings can be expressed by reference to a wide range of current images. Thus, the idea that manipulated objects are bigger, but unnatural and bizarre, can be articulated in terms of mad cows or a mouse with a human ear on its back. But individuals can also draw upon mythology and literature, invoking fantastic monsters or Frankenstein. The importance of local context is especially clear in the case of Britons citing mad cows and the Swedes tending to think in terms of the monster bull Belgian Blue, even though both groups of respondents were expressing their concern that biotechnology may bring about bizarre, weak and unwell animals. Considering the media attention BSE was receiving in Britain at the time, it is far from surprising that the association was made. The Belgian Blue cattle breed was an even more localised issue (with only a few Norwegian respondents also citing it), but since the import of twenty-five Belgian Blue bulls to Sweden in 1995 this topic had attracted a lot of attention in the Swedish media, which referred to them as 'monster bulls'. Likewise, a historical reference to Hitler can express the same fear of authoritarianism as quoting Huxley's *Brave New World*. Clearly, different images can stand for similar concerns; the concrete words and terminologies used can only be understood as being largely derived from the speaker's context.

The general semantics of biotechnology discourse

There seem to be two global representations of biotechnology in which the different images are organised: one recapitulates the theme of humankind's progress in overcoming natural obstacles and the other involves humans' unwarranted interference with nature.

Fighting enemies of mankind

Discourses that express a positive evaluation of biotechnology frequently focus on a war metaphor, which depicts the modern world as confronting a range of dangers that must be ameliorated. In this view, biotechnology is a weapon that helps in the 'fight' against diseases, such as cancer or Aids, as well as world hunger. This discourse emphasises progress, the creation of a better world. Nature represents a complex set of mechanisms that can be incrementally better understood, and whose more pernicious aspects can be overcome. Accordingly, the manipulated object receives little attention, and this discourse is able to draw upon few, if any, images.

Tailoring living nature: fake life

Biotechnology seems to offer the possibility of creating and fashioning a world according to human plans and ideas ('designer babies', 'made-to-measure humans'); humans are considered able to imitate the Divine ('man plays God' or 'interferes with Creation'). But respondents believe that it would be unacceptable for humankind to do so; humans, it is claimed, are not entitled to interfere with nature in this way. For this reason, comments on biotechnology are couched in terms of irresponsible 'playing', 'tampering', 'messing' and 'tinkering' with Nature. According to this view, the expected result must be an ultimate loss of control, with humankind being exposed to chaos and the suffering implicit in references to the Sorcerer's Apprentice and Pandora's box.

Notions of human interference in 'living nature' subsume several different themes.

- Interference with respect to upsetting the natural balance and harmony of ecological systems – this presupposes an anthropomorphic image of nature: since biotechnology is 'harming nature', it will eventually 'hit back' and 'take revenge'.
- Interference that can lead to humankind itself becoming 'de-naturalised' because of the possibility of 'unnatural birth' ('test-tube babies', 'artificial fertilisation', '*in vitro* fertilisation', 'designer babies', 'pig as loan mother', 'buying babies'); these technologies are seen as allowing 'unnatural lifetimes' ('eternal living'), made possible by the use of living 'spare parts', and this is perceived to imperil the uniqueness and value of being human ('producing life', 'mass production of babies', 'armies breed to fight', 'Brave New World').
- Biotechnological interference that represents a threat to 'natural' social orders and social justice, respectively (Hitler, Brave New World, living machines doing dirty jobs, subhumans and supermen).

From this perspective, there are a variety of potential outcomes of biotechnological intervention, and these relate to different conceptions of Nature. In all cases, however, modifying nature means that the result is, by definition, no longer 'natural', it becomes 'artificial'. But what does this nature 'made' by humans look like?

Fruit and vegetables are characterised as being 'bigger', 'more beautiful' and 'more durable' and as yielding 'more profits'; manipulated humans are supposed to be 'more beautiful', 'perfect', 'super-intelligent', 'geniuses' or 'supermen'. In this way, certain desirable attributes of living beings are magnified. Where tomatoes are just 'redder' and apples 'greener', the modification is perceived in a way analogous to the fairytale notion of the most tempting apple, for example, containing the poison. The manipulated food is depicted as aesthetically pleasing but artificial, unhealthy or toxic ('radiated food', 'chemical food', 'messing with hormones of animals'). Even where there is no evidence of the modification of food, respondents consider it possible that the food could have been insidiously altered; and concerns are expressed that it is impossible to determine whether or not a product is genetically manipulated ('gene food sneaking onto the shelves'). The tempting, desirable results of biotechnology bring with them invisible dangers, and the 'natural-looking unnatural' is viewed with either suspicion or ambivalence.

Where the results of manipulation are easy to perceive, i.e. where the results are bizarre or defective, respondents think in terms of the unintended and incalculable side-effects of biotechnology. The danger of inadvertently creating organisms (plants, animals and humans) that are degenerate, bizarre, weak or ill ('monster', 'Frankenstein', 'mad cows', 'monster bull', 'handicapped children') rises to the fore. This is also the case for applications of biotechnology in which normal traits are exaggerated and the attributes of different kinds of organism are combined ('mouse with human ear', 'pig with a cow's head', 'tomato with banana taste').

Biotechnology, and especially cloning, are viewed as enabling the 'production' of life by humans. Images such as 'artificial life', 'living machines', 'human robots', 'producing life' and animals as 'spare parts bin' all depict 'unnatural life', that is, a form of life that is simultaneously real and fake. Any modification of the nature of life raises a paradox, since the very essence of natural life is that it cannot be 'made'. The images employed express a feeling of unease, embodying the notion that, although it might appear to constitute life, life manufactured by humans cannot be real.

Conclusion

The philosopher Peter Sloterdijk (1987) views the modern world since Copernicus as characterised by not only a cosmological but also an epistemic shock. He argues that it is clear that what we see and what comprises reality can no longer be considered synonymous. The dilemma is that, when we believe we are seeing the sun rising, the sun is not actually rising, and the earth can no longer be viewed as the centre of the universe. The modern corollary of this is that it is necessary to suppose a world that is not as we perceive it; there is a disjunction between the world and our perceptions of it. More and more, our knowledge depends on information produced using technical apparatus and not on our own perceptions. Nowadays, we live in a world full of invisible radiation, ozone and toxic chemicals. This is our 'real' life. Equally there is no doubt in the public mind that genetic manipulation exists and that it alters the objects to which it is applied. The same certainty is conferred on this notion as was enjoyed by witchcraft in former times; in both cases, the arguments are scientifically justified (Luhmann, 1991). In this sense, biotechnology is just one of many scientific or technological facts that cannot be perceived directly but that must be dealt with.

Since Copernicus, modern life has been defined by scientific and industrial change capable of profoundly altering the world, as well as by the emergence of world-views in which it is impossible to take anything for granted (Sloterdijk, 1987: 63). The Copernican abolition of obviousness in modern life has provoked counter-reactions that seek to restore a sense of normality to the world. Thus, common sense is concerned with making available, not 'models of reality' but rather 'models for reality' (Geertz, 1973: 93). The invisibility of genetic manipulations that might, nonetheless, bring about massive change is, therefore, countered with attempts to impose upon it rigid classifications that enable us to come to terms cognitively with this potent new technology. Visualising biotechnology with the help of images brings about a socially shared truth in which it is depicted in a way that helps to resolve the ambiguities surrounding the new technology.

This understanding involves selecting descriptive categories. The central one, in the present context, is that of Nature and its meanings. However, 'Nature' is not a clear-cut but a rather ambiguous notion, employed by both supporters and opponents of biotechnology. As has been seen, according to the optimistic view Nature is a legitimate object for investigation that will gradually reveal all of its secrets to scientific researchers. A completely different understanding of Nature draws on the distinction between the natural and the unnatural, or between the natural and the artificial. This opposition underlies many of the images described in the context of moral evaluations and criticisms of biotechnology in which biotechnology is referred to as being in contradiction to Nature. In this case, 'Nature' provides the anchor for a diaphoric understanding of biotechnology. Contrary to the common uses of metaphors, the basic concept of Nature is not equated to the new phenomenon; rather, the phenomenon is likened to the source domain's – i.e. Nature's – negation.

The dominant images of biotechnology represent attempts to construct a clear distinction between the natural and the unnatural. But this drawing of boundaries goes further, and also provides a basis for the separation of the good from the bad. In this modern world, in which people know that 'the natural' (in the sense of being 'the original') can no longer exist, identifying the 'natural' with the 'good' provides a reassuringly solid foundation for morality. The normal (i.e. that which is taken for granted) is equated with the normative. Thus, because it is usual for humans to originate by way of sexual reproduction, this is seen as the way things must be. This restores stability to the world and its 'natural kinds'.

Although people in the six countries studied do not tend to use the term 'unnatural' when talking about medical applications, they do use it when discussing reproductive technologies. This is because the latter seem to challenge a form of procreation that is taken for granted. Being the result of the procreative act, and being oneself able to reproduce, have come to be part of the definition of being human. In a similar way, humans' bodily integrity is considered to be challenged by the transplantation of organs from other species. In both respects, biotechnology is seen to be capable of subverting both gender and generation relationships and the idea of the limited lifespan. Personal identity and the sense of uniqueness are endangered, and life is demystified and replaced by 'artefacts'. As with any artefact, there is a determined outcome; there is no variation and chance ceases to be an important factor. Thus, with biotechnology, science seems to be on the verge of abolishing the last 'miracles' of life. People fear a cold, sterile future in which everything that exists does so according to preordained but worldly plans, a life that is fake without fate. To repeat, the distinction between the natural and the unnatural is made synonymous with the divide between the good and the bad.

Potentially having the power to alter basic aspects of the material and social world also implies that we will be forced into making decisions (Beck, 1986). After all, which are really more desirable, blue or green eyes, fair or dark hair? We will need to establish the criteria upon which to base our decisions. However, being able to decide also implies a responsibility for that decision, ushering in new forms of guilt. In this future scenario, being ill or having some physical imperfection would no longer be conceptualised in terms of a destiny that must be accepted, but would be seen as an error of judgement.

Further, how can we be responsible for something whose consequences we do not know? Scientists, industrialists and politicians are not considered to be capable of bearing these responsibilities. Instead, they are perceived as 'playing God', striving to create a new world out of genes, the building blocks of life. The public just wonders what sort of world this 'second Genesis'[8] will bring about.

Methodological appendix

Method

ALCESTE investigates the distribution of vocabulary in data material consisting of text (in our case, answers to an open question and interview transcriptions). The method was introduced by Max Reinert (1983) and produces a descending hierarchical classification. That is, ALCESTE separates classes of specific vocabulary that empirically co-occur without regarding the meaning of the words. In a second step these semantic word classes must be interpreted. To be able to trace the vocabulary back to its original context, ALCESTE provides a list of statements associated with each class. This helps to check if the interpretation of the single words is correct.

ALCESTE is a French program that can be used for analysing data in any language using Latin letters. To yield more precise results it is advantageous to exclude words such as articles, prepositions, pronouns and conjunctions from analyses, as well as to reduce plurals and conjugations to the word's root form. For that purpose ALCESTE needs dictionaries of these suffixes and function words. Since dictionaries are available for only a few languages, the working files of ALCESTE were modified by us to exclude a number of words (articles, prepositions and the like) for all languages required.

In order to obtain reliable results, ALCESTE computes two analyses using statements of slightly differing length. Only those statements that can be classified in a stable manner in both analyses are used further on. A stability coefficient indicates the percentage of statements that can be classified in a stable manner.

Data sets

The results presented in the chapter are based on two separate sets of textual data for six European countries. These are single and focus group interviews, on the one hand, and responses to a survey question, on the other. The two data sets were collected as part of a European Concerted Action research project, 'Biotechnology and the European Public'.

The Eurobarometer survey 46.1, which was conducted in each European Union country in 1996, was based on multi-stage random sampling methodology. It provides a representative sample of individuals aged 15 and over. In a face-to-face interview, respondents were asked the following question: 'Now I would like to ask you what comes to your mind when you think about modern biotechnology in a broad sense, that is including genetic engineering.'

Interviews with individuals were conducted in Austria and France (data from the latter country are integrated only to a small extent in this chapter); focus group interviews took place in Austria, the UK and Sweden. It was intended not to obtain a representative sample but to cover a wide range of perspectives on the topic of biotechnology. For that reason, persons associated with different socio-demographic categories were recruited. The interview questions covered the topics of medical applications, agricultural applications, genetically manipulated food, transgenic animals, reproductive genetic technology, control, risks

Table 9.2 *Main characteristics of data sets and results of analyses*

	France	UK	Sweden	Germany	Austria	Norway
Total number of responses (open question)	1,004	1,074	1,008	1,990	1,009	966
Number of analysed responses to open question (statements indicating 'I don't know' excluded)	812	973	836	1,990	814	730
Percentage of stable classified responses: open question	83	76	77	76	71	70
Number of classes	10	6	7	6	4	7
Number of interviews	20	–	–	–	18	–
Number of focus group interviews	–	5	2	–	7	–
Percentage of stable classified responses: interviews and focus groups	–	83	–	–	75	–

and consequences of biotechnological applications. Data from Austria and the UK were analysed by using ALCESTE, whereas data from Sweden were analysed according to the grounded theory approach (Glaser and Strauss, 1967).

Table 9.2 shows the number of respondents to the open question for each country, the number of responses analysed (for most countries, answers indicating 'I don't know' were excluded from the ALCESTE analysis) and the number of responses classified in a stable manner in the ALCESTE analysis. The stability of the results – ranging from 70 per cent to 83 per cent stable classified statements – can be considered as satisfactory for all countries. Table 9.2 furthermore indicates the number of interviews and focus groups conducted in each country, as well as the stability coefficient for the ALCESTE analyses of the British and Austrian interview data.

NOTES

1. Note that France was not included in the analysis of categorisations owing to the unavailability of data, but will be included in the following sections. Likewise, the Netherlands, Finland and Greece were included in the categorisation analysis, but will not be considered in the following section owing to missing content analyses.

2. One has to remember that we are dealing with different European languages here; differing vocabulary and language use can have an impact on the construction of the word classes.
3. Note that the survey took place before intensive media coverage about cloning Dolly the sheep started in 1997.
4. Throughout the text, literal quotations of responses to the open question are marked by the country code and the respondent's number. Statements from interviews and focus groups are marked accordingly. The codes are: AT = Austria, FR = France, GE = Germany, NW = Norway, SE = Sweden and UK = United Kingdom.
5. These cattle originated in central and upper Belgium and the breed was established in the early twentieth century. There is a large proportion of muscle hypertrophy in Belgian Blue, which is genetically inherited; this means that the animals are born with double thigh muscles, which give more meat and better productivity. Muscle hypertrophy also makes the internal organs of the animal smaller and caesarean delivery is frequently called for.
6. Even though the mouse bearing the human ear was not a product of xenotransplantation.
7. It is interesting to note that many of the images described in this chapter also emerge as part of the media discourse about Dolly the cloned sheep. Since the survey and interview data were collected before coverage about Dolly started, it can be supposed that media and public discourses are mutually determined. Media discourses equally reflect and influence public opinions (Wagner and Kronberger, 2001).
8. This was the title of an Austrian radio transmission about biotechnology.

REFERENCES

Bauer, M.W. and G. Gaskell (1999) 'Towards a Paradigm for Research on Social Representations', *Journal for the Theory of Social Behaviour* 29: 163–86.

Beck, U. (1986) *Risikogesellschaft. Auf dem Weg in eine andere Moderne*, Frankfurt am Main: Suhrkamp.

Geertz, C. (1973) *The Interpretation of Cultures*, New York: Basic Books.

Glaser, B.G. and A.L. Strauss (1967) *The Discovery of Grounded Theory. Strategies for Qualitative Research*, New York: Aldine.

Herzlich, C. (1973) *Health and Illness: A Social Psychological Analysis*, London: Academic Press.

Jodelet, D. (1989) 'Représentations sociales: un domaine en expansion', in D. Jodelet (ed.), *Les Représentations sociales*, Paris: Presses Universitaires de France.

Johnson, M. (1987) *The Body in the Mind. The Bodily Basis of Meaning, Imagination and Reason*, Chicago: University of Chicago Press.

Kronberger, N. and W. Wagner (2000) 'The Statistical Analysis of Text and Open-Ended Responses', in G. Gaskell and M. Bauer (eds.), *Methods for Qualitative Analysis*, London: Sage.

Lakoff, G. and M. Johnson (1980) *Metaphors We Live By*, Chicago: University of Chicago Press.

Luhmann, N. (1991) *Soziologie des Risikos*, Berlin: de Gruyter.
Moscovici, S. (1984) 'The Phenomenon of Social Representations', in R.M. Farr and S. Moscovici (eds.), *Social Representations*, Cambridge: Cambridge University Press, pp. 3–69.
Reinert, M. (1983) 'Une méthode de classification descendante hiérarchique: application à l'analyse lexicale par contexte', *Les Cahiers de l'Analyse des Données* 8(2): 187–98.
(1990) 'ALCESTE. Une méthodologie d'analyse des données textuelles et une application: Aurélia de Gérard de Nerval', *Bulletin de méthodologie sociologique* 26: 24–54.
(1998) 'Manuel du logiciel ALCESTE (Version 3.2) [computer program]', Toulouse: IMAGE (CNRS-UMR 5610).
Sloterdijk, P. (1987) *Kopernikanische Mobilmachung und ptolemäische Abrüstung*, Frankfurt am Main: Suhrkamp.
Wagner, W. (1998) 'Social Representations and Beyond: Brute Facts, Symbolic Coping and Domesticated Worlds', *Culture and Psychology* 4(3): 297–329.
Wagner, W. and N. Kronberger (2001) 'Killer Tomatoes! Collective Symbolic Coping with Biotechnology', in K. Deaux and G. Philogene (eds.), *Social Representation: Introduction and Exploration*, Oxford: Blackwell.
Wagner, W., F. Elejabarrieta and I. Lahnsteiner (1995) 'How the Sperm Dominates the Ovum – Objectification by Metaphor in the Social Representation of Conception', *European Journal of Social Psychology* 25: 671–88.
Wagner, W., G. Duveen, R. Farr, S. Jovchelovitch, F. Lorenzi-Cioldi, I. Marková and D. Rose (1999) 'Theory and Method of Social Representations', *Asian Journal of Social Psychology* 2: 95–125.

Part III

The watershed years 1996/97: two case studies

10 Testing times – the reception of Roundup Ready soya in Europe

Jesper Lassen, Agnes Allansdottir, Miltos Liakopoulos, Arne Thing Mortensen and Anna Olofsson

In the autumn of 1996, ships carrying the annual harvest of soybeans for the European Union (EU) market were sailing from US harbours. This soya was intended partly for the European food industry, which uses soya as a raw material in the production of additives, food products and livestock feed. But, for the first time, the ships' holds contained more than traditionally bred soya. These shipments were mixtures in which up to 2 per cent comprised a genetically manipulated strain of soya. The new strain was called 'Roundup Ready'.

The soybean was the first genetically manipulated food to be marketed on a large scale on the European market. It was the culmination not only of many years of biotechnological research, but of the improving efforts of the agricultural industry as a whole, and of Monsanto (the US company responsible) in particular. The novelty of 'Roundup Ready' lay in the genetic modifications that rendered it resistant to Glyphosate, the active ingredient in 'Round-up', a herbicide produced by Monsanto and one of the most widely used agricultural chemicals in the world.

Thus, apart from being the first genetically manipulated product intended for consumption by the European market, the soybeans represented a bio-agricultural strategy with considerable economic and practical implications. This was also a technology that had already been heavily criticised for its potential environmental and social consequences (see, for example, Kloppenburg, 1988; Goldburg et al., 1990; Busch et al., 1991). As a result, the ships approaching Europe were not only transporting the first genetically altered food, they were also carrying with them the potential for serious conflicts – controversies that would not be confined to environmental, social or food-safety issues. As the controversy over 'Roundup Ready' soya escalated, critics began to argue that businesses such as Monsanto are jeopardising democratic processes. In this context, Monsanto's Roundup Ready soybean emerged as the European test-case for the food-gene technology project, and became the focus for more than a decade of public debates about the pros and cons of genetic manipulation in food and agriculture.

The introduction of a product such as Round-up Ready soya (or any other genetically manipulated food product) into the European Union was formally regulated by the Directive on Deliberate Release enforced in 1991 (EEC, 1990). This directive required manipulated organisms intended for release in the environment to acquire official approval and undergo risk assessment studies prior to import, marketing or release. Accordingly, Monsanto submitted an application to have Roundup Ready sanctioned by the UK in the winter of 1994–5. Approval from the British Advisory Committee on Novel Foods and Processes was obtained for the use of modified soya in food in February 1995 and as livestock feed in May 1995. Thereupon, in early spring 1996, an application for approval was sent to the European Commission. Subsequently the approval was considered by the Commission, submitted to hearing among the member states, and finally passed by the Commission in April 1996. Yet, despite this approval, no further progress was made with Roundup Ready soya. This was partly because the question of labelling was left open for interpretation by the national authorities, and partly because the member states were still able to demand that approval of food products containing genetically modified organisms (GMOs) could be dealt with at a national level. This situation was the result of an ongoing debate among the member states and the institutions in the EU on the so-called Novel Food Regulation – a set of rules intended to regulate novelties in the food area, including products containing or derived from GMOs.

The Commission had initiated work on novel food regulation in 1992, but resolutions were still pending in October 1996 when the soya controversy was heating up. This delay was due to disagreements among the member states in the Council, as well as among such EU institutions as the European Parliament, the Commission and the Economic and Social Committee (Behrens et al., 1997: 105ff.). Conflict centred on two fundamental issues. First, and most importantly, the question of labelling had been left open for interpretation by national authorities in the Directive on Deliberate Release. Secondly, debate focused on whether regulation should be based on an additional notification and/or approval procedure. Clearly, until the final version of the Novel Food Regulation was adopted by the Commission (on 27 January 1997), there was no clear perception of what a common policy on the marketing of genetically manipulated food products would involve (EC, 1997). Four months later, however, this new directive was put into force. Crucially, as a consequence, the door was left open for the formulation of national policies concerning the marketing of the products based on 'Roundup Ready' soya.

Seen at a societal level, it is interesting to observe how differently the soya story developed in the various EU countries during the four-month gap before the Novel Food Regulation was decided upon. At one end of the spectrum, the soya issue became a 'hot' issue in countries such as Denmark and Austria, where both public attention and political awareness were intense. At the other end of the spectrum were countries such as Greece and Italy, in which neither the public nor the political system paid the issue more than the most perfunctory attention during these months. In between these extremes were countries such as Sweden and the UK, with a moderate degree of attention directed at the soya issue.

The prospect of the introduction of modified soya struck all EU countries simultaneously, yet it elicited widely different responses in terms of both public attention and political action. This case therefore offers a unique opportunity to study the development of a controversy and political reactions to it. The two main questions this raises are, first, why did Roundup Ready spark controversy in some countries while being almost totally neglected in others; secondly, how we can explain the differences in how the political outcomes were shaped in those countries where soya *did* become an issue?

This chapter will offer some answers to these questions through an analysis of the soya issue in Denmark, Sweden, Italy and the UK. We do not claim that these four countries provide a full picture of the reception of the genetically manipulated soya in the EU. Rather, we contend that the fate of a social issue and the political reactions elicited can be understood only if we proceed from an examination of different national contexts. Explanations *must* focus upon the actors engaged in the conflict and the different material, cultural and political structures and traditions within which the potential issue evolves. These four countries have been selected in order to illustrate the variability in reception across the EU. Each country represents and embodies a different political culture; different agricultural interests in the biotechnology issue; and different public attitudes towards biotechnology in general, and food biotechnology in particular.[1]

The analytical background: the development and political impact of social issues

There is a long tradition of research into the question of how social issues develop and then disappear. It has been suggested that the evolution of social conflicts follows a systematic 'issue attention cycle' in which the social problem lasts for a relatively limited period of time, during which it passes through different stages until it eventually disappears from public view (for example, Downs, 1972). A social problem fading from view, it is claimed, will either leave a 'trace' in the shape of political action or simply melt back into oblivion. Such 'natural history' models are, however, open to criticism: first, because they lack the ability to explain the dynamics that guide the selection of potential issues and those that gain public attention and thereby become true public issues see for example, Hilgartner and Bosk, 1988); and, secondly, because their stipulation that social issues last for a relatively short period of time is arguably erroneous (Jasper, 1988).

Consequently, in attempting to explain the varying degrees of attention focused on the soya issue in different countries, this chapter will not proceed along the route preferred by 'natural historians', but will instead build on research into the development of social issues that stresses the importance of context and of actors asserting different positions (for example, Hilgartner and Bosk, 1998; Jasper, 1988). Following this line of research in explicating the development of social conflicts, the following core issues must be addressed: the discursive environment, the material setting, the political culture and the actors and their strength.

The discursive environment Here the question concerns how the new social issue relates to present fields of discourse. Those that are consistent with existing discourses are likely to benefit from this relationship and prosper at the expense of other issues. Thus, the general level and direction of the science and technology discourse will be of importance with respect to the controversy over Roundup Ready soya, as will be the more specific characteristics of the biotechnology debate. Closely related to this is the question of how the issue is defined or interpreted. Some interpretations will relate to and draw nourishment from existing discourses within the science and technology area, or even wider discourses.

The material setting This concerns the relationship between particular issues and wider economic structures. On the one hand, the general standard of living and welfare provision will be of importance in determining the amount of attention assigned to an issue, considered as a simple marginal utility relation. This view has been put forward as one explanation of different attitudes towards biotechnology in general in the EU (Nielsen, n.d.). On the other hand, the relative strengths of different national business sectors will affect the uptake of, and responses to, new social issues. In the present case, countries with powerful food and agricultural sectors are likely to produce different conditions for the emergence of the soya issue than countries with no such industry.

The political culture The development of an issue, once it has emerged as a prominent public debate, will be profoundly affected by a country's participatory tradition, as well as by its openness and, therefore, the public's access to information. Issues developing within a political culture with participatory traditions, where the views of different actors are incorporated in the policy process, are thus more likely to reflect these different views than are issues developing within countries with a more technocratic culture.

The actors and their strength Issues do not enter the public arena by themselves; they require 'social carriers' both to introduce them and to maintain their position on the public agenda. Issues will also encounter opposing actors, groups or individuals who will try to interpret them differently or even to remove them (Lukes, 1974; Bachrach and Baratz, 1962). Thus, the life of an issue depends not only on the resources (economic, knowledge, organisational, etc.) of the actors involved but also on their relations to the media and the political system. Closely related to the question of the actors is the question of 'problem definition'. Actors may use their resources to influence the way in which a certain issue is debated and thereby bring it into line with dominant discourses that either promote or reduce the longevity of the issue. Two of the opposing strategies within biotechnology have been to relate the industry to a positive image or a discourse of welfare, which includes industrial progress and prosperity, or conversely to implicate biotechnology in the discourses of social, environmental and health risks. These framing processes impact not only upon the ability of an issue to reach, and remain upon, the public agenda, but they may also influence the political reactions the issue evokes.

Of course, an analysis of the dynamics affecting the public profile achieved by the soya issue in different countries must take into account the political process targeting 'Roundup Ready' soya. On the one hand, actors may attempt to steer the political process as a means of achieving a congenial political outcome. Therefore the established political system (the government, the parliament and the bureaucracy) may be the main arena for the development of the issue, both because actors in the early stages of the conflict may want to persuade the political system to take up the issue and because, in the more mature stages, they may lobby for particular decisions to be made. On the other hand, the political system can become a dominant player in the development of the issue. This may happen either because outside pressure shifts the issue from the public into the political sphere, or because the government takes up the issue on its own initiative.

In explaining the outcome of a conflict, the first question to address, therefore, is what the political outcome was. The outcome is important because it represents the formal trace of the conflict: once the contest is over, its effects – new or altered regulation and new procedures (institutional learning) – will remain. A conflict may also leave a trace in the form of changes in everyday language, as well as by becoming enshrined in the stories or myths of the public and political spheres. In the long term, such traces will contribute to the construction of a shared frame of reference within the nation, and will thereby influence the trajectory of new controversies and political debates.

In this chapter, attention will largely be focused upon the formal traces of the 'Roundup Ready' soya issue; and particular attention will be devoted to the wording of the laws or political resolutions passed, and an investigation of whose interests these decisions advanced. This will involve, in accordance with the pluralists' tradition of political analysis, a study of the visible conflict and an identification of its winners (for example, Dahl, 1986). In addition, we will attempt to identify the processes involved in producing this actual outcome. For this reason, the analysis will also encompass an examination of the ways power was exerted in shaping the political result of the conflict. Here we make use of the concept of non-decision-making (Bachrach and Baratz, 1962, 1963), or how the interests of certain groups may be excluded by powerful actors controlling social values, political values and institutional practices in order to secure policies that minimise harm to their own interests. At the same time, we recognise that the outcome is also shaped by the more subtle and covert use of power, as manipulation or the exercise of authority, involving the deliberate shaping of the counterpart's wishes and interests (Lukes, 1974: 23ff.).

The media play a key role in the processes described by Lukes, Bachrach and Baratz, since they constitute an important stage not only in the marketing of counter-arguments but also in the attempts to control values, manipulate public opinion and demonstrate authority. Taking this a step further, we will analyse the relationships between the formulation of public policy and the activities of the media. In doing so, we accept in part the social constructionist view that a problem does not exist until it has been socially defined. This raises the central question of who determines how a problem is defined. As Rochefort and Cobb (1994) note, the definition and redefinition of problems serve as tools employed by opposing parties in order to gain advantage in the political conflict. By way of example, they claim that some actors may be inclined to define issues in narrow and technical terms since this might serve to confine the debate to just a few parties. Conversely, others may prefer a wider definition of the problem – as a matter of democracy or liberty – with the effect of involving more, potentially supportive, groups of actors.

During this process of definition, the media play a central role. This is partly because they provide important arenas for the development of the issue itself, but also because their control of these arenas enables them to influence the debate as, in effect, indirect actors. Problem definition invariably becomes a key issue for those social actors seeking to exert influence on the policy process. After all, certain definitions of the problem will significantly reduce the range of political actions considered appropriate; in this way, controlling the process of definition confers considerable influence over the political reactions to the problem. Pline (1991), for example, showed that proponents of biotechnology in the USA have successfully transformed the public image of their industry, generating a positive image that links biotechnology with economic progress. This has provided a secure platform for fighting off opponents on the political scene. Von Schomberg (1993) gives another illustration of this phenomenon in his analysis of political decision-making in relation to the release of genetically engineered organisms. He demonstrates that ecologists and biotechnologists are advancing two different definitions related to the issue of deliberate release. Ecologists maintain that 'release' is a new issue, and one that therefore demands that that new knowledge (i.e. test systems) is established before any major decisions are taken. Biotechnologists, in contrast, argue that there is nothing novel about genetically engineered organisms, that they are comparable to organisms produced by traditional methods, and that existing knowledge and methods of risk assessment should for this reason be deemed sufficient.

From the example put forward by von Schomberg, one would expect the biotechnologists to prefer issue definition to take place in a restricted technical arena. In contrast, ecologists will gain from moving the issue into the public arena via the mass media, partly because they could benefit from what Pline (1991) calls issue association (i.e. linking the issue to prominent discourses), and partly because in this way they might be able to draw other organised interests, or even public opinion, onto their side. This emphasises the key role the mass media play in issue definition as part of the democratic process, and raises the further question of the extent to which the media themselves reflect public opinion. The growing influence of the media in the political process in general highlights the importance of access to the media for those desirous of securing a particular political outcome. Moreover, as the media tend to become more concentrated, ordinary citizens are less and less likely to gain access to media expressing their views. Of course, public opinion will, to some extent, affect the political process through the activities of different non-governmental organisations (NGOs) representing the different sides of the issue. Following Schattschneider (1975), however, this raises the question of the NGOs' legitimacy. It is arguable that the media do not represent the public because they instead tend to report the attitudes of large collective actors (Schattschneider, 1975). Following this argument, any assessment of the effectiveness with which public opinions influence the political process must determine the extent to which the NGOs themselves represent the public.

This chapter does not aim to provide complete answers to all these questions. Instead, we will focus on the role of the media in this debate between early autumn and the settlement of the Novel Food Regulation. More specifically, we will show how the arguments related to the soya issue put forward in the media can be interpreted as a process of issue definition. In doing so, the media will be shown to be a basic part of the political process.

The reception of Roundup Ready soya in Denmark

Aside from occasional references to GM soya as one of the most advanced applications of biotechnology, soya was not an issue in Denmark until Monsanto delivered its application for approval to EU member states in 1995. This application prompted two questions to be asked in the parliament from the left-wing opposition Socialist People's Party (Socialistisk Folkeparti or SF) during the spring of 1995, but the story never provoked more than a few media articles. At this stage, however, the soya question became part of an attempt by SF and Enhedslisten (EL), another left-wing opposition party, to attack Danish policy regarding the Novel Food Regulation. They ensured that the government endorsed the unanimous parliamentary decision of 1994 (Folketinget, 1994) that, among other things, stated that it was Danish policy to demand, in both Denmark and the EU, the labelling of food products containing and/or produced with the aid of GMOs.

When Monsanto's application was sent for its hearing among the member states in the spring of 1996, for the first time soya became an issue – albeit a minor one – in the media as well as in the parliament. And the story was conjoined with that of Ciba Geigy's manipulated maize, for which Novel Food Regulation was also pending. Again the debate was led by EL and SF, and again the main concerns were to ensure that all soya products (as well as all other products based on GMOs) are labelled according to the parliamentary decision of 1994, and that organic products are kept free of GM organisms at all stages of production. These debates were also represented in the media during the spring and summer of 1996. Actors presented the pros and cons, and the possibility of banning genetically manipulated soya was also raised by representatives from EL and Greenpeace.

On 14 March 1996, the Danish government voted against Monsanto's EU application. Its decision was largely prompted by the absence of labelling requirements, but, in addition, the government drew attention to the fact that Danish regulations require that the Minister of Health issue an approval before such products can be released in Denmark. Monsanto applied for Roundup Ready soya to be approved by Denmark in May 1996, and permission was granted on 18 November, following the Levnedsmiddelstyrelsen's (the National Food Agency) insistence that the manipulated soybean should not be considered to be different from the traditional beans. However, the conditions of approval stated that imported beans must be accompanied by information about the content/or possible content of GM soybeans, so as to permit labelling 'in a suitable way' (Levnedsmiddelstyrelsen, 1996a).

The soya issue in Denmark: November 1996–January 1997

The Danish media closely followed the soya issue in the period analysed.[2] The media focused on basic arguments and events that promised to make good stories, and succeeded in making Roundup Ready soya one of the most prominent social issues of autumn 1996. In the first half of December, the newspaper *Politiken* alone carried seventeen longer articles, fourteen letters from readers and two leading articles dealing with Roundup Ready soya. Moreover, the issue was addressed in ten questions and one enquiry in parliament between mid-November and mid-December.

The controversy in Denmark really began, then, in mid-November, when it became known to the media that cargo ships were crossing the Atlantic carrying modified soybeans. They learned that one of them, *Hanjin Tampa*, was destined for Aarhus, Denmark. This ship was transporting 23,000 tons of soybeans to the Danish company Central Soya, a subdivision of the French company Eridania Beghin-Say, owned by the Italian Montedison. Crucially, 0.5 per cent of this cargo was derived from GM plants. It had been known since the spring of 1996 that genetically manipulated US soybeans were likely to reach European – including Danish – markets during the autumn. Yet these approaching ships were to assume immense symbolic significance. And it was this event, above all, that sparked the major debates in the media as well as in parliament. The parliamentary soybean debate ceased only after the Minister of Health issued guidelines for labelling GM soybeans on 6 December; and a parliamentary decision on GMOs and organic products was reached on

12 December. Nonetheless, the soya issue continued to appear in the media until the spring of 1997.

By mid-November, the ships were nearing European ports and GM soya was emerging as a major media and policy issue. But though GMOs had been approved in Denmark, the necessary guidelines had not been put into place. In effect, therefore, the Minister of Health, Yvonne Herløv Andersen, had been ignoring the 30 October recommendations made by her own administration (Levnedsmiddelstyrelsen, 1996b), as well as the parliamentary decisions of 1994, which had demanded the detailed labelling of GM products.

To resolve this uncertainty, the minister was called in for consultation by the Parliamentary Committee on Health on 28 November. At this consultation, she expressed her view that the labelling requirements stipulated by parliament in 1994 had been met sufficiently by the display of signs on shop shelves. She repeated this statement during a debate in parliament on 4 December (Folketinget, 1996), which resulted in both the left and the right insisting that the 1994 decision was quite unequivocal in its demand for each product to be labelled. Alongside this debate, the left-wing opposition repeatedly raised broader issues concerning the relative merits of GM organisms and organic products.

At about the same time, towards the end of November 1996, the EU parliamentary working group on the Novel Food Regulation met with the fifteen national Ministers of Health and settled upon a preliminary agreement on the labelling matter. In the process, the official Danish policy on labelling suffered a defeat. The preliminary agreement still required the approval of the EU Parliament and the Council, but demanded labelling only of products that contain live GMOs. Where appropriate, the agreement also permitted the use of labels informing consumers that a product contains no ingredients produced by gene technology. From this moment it became clear that, within a few months, Danish policy would be replaced by a rather more relaxed set of EU regulations.

Despite this development in Brussels, the Danish Social Democrats (the largest of the three parties in the minority government) proposed the adoption of a policy analogous to that agreed in the Danish Parliament of 1994. This would have made mandatory the individual labelling of all products, or ingredients thereof, containing GM organisms. This move, combined with opposition pressure on 4 December, caused the Minister of Health (from the small centrist coalition party CD) to make a U-turn when issuing labelling guidelines on 6 December. The guidelines now stated that 'Food and food ingredients produced on the basis of genetically manipulated soybeans, intended for sale to consumers, must be labelled accordingly' (Levnedsmiddelstyrelsen, 1996c). They also specified that labels must appear on the product, and further stipulated that additives, such as lecithin, are exempt from this labelling requirement. Finally, the guidelines allowed for the positive labelling of products produced without manipulated soybeans.

In formulating these guidelines, the policy-makers first produced regulations. However, subsequent controversy, and particularly the lengthy debate in parliament on 12 December, showed that there were outstanding questions demanding resolution. One question involved keeping organic products free from GM soybeans, or, for that matter, any other GMOs. Another issue concerned the likelihood of imported soya entering the food system *via* livestock feed and meat consumption: even though the greater part of soya imports were used in animal feed, the implications of this fact were not covered by the guidelines.

Meanwhile, *Hanjin Tampa* arrived in Aarhus on 11 December to be met by thirty to one hundred activists representing the newly founded organisations Økofolk Imod Gensplejsning (Ecopeople against genetic engineering) and Danmarks Aktive Forbrugere (Danish active consumers), who successfully prevented the unloading of the cargo.

The 12 December parliamentary enquiry had been scheduled by Enhedslisten on 15 November, before the soya issue had escalated. It was due to address general issues related to GMOs and to discuss the possibility of an outright ban. However, in the event, soya became one of the chief bones of contention, and the result of the enquiry was a rejection of the Enhedslisten agenda. In its place there emerged an alternative agenda, championed by the government parties and the Socialist People's Party, which called for the adoption of rules to ensure the availability of organic products uncontaminated by GM organisms at every stage of the production chain. In this way, two of the main themes of the debate were settled, while the use of GM organisms in livestock feed was unaffected and the question of a ban was rejected by most parties. The day after the enquiry there was a meeting of the parliamentary EU Committee, where all except the two socialist parties (SF and Enhedslisten) granted the government a mandate to vote for the proposed version of the Novel Food Regulation. On the morning of the same day, police moved in and broke up the blockade at Aarhus harbour, the off-loading of the cargo began and soya ceased to figure on the public agenda as a single issue.

Outside parliament, the demand for a ban was supported by a number of NGOs. After Greenpeace International had taken up biotechnology as an issue, Greenpeace Denmark became one of the most prominent NGO participants. Greenpeace's resistance is based partly on environmental considerations, and partly on health grounds. The environmental argument for securing a ban focuses on the increased use of pesticides; and their contentions were strengthened after Round-up breakdown products were found in UK, German and Dutch groundwater. On the basis of this argument, Greenpeace rejected and criticised the proposed introduction of labelling. Similar arguments were put forward by most NGOs. The only significant exception was the rather more positive National Consumers' Association, which supported the labelling solution.

The dynamics of the soya controversy in Denmark

The dramatic and newsworthy theme of a ship, laden with a highly controversial cargo, approaching Denmark was noted by all newspapers. But this never constituted the main story. At the risk of oversimplification, we argue that the main issue considered in newspaper stories was how the Danish labelling decision could possibly be upheld in the face of both direct attack from Monsanto and the introduction of EU regulations concerning novel foods. The dramatic tension in the stories was related to themes or discourses that aroused the readers' interest, such as decreasing food quality, national versus European regulation, assaults on democratic rights or the failure of politicians to solve the nation's basic problems.

There can be little doubt that the intensity of this media coverage can in part be attributed to the fact that the soya issue was framed as a question of labelling. It therefore tapped into the discourses about food quality that were so prominent during the first half of the 1990s. This food discourse arose from the prevalent tendency to perceive food quality materially as a matter of taste, safety, texture, and so on, but also as including the ethical, environmental and social aspects of the way food is produced (Lassen, 1993). In the 1990s, this was seen in the near-explosive growth in the purchase of organic products and, after the controversy over Brent Spar and the French nuclear experiments, with the emergence of a 'political consumer', prepared to express his or her politics through purchasing decisions. These historical circumstances both serve as a background for understanding why soya became a contentious issue, and also help explain why labelling per se became the dominant way in which the issue was framed both in the media and in the parliamentary debates. The ability of the consumer to identify the right commodity depends on information about the product itself. Therefore labelling is important and becomes a key issue for the political consumer.

Opposition to the unregulated introduction of GM soya also drew support from an environmentalist sub-discourse that had been prominent in Denmark for more than a decade. This sub-discourse related to concerns over the quality of groundwater. This was an especially sensitive issue in Denmark because the country is highly dependent upon this source of water, and until recently it had had a reputation for being clean and pollution free.

Behind the media stories, though not explicitly elaborated, lay a further theme: the discrepancy between the moderately positive attitudes to biotechnology held by most politicians and opinion-leading media personalities, and the generally negative attitudes of – perhaps – a majority of the population. An opinion poll initiated by *Berlingske Tidende*, one of the three leading national newspapers, confirmed the existence of this disparity: 68 per cent of respondents expressed the belief that GM food products should be prohibited by law (Gallup, 1996). This figure was not stressed by the newspapers (probably because it was considered unreliable) and was not taken up by the rest of the media, although the majority of the population unequivocally wanted a ban. None of the papers supported an outright ban, but all lodged strong support for consumers' right to choose and to be informed about possible GM ingredients in food products. This right was generally accepted and not disputed, which further explains the importance attached to the issue of labelling. In all interviews, the importation of Monsanto's soybeans was seen as a provocation and the importer was, for obvious reasons, placed in a defensive position.

The prerogative of the EU in regulating national labelling policy was not disputed. As a result, deeply rooted conflicts over the role of the EU were not explicitly aired in the media coverage. This was partly because of a widespread belief that the coming 'novel food' directive would in practice be equivalent to existing Danish labelling regulations. Alternatively, it was held that, even if the directive did not reproduce Danish regulations regarding soybeans, a strict reading of the EU decision could be imposed that would not significantly alter the Danish regulations. Where the delusional nature of both positions was remarked upon, criticism was diverted into another popular theme: the shortcomings of Danish politics.

The media debate shows two peaks in the level of coverage: between 3 and 6 December and between 12 and 14 December. On the whole, the problems related to the Danish political scene were dominant in the media coverage of the soybean case. Moreover, compared with other countries, these problems were of a special kind and serious enough to account for their prominence in the Danish media.

In November, the Minister of Health had approved Monsanto's application to begin marketing Roundup Ready soybeans as food, a decision she could hardly have avoided considering the recommendations of Danish scientific authorities. However, this meant that she was obliged to issue guidelines explaining how the earlier parliamentary decision about labelling should be implemented. Monsanto's refusal to separate genetically manipulated soybeans from traditionally bred beans made that practically impossible without her bending the formulation of the parliamentary decision. And this is what she tried first. Doing so proved to be a political miscalculation, and she was forced to replace her first guideline with a stricter version that interpreted the earlier decision in a literal fashion. This happened in the period around 4 and 6 December and produced a storm of media interest, with the event being covered by most newspapers as well as the electronic media. One week later, the arrival of the soya ship and the beginning of the parliamentary debate requested by Enhedslisten one month earlier happened to coincide; the media coverage reached its peak during 13–14 December.

What can be seen as remarkable in the representation of these two events is the difference in focus between the media and parliament. The media did not pay much attention to the fact that the parliamentary decision about organic food labelling had been almost unanimously accepted and was considered by all, even by Enhedslisten (the left-wing party that had initiated the debate), to be a demonstration of constructive political resolution. In the media, the focus was still on the labelling problem, and addressed such questions as: how can the decision be implemented? how could it be made compatible with the coming EU regulation? how might it benefit consumers? Questions such as these emerged during the following weeks as attention shifted from Denmark to Brussels, and perhaps also when consumers began wondering why they never saw any labels in the shops.

One of the remarkable results of the controversy over Roundup Ready soya in Denmark was a restructuring of organised resistance to genetic engineering. Previously, in the 1980s, organised resistance had been virtually monopolised by NOAH, the Danish Friends of the Earth. Then, as biotechnology declined in prominence after 1990/1, NOAH had also disappeared, and organised resistance with it. Yet, during the winter of 1995/6, the soya conflict dramatically altered this state of affairs. NOAH reappeared as a critic of the new biotechnologies, although not at the same level as earlier and this time sharing the spotlight with several new NGOs and with some older organisations such as Greenpeace and the Danish Society for the Protection of Nature. Despite this diversification in organised resistance, late in 1997 when eighteen different NGOs, organised by NOAH, joined together to collect signatures against genetically manipulated food it became apparent that they had common interests. In the wake of this development, for the first time in Denmark biotechnology was challenged by activists prepared to take direct action, and they sought to prevent the unloading of soybeans in Aarhus harbour. Until this episode, organised resistance had 'played by the rules' and had mainly focused upon raising public awareness.

The reception of Round-up Ready soya in the UK

After the approval of Round-up Ready soya by the US Food and Drug Administration (FDA) in September 1994, Monsanto applied to the Ministry of Agriculture, Fisheries and Food (MAFF) in the UK. This was the first in a series of applications that soon included Canada, Mexico, Argentina, Japan, the Netherlands, Denmark and eventually the European Union. The UK had already been the first country to approve another GM food, a tomato paste, which had proved to be commercially successful.

In February and May of 1995, Roundup Ready soya obtained official approval for food and feed respectively. This approval was based on recommendations from the Advisory Committee on Novel Foods and Processes, an independent scientific committee established in 1988 with the mandate to advise the government on the appropriateness of new foods for human consumption. The committee recommended that the new soya be approved without caveats. The application then moved to the European Commission for approval and further clarification on aspects such as labelling.

During this time there was no official debate in Parliament on the issue, and there were only fifteen articles in the press referring to it. However, the scene changed dramatically when it became clear that Monsanto was shipping the new strain of soya to Europe mixed in with the ordinary crop. When this news broke, various NGOs in the UK were active in trying to stop the shipment, instigating a public debate that was taken up in the media and in Parliament.

The soya issue in the UK: September 1996–February 1997

Following news of the Monsanto shipment, the debate was lively in parts of the media. During the peak period of the debate, a total of one hundred articles were written in the British press concerning the soya issue. There is an obvious difference between the coverage of the tabloid and the broadsheet daily national press. Only the broadsheet or 'quality' press – the *Daily Telegraph*, the *Financial Times*, the *Guardian*, *The Independent*, and *The Times* – tackled the soya issue, providing 81 per cent of the total press coverage. The tabloid press had a minimal input into the debate, amounting to only 5 per cent of this total; the rest of the articles (14 per cent of the total) appeared in the international press and magazines. The press covered a wide spectrum of arguments for and against the introduction of GM soya. It also represented the voices of all the main actors in the debate: industry, NGOs (Greenpeace, the Green Alliance) and the scientists.

The industry viewed Round-up Ready soya primarily as a safe, economical product and an unmistakable step towards the elimination of world hunger. The official regulatory stance over Roundup Ready soya, as stated in the reasoning points on which the FDA had approved the soya, was used to warrant the claims over its safety. The assertion that GM soya is safe was justified on similar grounds to those invoked by the FDA, with particular emphasis being placed on the argument that there is no substantive difference between GM and ordinary strains. That GM foods had acquired a negative image was acknowledged, but this was attributed to ignorance about the true status of the technology and misinformation given to the public by interest groups.

The industry clearly saw winning consumer acceptance as its objective, and it clearly acknowledged the need for an information and image campaign. Indeed, following this line of thinking, the European biotechnology industry has recently launched just such a campaign.

Moreover, it is significant that, apparently taken by surprise by the public reaction to the Roundup Ready shipments, the industry's arguments and claims changed over time. The argumentation of the early days of the debate showed considerable confidence as it evolved around the safety issue of the Roundup Ready soya, with the sole backing coming from its acceptance by regulators around the world. Yet it soon became apparent that GM soya raised doubts beyond the specific regulatory issues, and could not be separated from debates over the acceptance of biotechnology as a whole. As the focus of the dispute shifted from the specific product to the technology in general, questions of morality entered the debate. At that stage, the industry changed its main line of argument: rather than emphasising safety issues relating to Roundup Ready soya, it began to herald biotechnology as the solution to global problems.

Scientists, another main actor, approached the debate from a technical point of view. Their arguments concentrated on specific technical aspects of the Roundup Ready soya case, including the regulatory safety check procedures and past genetic research. They also questioned the naturalness of genetic engineering technology, the integrity of the regulatory procedures for the acceptance of biotechnology products, and even the ethical credentials of the scientific research itself.

Furthermore, the scientists directly challenged the official line of the regulatory authorities. In particular, on the basis of a recent discovery that foreign genes can pass into human intestinal cells, they strongly criticised claims that Roundup Ready soya is fundamentally the same as ordinary soya and therefore unlikely to produce any negative side-effects upon human consumption.

On the whole, the scientific community adopted a critical stance on biotechnology. Of course, scientists were not unanimously sceptical of

GM soya; after all, the regulators who had approved the soya also based their decision on scientific arguments. But most of the scientific arguments to reach the media, and therefore that contributed to the public debate, were strongly critical.

NGOs (environmental groups and consumer associations) took a strong and unambiguous stance in this debate. They advanced arguments at three different levels. First, at the scientific level they recapitulated the arguments put forward by sceptical scientists, especially those who called into question the safety of GM foods for human consumption. Secondly, at the level of the status of biotechnology as a whole, they portrayed the technology as unnatural and invoked technical procedures such as those involving the transferring of genes from animals to plants in attempting to evince this. Thirdly they questioned the morality of the political decision-making process, especially with respect to the labelling issue (i.e. the idea that all foods containing genetically manipulated ingredients should be clearly labelled). Labelling was at the heart of the NGOs' position during the Roundup Ready soya controversy; and this echoed the will of the majority of the public (82 per cent), who wanted genetically manipulated foodstuffs to be labelled (European Commission, 1996).

Between September 1996 and February 1997, the soya issue was debated on ten occasions by Members of Parliament in the House of Commons. The debate was mainly between members of the Opposition (at that time the Labour Party) and the Minister of Agriculture, Fisheries and Food (of the then Conservative government), whose ministry was responsible for overseeing the introduction of new foods into the food chain. In Parliament, the Opposition's case was clearly based upon the environmental risks stressed in the media debate. Accordingly, it laid particular emphasis on long-term risk assessment and the question of labelling. The Opposition was especially mistrustful of existing risk assessment schemes, and used the analogy of BSE to claim that the consumption of GM foods might have unforeseen long-term effects on health. This was reinforced by the scientists' argument that the consumption of antibiotic markers in GM foods might lead to a resistance to antibiotics. Further, the Opposition claimed that foods containing Roundup Ready soya should be labelled in view of the public's mistrust of the industry and the technology itself. Overall, the Opposition assumed an unambiguous position towards agricultural applications of biotechnology. It also displayed considerable sensitivity to lay fears, openly acknowledging the arguments advanced by organised bodies, for example religious groups, retailers and consumer organisations.

Largely in response to the Opposition's attacks, the government itself also focused its attention on the issues of risk assessment and labelling.[3] The Conservative government had set up the existing mechanisms of risk assessment, and unsurprisingly, therefore, expressed considerable confidence in their efficacy. The government dealt with the issues of labelling and GM contamination with apparent resignation. Although expressing sympathy for the critics of existing labelling policy, the government claimed that it was unable to take any action because of European and international regulations. But it is also clear that the government was unwilling to inconvenience industry by introducing constraints on its activities. Aside from expressing the government's support for industry, those government spokesmen involved in the debate adopted the biotechnology industry's preferred argument: that by increasing crop yields GM foods would help alleviate world hunger.

The dynamics of the soya controversy in the UK

In conclusion, we see that the media and parliamentary debates concerning the introduction of Roundup Ready soya in the UK followed parallel courses. As the debate developed, it was framed in two main ways: GM soya as a question of food quality and as a matter of progress and economics.

The soya issue was framed as a matter of food quality primarily by the parliamentary Opposition, and, in the media debate, by the NGOs. This framing, one that stressed safety issues and labelling, was strongly influenced by the prevalent UK discourse concerning BSE. The way in which the British government had handled the crisis over infected British beef led to a deep public mistrust of the government's competence to form reliable judgements on questions of food safety. Further, this dissatisfaction only accentuated existing discourses on food safety provoked by such issues as salmonella-infected eggs. Thus, the BSE crisis provided a point of reference against which every new issue of food quality was judged and developed, including GM soya. The question of BSE and food quality was consistently promoted as a matter of science and technology, and the discussion did not touch on ethical concerns. For this reason, the controversies surrounding GM soya were exclusively concerned with future health risks that might emerge.

Moreover, the Opposition repeatedly accused the government of acting in an irresponsible and underhand manner during the BSE crisis, on the grounds that it had ignored vital scientific reservations and withheld such information from the public. The GM soya debate was set against this background, with the parliamentary Opposition insinuating that the public no longer trusted either the government or industry.

The overall result of this was to relocate the debate's chief point of reference from the sphere of expert opinion to that of public choice. The very considerable support shown by the government for the labelling of *all* products that contain (or might contain) manipulated foodstuffs indicates a will both to regain public trust and provide a means of public regulation in the GM foods issue.

The second way in which the debate was framed, in this case mainly by the industry and government, was related to technological progress and the economic competitiveness biotechnology might bring to the UK. Here an important point of departure was the fact that the UK has the second-largest biotechnological industry in the world, with 250 active companies in 1998. In this context, industry and government saw a serious danger in that any decision that threatened the production and/or marketing of manipulated soya would threaten the viability of this commercial sector.

Having this material setting in mind, it is not surprising that the government consistently stressed the positive aspects of this technology. In a similar fashion, the industry itself focused on the alleged ability of GM crops to produce more and cheaper food. Framing manipulated soya as an economic issue limits the range of alternative arguments and solutions, because it makes it clear that any decision will have consequences for the economic competitiveness of the country – and, taking it to its logical conclusion, also for the welfare state.

In addition to these two framing devices, the issue of GM soya was less frequently framed and debated in terms of the relationship between national and European government and of notions of trust in public authorities. In the former case, the inability of the government to act on the labelling issue was attributed to the stifling effect of European legislation. Allusions to the limits placed upon national independence by the UK's membership of the European Union may be identified, but this never developed into a full-blown debate either in Parliament or in the media. The issue of how much trust people should have in public authorities, especially after the BSE crisis, was also brought into the debate. Both the media and parliamentary debates contained references to the public mistrust of official decision-making procedures. Again, however, these never escalated into major issues.

The reception of Roundup Ready soya in Sweden

In Sweden, GM soya first emerged as a public issue at the beginning of 1995, when Monsanto submitted its application for approval to MAFF in the UK. The Swedish Board of Agriculture handled the application during the hearing phase, and concluded that the import of GM soya should not be restrained by Swedish legislation. In the spring of 1996, the Board handled the application for the import of GM soya prior to the formal voting procedure of the EU member states. Once again the Board of Agriculture assumed a positive stance. The Ministry of Agriculture, which was responsible for dealing with GM plants, did, however, ensure that the matter was discussed by other relevant ministries. The Ministry of Agriculture suggested that the proposal by the Swedish Board of Agriculture be followed, but the Ministry of the Environment registered its opposition on the grounds of contrary public opinion and the potential for negative environmental consequences. Conversely, keen to avoid jeopardising trade relations with the USA, the Ministry for Foreign Affairs endorsed the Board's proposal. Yet, as a result of this round of consultations, three days later the proposed approval was changed into a rejection. The formal outcome of this was that Sweden voted against the EU proposal to approve GM soya, even though Sweden had not objected to the initial application.

This rejection was based mainly on political arguments, and was not clearly supported by scientific evidence of risks, as required in the EC Directive on Deliberate Release (EEC, 1990). An important consequence of this first stage of the issue was that the government decided to change the Swedish statute that dealt with applications for approving the release of GMOs. From then on, the government, rather than the competent authorities, was to express its views as early as the 'objection phase', and not simply at the formal voting stage. Later on, when the EU had approved the import of genetically manipulated soya, Sweden accepted this and made no further moves to prevent its import.

Roundup Ready soya shocked the Swedish into thinking more seriously about GMOs. So far, the debate over gene technology had focused mainly on issues such as medical applications, formal regulation and ethical issues, while GM food had been only a minor subject. From the beginning of 1996, however, the debate escalated and came to focus largely on GM soya. The Centre Party (former Agriculture Party) was clearly hostile to the planned import of GM soya into the EU. At this point, the Swedish food distributors and retailers entered the fray and they too expressed themselves in opposition to the import of GM food. Furthermore, on 23 October 1996, fifteen NGO spokespersons issued an appeal in the large-circulation national quality newspaper *Dagens Nyheter* (Ström et al., 1996), urging the food industry and distributors not to buy GM soya from the USA.

The soya issue in Sweden: October 1996–January 1997

After the appeal issued by the NGOs, the issue received greater attention in the media as well as in parliament. Between October and the end of 1996, about 300 articles addressing the question of GM food were published in all kinds of newspapers and magazines. Most of these, however, dealt with the EU's decision, a couple of days before Christmas, to approve the Ciba-Geigy GM maize. In parliament, there were a number of questions on this subject and MPs' bills were prepared, mainly by the Green Party, the Centre Party, the Christian Democrats and the Left Party. All of these expressed a distinctly negative attitude towards GM food, and GM soya in particular. Indeed, a majority of MPs argued for labelling regulations.

As a result of the interventions by the Centre Party and the Green Party, a short debate was held in parliament on 29 November 1996. Opponents of GM soya argued that Sweden should prevent its import by invoking article 16 in the EC Directive on Deliberate Release (EEC, 1990), on the grounds that the potential consequences for the environment and possible health risks were unknown. As an illustration of the potential hazards of GM food, opponents offered the example of the introduction of the Nile perch in Lake Victoria. The very same example had been used, a few days earlier, by a representative from Greenpeace in *Göteborgs-Posten*, one of the largest-circulation quality newspapers (Berg, 1996). This camp also claimed that, if the use of GM soya were permitted, then other products – such as GM oilseed rape and GM sugar beet – that are not grown in Sweden should also be permitted. Furthermore, it was argued, if it is impossible to prevent the import of GM soya, a proper labelling policy had to be implemented, giving consumers the option of choosing whether or not to buy GM food. Parliamentary opponents insisted that they identified with, and were representing, the consumer's position. This is quite clear from the statements of critics such as Gudrun Lindvall (MP): 'We consumers are not moderately restrictive in this case; we don't want this kind of crap' (Lindvall, 1996).

The Agriculture Minister, Annika Åhnberg, explained the official Swedish position on genetic engineering in general, and GM soya in particular (Åhnberg, 1996). She did not oppose the EU decision, and claimed that article 16 could not be used because the potential risks of releasing GM soya were insufficiently serious and because growing soya in Sweden is impossible. Her view, consistent with her previous statements, was that Sweden should be restrictive towards genetic engineering without rejecting the possible benefits it might bring. Indeed, she hoped that Swedish industry would successfully adopt this new technology. On the whole, Annika Åhnberg gave the impression of being well informed and quite positive about the future uses of genetic engineering.

An analysis of the media debate in Sweden reveals that thirty-nine articles addressed this issue.[4] Thus, GM soya and the US shipments to Europe constituted a media 'event', but not a very significant one. It is noteworthy that the majority of the articles appeared towards the end of December, and were triggered by the more intense debate over the EU's approval of Ciba-Geigy's GM maize. Looking at the actors and how their positions were represented in the media, a pattern similar to that of the preceding stage is discernible. The retailers promised to boycott food containing GM ingredients and NGOs such as Greenpeace and the LRF (the National Organisation for Farmers) expressed criticism. However, the LRF and the retailers were mostly preoccupied with consumer reactions, whereas Greenpeace was more concerned with environmental and health risks. Conversely, the politicians were fairly positive and saw an opportunity for enhancing food quality and improving Sweden's future economic performance. Actors such as scientists and industry hardly participated in the media debate at all.

Taking a closer look at the arguments put forward by the different actors, it appears that the representatives of NGOs sought to prohibit the selling of GM food on environmental and health grounds. For instance, they stressed the potential for threats to biodiversity and the unpredictable long-term consequences of an increased use of pesticides. Furthermore, GM soya was not perceived as something from which consumers could benefit. On the contrary, it was seen in a negative light because of the risk of increased allergenic reactions and the spread of antibiotic resistance.

Critics also demanded labelling as a means of protecting consumer interests. Likewise, the retailers argued that GM soya should be labelled, though they did so on the presumption that consumers do not wish to buy GM products. The retailers also perceived the need for information campaigns to give people a better foundation upon which to make well-reasoned decisions.

Journalists also appeared as actors. Most of them articulated sceptical and negative attitudes, insisting that GM soya should not be imported into Sweden because of risks to health such as antibiotic resistance and allergenic reactions. Journalists also reiterated the claim that more information needed to be supplied to the public, on the assumption that the populace was too ignorant to make its own proper judgements.

Representatives of the government (most often Annika Åhnberg) saw genetic engineering, in general, as a progressive field encompassing many opportunities in terms of both economics and increased food quality. The risks were viewed as exaggerated and the opponents of GM as reactionary. Moreover, the government stressed the fact that a common agreement in the EU on labelling should be given the highest priority since the issue was not confined to Sweden.

Many of the published articles comprised small news items or short articles from news agencies that did not tend to focus on the different actors. It is interesting to note that, in this kind of article, genetic engineering is by definition considered to be something negative. This hostile climate is exemplified by an account in *SydSvenska Dagbladet* of the NGOs' blockade in Aarhus, Denmark, which does not give the reader an explanation of why the demonstration occurred. The author baldly states, 'The ship contains 23,000 tons of soya beans, of which some are genetically manipulated' (Nilsson, 1996), and feels able to assume that the reader will understand the justification for the demonstration.

There was almost no significant reaction after the adoption of the Novel Food Regulation by the Commission on 27 January 1997. On 7 April 1997, the food industry and the retailers voluntarily agreed upon labelling certain products that contained GM soya. At this point, the debate faded to some extent. However, a new issue had now been placed on the public agenda, and from then on the public was aware of what had been dubbed 'gene food'.

The dynamics of the soya controversy in Sweden

Overall, the amount of attention paid in Sweden to the issue of GM soya has been rather modest, especially at the political level. This relatively low level of debate can partly be explained by the fact that none of the ships sailing from the USA were bound for Sweden. In addition, for simple biological reasons, soya cannot be grown in Sweden.

When the issue was discussed by actors other than the government, debate was dominated by Greenpeace, the Green Party and the Centre Party. As demonstrated above, environmentalist and consumer groups did show a fair amount of interest in the issue. Their comments were solicited during the early phase of the debate, but their overall impact on the political process seems to have been marginal. The actors who opposed GM soya did so for different reasons: retailers, politicians and NGOs (such as LRF) were sensitive to the views of consumers and concluded that the public does not want GM food; journalists and NGOs such as consumer organisations and Greenpeace, on the other hand, perceived the potential for environmental and health hazards and pointed to the fact that consumers have nothing to gain from the import of GM soya.

The dominant framing of the issue in parliament and in the media seems to have been very similar, with GM soya being identified in both arenas as a possible threat to the environment and, to some extent, to public health. As in Denmark, this way of framing the issue, although dominant, did not result in a rejection of GM soya in the end, when Sweden accepted the EU decision. This is probably because these arguments were always countered by a more positive discourse in which GM soya was framed as a source of prosperity. At the political level, this is reflected in the fact that restrictive policies were preferred to the introduction of prohibitive measures. Thus, existing legislation was retained and the EU's decisions were accepted. At a more practical level, the labelling of products containing GM ingredients was debated, and then implemented, by the food industry and retailers in late spring 1997.

The framing of the issue as an environmental question must be seen against the background of the pronounced environmental discourse in Sweden. As in most other northern European countries, the environmental question has ranked high on both the public and the political agendas. In debates concerning food products, this discourse has been expressed through a general conviction that, in general, Swedish products and farmers are environmental friendly. In relation to the more general debate over genetic engineering, the environmental discourse fostered the view that genetic engineering implies (unacceptable) interference with nature.

As indicated earlier, the positive, mainly economic, framing of the soya issue cannot simply be explained in terms of the interests of the Swedish food industry. The domestic food industry is of minor national importance and mainly supplies the home market. A possible explanation for the promotion of the positive framing of GM soya could, on the one hand, simply be a general optimism about technology among Swedish politicians. But the explanation could, on the other hand, also be a fear about the indirect consequences the soya issue might have for uses of biotechnology within, for example, the pharmaceutical sector. Were the application of GMOs within the food sector questioned, this could have the potential of reopening more general debates over the industrial use of biotechnology. This might threaten the pharmaceutical industry and other biotechnological industries. In addition, the importance of the USA as a market for Swedish industrial products seems to have been a consideration. Sweden is highly dependent on exports from its large and successful industrial sector and was therefore reluctant to harm relations with the USA.

Another element, which adds to our understanding of the relative success of the critical framing of the soya issue, is of a cultural nature. Economic prospects and profits are not always seen as something positive, especially not if those benefiting are (foreign) multinational companies. In the media debate, the view that only multinational industry – and not consumers – would benefit from the introduction of manipulated soya predominated.

The reception of Roundup Ready soya in Italy

GM food has traditionally been an issue of only minor importance in Italy and, symptomatic of this, the Monsanto Round-up Ready soya did not generate public concern in Italy prior to the arrival of the first cargoes containing the product.

The soya issue in Italy: October 1996–January 1997

The soya story was relatively slow to take off in Italy. In late October, representatives of NGOs (Greenpeace Italia and Lega Ambiente) raised questions with the Interministerial Commission for Biotechnology (CIB) about the actual levels of residual pesticides in the Monsanto soya, and in November Greenpeace called upon the Minister of Health to block the use of GM soya and take the matter up at the European level. This appeal did not lead to any direct policy responses. In mid-November, the media coverage of naked women protesting against Monsanto soya at the FAO summit in Rome caught the public attention for a fleeting moment. At this point, the issue of GM soybeans was clearly not a question for public debate in Italy. But, by the second half of November it had become practically impossible to separate the issue of GM soya from that of the imminent importation of the Ciba-Geigy maize.

The Novel Food Regulation elicited a very critical reaction among the media, which criticised the EU for giving in to pressure from the biotechnology industry at the expense of the well-being of the public. The new regulation was seen as a compromise that could jeopardise public safety. In the time-frame under study here, there is little evidence of any kind of policy activity regarding the soya case per se. This perhaps reflects a tendency to delegate responsibility in such matters to European rather than national authorities. In fact, there were no discussions in parliament over the issue until March 1997 when news of Dolly, the cloned sheep, rekindled the whole debate about biotechnology in Italy.

Within our time-frame, the Italian media coverage of the arrival of the Monsanto soybean hardly constitutes a debate. The intensity of the media's coverage of the issue was weak, and would be more aptly described as news reporting in response to events that for the most part happened far away. Articles that dealt exclusively with the soya issue did not amount to more than a handful in the whole of the national press. However, from the second half of November, particularly after the decision in December to allow the importation of GM maize, the soya story became fused with the maize issue. The approval of GM maize was covered by all the leading national papers. The real peak in media coverage of the soybean, and other GM foodstuffs, was in the following spring after Dolly the cloned sheep lowered the threshold of media attention to all aspects of biotechnology. With the exception of the financial paper, the tone of the press coverage started out sceptical and ended up being generally negative.

The main actors with a voice in the debate or in media coverage were NGOs (Greenpeace and Lega Ambiente), the Italian biotechnology industry and a number of leading scientists in the field. Some of the main actors – or rather standard commentators – involved in the debate at the global level were notably absent in Italy, as were governmental actors. The three main actors involved advanced differing arguments and had a varying influence upon the debate.

Overall, the arguments advanced by the NGOs were clear and consistent throughout the three months under study. Their general argument was concerned with the long-term and indeterminate risks to human health, and they questioned the ability of government institutions to ensure the safety of GM products. In this context they referred repeatedly to the well-known BSE crisis. This argument is closely related to the scientists' concerns regarding the impossibility of guaranteeing a product to be risk free. The political, or perhaps moral, dimension to this argument was linked to appeals for an adequate labelling of GM products that would allow the public, as customers, a choice. The environmental movement also put forward a more specific technical argument that found its way into the media. The actual level of residual pesticide in the soybeans, they claimed, was too high and imports should therefore be banned.

The initial arguments put forward by industry mostly concerned the safety of the product. The industry claimed that GM food is safe on the grounds that it is not substantially different from ordinary strains. In this context, the earlier FDA approval was widely cited. Industry also advanced a more global argument in favour of the application of biotechnology, insisting that GM crops represent the possibility of a new green revolution that promises to benefit the environment and reduce world hunger.

The dynamics of the soya controversy in Italy

The Roundup Ready soybean on its own did not instigate a public debate in Italy. Media attention to this issue was relatively weak in intensity, while the tone of coverage was clearly negative. Furthermore, no policy initiatives emerged that were directly related to the soya case or the issue of labelling GM products.

The issue of GM plants and their use in foodstuffs was hard to reconcile with the way in which the biotechnology debate was conventionally framed in Italy – the benefits of medical progress versus a number of ethical concerns. Initially, the question of GM soya was put on the agenda by environmental groups as a technical regulatory issue focusing on the unacceptably high levels of pesticide in the soya, and not as an issue within the wider biotechnology debate. In 1996, advanced biotechnology was not a major issue in the Italian public arena and the intensity of media coverage was even lower than in preceding years. In the Italian context, environmental impacts and risk assessment issues traditionally rank second to medical and ethical questions. This is reflected in the fact that the issues that tend to trigger public debate over biotechnology in Italy concern not plants but the delicate connections between biotechnology and reproductive technologies, or any intervention in human life.

As time went on, the soya issue became increasingly linked to other discourses. By mid-November, attitudes towards GM soya had become inseparable from the sceptical reactions to the modified maize. At this point, the soya and maize discourses became part of an EU discourse. This introduced an anomalous theme into the Italian political handling of biotechnology. Following the EU decision to allow the import of GM maize, the issue focused on the claim that the EU had yielded to pressure from the industry, placing the interests of North American or multinational companies above the well-being of its citizens.

A discourse of diminishing trust in scientific institutions adequately to assess risks also appeared. Frequently linked to the BSE crises, this argument was discussed only with reference to other countries, but it stood as a lesson on the dangers of the authorities not responding adequately where public health was potentially at risk.

The issue of labelling was discussed only after this. In Italy, food safety was not a sufficiently prevalent issue in 1996 for it to be strongly implicated in debates concerning GM foods. Hence, the parliament did not discuss this question until the Novel Food Regulation was in place. Whereas the soya issue failed to provoke a public debate, let alone a political reaction on its own, the issue was catapulted back onto the political agenda in the wake of the controversy over Dolly, the cloned sheep.

The principal framing of biotechnology was initially that of economic progress and development. The application of biotechnology to agriculture was seen as a positive step forward, and any opportunity to further the reportedly weak Italian biotechnology industry was welcomed by the authorities. This framing was particularly evident in the financial press. In the media coverage as a whole, in contrast, the issue was most commonly framed in terms of safety and risk, and GM food became increasingly associated with the theme of trespassing on the boundaries of nature with unknown consequences. The third main framing strategy was that of regulation and public accountability. This became dominant after the decision to allow the import of the Ciba-Geigy maize in December 1996. This discourse stressed the allegation that the EU had betrayed its citizens by yielding to pressures from industry and multinationals. This story took a further turn when the Novel Food Regulation was introduced; this was depicted as an inadequate response and a further betrayal of the public.

A knowledge of the political culture in Italy enhances our understanding of the fate of the soya issue. When soya was on its way to Europe, Italian political life was preoccupied with economic and fiscal reform. Issues such as food safety, consumer rights and biotechnology were not of prime concern in the midst of an already hectic political agenda. Questions of this kind, defined as technical in nature, are typically handled behind closed doors by appointed governmental experts. Moreover, the issue was seen as an EU responsibility, on which little national policy activity was called for. In all matters concerning the regulation of modern biotechnology, except those considered to be of an ethical nature, Italy has conformed to European-level decisions.

The main actors in the soya debate in Italy were industry, scientists and environmental NGOs (mainly Lega Ambiente and Greenpeace Italia). Relations between the environmental NGOs and their European sister organisations had some impact on the debate in Italy, but international NGOs more directly influenced the Italian debate, Greenpeace International being the first to stimulate concern and then to organise opposition.

In contrast, the industry emphasised economic aspects and the potential benefits to be derived from the use of biotechnology in modern agriculture and food production. Scientists, mostly governmental experts, also played an important role in the debate. Scientific, or rather technical, arguments were used by all actors but, considering the unknown risks involved with GMOs, the scientists remained somewhat sceptical of the use of the new biotechnologies.

All in all, the case of Roundup Ready soya in Italy turned into a story about the general acceptability of GM foods, and in particular the competence with which the issue had been handled at both a European and a national level. The most interesting feature of the Italian reaction to the Monsanto soybean is how slowly and late it came. The debate really started after the event, indicating a low prior awareness of these issues among actors. Indeed, in the policy arena, most of the action took place at a very late stage.

Conclusion

Impacts on the discursive environment

Roundup Ready soya served to bring biotechnology back onto the public agenda, with the possible exception of Italy. Yet it played different roles in the disparate discursive environments of the four countries studied.

Whereas its effect in Denmark and the UK was largely to reopen dormant debates on biotechnology and cause qualitative developments within these discussions, the impact in Sweden and Italy was more profound, with the controversy introducing new discourses. In Sweden, a new and critical discourse concerning GM foods emerged from both the soya and the maize issues, where the dominant biotechnology-related discourses had previously been medical, ethical and regulatory. In Italy, the pattern is somewhat different, because the soya issue itself was of only minor consequence. However, combined with the maize issue, it has had a more or less similar effect to that in Sweden. In addition to this, there are subtle traces of a change in the Italian EU discourse, from a rather positive attitude towards a more critical position triggered by the GM maize and soya controversies. Thus, Italy differs from the three other countries, where EU scepticism was an existing and viable discourse that the soya issue could benefit from being associated with.

USA–Europe relations

The USA's reaction to the EU's critical reception of soya largely comprised allegations that the EU was using the soya issue as a pretext for setting up protective walls around itself with the aim of prohibiting US imports. The USA's interpretation of the soya issue is therefore consistent with Pline's finding that biotechnology was defined as an economic issue (1991). However, the analysis of these four countries indicates that this was not the case. There is no evidence that the GM soya issue was framed in terms of US imports threatening the European food and agricultural sectors. Such a framing would have been part of an economic argument for protectionism, which does not seem to have any pertinence here. Only in the UK was an economic frame dominant. But even in this case, as in the USA, economic arguments stressed the positive economic rewards biotechnology might bring and emphasised the need for a low level of restrictions.

Furthermore, for there to have been a protectionist discourse in the EU, European countries would have to have been producers of soya, or of crops with similar functional characteristics. There is, however, no European soya production to protect, nor is there a European surplus of plant proteins from other sources that could be used in feed production and thus be endangered by US exports.

It should be noted that critiques of the way Monsanto was handling the soya issue were developed in all four countries. But this was not so much a question of EU/US relations as a critique of the way in which multinational companies attempt to dictate politics and bypass fundamental consumer rights, such as the freedom of choice. This was, in other words, a moral statement and not a rejection of Monsanto's right to sell products on the European market per se.

The Eurobarometer survey in 1996 (see chapter 7 in this volume; Durant et al., 1998) indicates a widespread negative attitude towards GM food in the four countries studied here. Around 60 per cent of the respondents in Denmark and Sweden disagreed or tended to disagree with the statement that the genetic manipulation of crops should be encouraged; the percentage in Italy and the UK was around 45 per cent. Unfortunately, Denmark is the only country in which a survey was carried out during the controversy. However, the survey's results are consistent with those of the Eurobarometer survey, showing that as many as 68 per cent of respondents wanted a ban on GM foods. It is reasonable to expect that the same picture would have been found in the other three countries. The political consequence of accepting the public's reluctance to support or encourage biotechnology – in this case, soya – would be to impose a ban. Yet none of the countries has taken this step. This raises the issue of the extent to which political decision-makers should follow public opinion. Ultimately this is a question of populism. Closely related to this, it raises a more fundamental question of democratic influence on decision-making.

Political decisions that do not reflect the desires of large sections, or even the majority, of a population should not necessarily be interpreted as an expression of an undemocratic political process. So long as the different arguments present within the population are given an opportunity to influence the decision-making process, democracy is not subverted. The problem in the soya case, however, is that the critical argument and its ultimate consequence – a ban – are very weakly represented in the political process. Denmark provides perhaps the strongest evidence of this, since the survey showed that around two-thirds of the population wanted a ban. And, despite its persistence, the only political party prepared to promote this position (Enhedslisten) never really succeeded in getting the issue of a ban onto the political agenda. The result is that the question of 'if soya' has never been allowed to enter the process of decision-making; instead, the question has become one of 'how soya'. This provides a concrete example of the observation, made in chapter 2 of this book, that the political process has been directed at the goal of 'making biotechnology happen' and not seriously at questioning its application.

The same pattern is observable in Sweden and the UK, although the Swedish parliament saw rather stronger representations of the critical attitude. Conversely, in Italy, GM soya as a single issue never managed to provoke a political process in parliament.

A logical consequence of framing the GM soya issue as a question of 'how' is that labelling becomes a key issue in the debate. The problem is thereby transferred to the individual sphere and the decisions consumers make in the supermarket. Individuals can choose not to buy products containing Roundup Ready soya if they are labelled properly. From this perspective, the industry is likely to have gained most from the struggle over how the GM soya issue was framed. Crucially, it avoided what was, for it, the worst-case scenario: the discussion of a total ban of food biotechnology.

At present, consumers are left with the role of individuals in the marketplace. As a group they are heterogeneous and weak, and will remain so until consumer organisations succeed in taking up the issue and encouraging consumers to act collectively in the form of a boycott; or until retailers decide to obey the voice of consumers and remove GM products from their shelves, as major European supermarkets tried to do during the spring of 1999.

If no collective action is taken by organisations or retailers representing consumers, Roundup Ready soya will continue to be produced. To put it crudely, in such circumstances, individual consumers will be able to gain no more from proper labelling than a clear conscience and a certainty that their health will not be adversely affected by eating manipulated soya.

NOTES

1. For a detailed account of the different approaches towards the new biotechnologies in Denmark, see Jelsøe et al. (1998); for Sweden, see Fjælstad et al. (1998); for The United Kingdom, see Bauer et al. (1998); and, for Italy, see Allansdottir et al. (1998).
2. The analysis covered the period between 1 November 1996 and 31 January 1997. Most national newspapers in Denmark were included (*Politiken*, *Berlingske Tidende*, *Aktuelt*, *Information*, *Kristeligt Dagblad*, *Børsen*, *Ekstrabladet* and *Weekendavisen*). The total number of articles addressing soya in this period was 208, including 10 editorials, 138 longer articles, 25 short notices and 25 letters to the editor.
3. The political landscape in the UK changed dramatically after the elections of May 1997. The Opposition (Labour Party) won with a large majority of parliamentary seats. As a government, Labour has shown a willingness to stand by the attitude towards GM soya it adopted when in opposition, and it formed a Select Committee on Agriculture to examine the GMO issue. The committee recognised previous mistakes concerning food-quality issues and suggested more openness to the public and the introduction of consumer-friendly legislation. Indicative of the change in government policy is the wording of the committee's report: 'It is our view that consumers have the right to know if foods contain genetically-modified organisms, or if there is a possibility that they may contain them, and we fully support the Government's labelling policy for such

foods. GMOs also have potential environmental consequences, but these matters lie outside the scope of the Report. We strongly support Dr Cunningham's [Minister for Agriculture, Fisheries and Food] call for continued vigilance both on the food safety and environmental consequences of GMOs' (22/04/98).

4. The analysis covered the period between 21 October and 31 December 1996. All newspapers in Sweden with a circulation greater than 90,000 per day were included.

REFERENCES

Åhnberg, A. (1996) 'Svar på interpellation 1996/97:72 "om genmanipulerad soya", 29 November 1996'. Rixlex, http://rixlex.riksdagen.se.

Allansdottir, A., F. Pammolli and S. Bagnara (1998) 'Italy', in J. Durant, M.W. Bauer and G. Gaskell (eds.), *Biotechnology in the Public Sphere. A European Sourcebook*, London: Science Museum, pp. 89–102.

Bachrach, P. and M.S. Baratz (1962) 'Two Faces of Power', *American Political Science Review* 56.

(1963) 'Decisions and Nondecisions: an Analytical Framework', *American Political Science Review* 57.

Bauer, M.W., J. Durant, G. Gaskell, M. Liakopoulos and E. Bridgman (1998) 'United Kingdom', in J. Durant, M.W. Bauer and G. Gaskell (eds.), *Biotechnology in the Public Sphere. A European Sourcebook*, London: Science Museum, pp. 162–76.

Behrens, M., S. Meyer-Stumborg and G. Simonis (1997) *Gen food. Einführung und Verbreitung, Konflikte und Gestaltungsmöglichkeiten*, Berlin: Edition Sigma, Rainer Bohn Verlag.

Berg, A. (1996) 'Den nya maten. Gentrixandet – katastrof eller framtidshopp?' *Göteborgs-Posten*, 18 November.

Busch, L., W.B. Lacy, J. Burkhardt and L.R. Lacy (1991), *Plants Power and Profit. Social, Economic, and Ethical Consequences of New Biotechnology*, Oxford: Basil Blackwell.

Dahl, R. (1986) 'Power as Control of Behaviour', in S. Lukes (ed.), *Power*, Oxford: Basil Blackwell, pp. 37–58.

Downs, A. (1972) 'Up and Down with Ecology: the Issue Attention Circle', *The Public Interest* (New York), No. 28: 38–50.

Durant, J., M.W. Bauer and G. Gaskell (eds.) (1998) *Biotechnology in the Public Sphere. A European Sourcebook*, London: Science Museum.

EC (1997) Regulation of the European Parliament and of the Council of 27 January 1997 Concerning Novel Foods and Food Ingredients, 97/258/EC, *Official Journal of the European Communities* L43, 14.2, p. 1.

EEC (1990) Council Directive 90/220/EEC of 23 April 1990 on the Deliberate Release into the Environment of Genetically Modified Organisms, *Official Journal of the European Communities* L117, 8 May, p. 15.

European Commission (1996) 'Proposal for a Parliament and Council Regulation on Novel Foods and Novel Foods Ingredients', *Bulletin* EU 11-1996, point 1.3.30.

Fjæstad, B., S. Olsson, A. Olofsson and M.-L. von Bergmann-Winberg (1998) 'Sweden', in J. Durant, M.W. Bauer and G. Gaskell (eds.), *Biotechnology*

in the Public Sphere. A European Sourcebook, London: Science Museum, pp. 130–43.
Folketinget (1994) 'Forespørgselsdebat nr. F13: Forespørgselsdebat til Sundhedsministeren', *Folketingets Tidende*, p. 1440.
(1996) Spørgsmål nr. 707, 4 December 1996, *Folketingets Tidende*, pp. 2061 ff.
Gallup (1996) 'Gallup Instituttet for Berlingske Tidende: Opinionsundersøgelse. Forbrugernes holdninger til gensplejsede fødevarer', Projekt 9363.
Goldburg, R., J. Rissler, H. Shand and C. Hassebrook (1990) 'Biotechnology's Bitter Harvest. Herbicide-Tolerant Crops and the Threat to Sustainable Agriculture', a report of the Biotechnology Working Group, New York, March.
Hilgartner, S. and C.L. Bosk (1998) 'The Rise and Fall of Social Problems: a Public Arenas Model', *American Journal of Sociology* 94(1): 53–78.
Jasper, J.M. (1988) 'The Political Life Cycle of Technological Controversies', *Social Forces* 67(2): 357–77.
Jelsøe, E., J. Lassen, A.T. Mortensen, H. Frederiksen and A.W. Kamara (1998) 'Denmark', in J. Durant, M.W. Bauer and G. Gaskell (eds.), *Biotechnology in the Public Sphere. A European Sourcebook*, London: Science Museum, 9 pp. 29–42.
Kloppenberg, J.R. (1988) *First the Seed. The Political Economy of Plant Biotechnology*, Cambridge: Cambridge University Press.
Lassen, J. (1993) 'Food Quality and the Consumers', MAPP Working Paper No. 8, Aarhus School of Business.
Levnedsmiddelstyrelsen (1996a) 'Letter of approval from National Food Agency to Monsanto', J.no. 571.1065-0001, 18 November.
(1996b) 'Notat til Folketingets Sundhedsudvalg om godkendelsen i henhold til §11 i lov om miljø og genteknologi af genetisk modificeret soja til anvendelse i levnedsmidler', J.no. 571.1065-0001, 30 October.
(1996c) 'Vejledning om mærkning af levnedsmidler og levnedsmiddelingredienser fermstillet på grundlag af gensplejsede sojabønner', Sundhedsministeriet.
Lindvall, G. (1996) 'Interpellation 1996/97:72 om genmanipulerad soya (29 November 1996)', Rixlex, http://rixlex.riksdagen.se.
Lukes, S. (1974) *Power: a Radical View*, London: Macmillan.
Nielsen, T.H. (n.d.) 'Bioteknologi i "Anderledeslandet". Politik, etik og opinion i norsk lovgivning om bioteknologi'.
Nilsson, G. (1996) 'Sojademonstranter kördes bort av polisen', *SydSvenska Dagbladet*, 20 December.
Pline, L.C. (1991) 'Popularizing Biotechnology: the Influence of Issue Definition', *Science Technology and Human Values* 16(4): 474–90.
Rochefort, D.A. and R.W. Cobb (1994) 'Problem Definition: an Emerging Perspective', in D.A. Rochefort and R.W. Cobb (eds.), *The Politics of Problem Definition – Shaping the Political Agenda*, Kansas City: University Press of Kansas, pp. 1–31.
Schattschneider, E.E. (1975) *The Semisovereign People. A Realist's View of Democracy in America*, Hinsdale, IL: Dryden Press; first published 1960.
Schomberg, R. von (1993) 'Political Decision-Making in Science and Technology. A Controversy about the Release of Genetically Engineered Organisms', *Technology in Society* 15: 371–81.

Ström, T., B. Jonsson, G. Axell, B. Thunberg, L. Andreasson, B. Bergnér, M. Persson, C. Elmstedt, U. Paulsson, C. Axelsson, E. Lundmark, K. Nilsson, N. Carlshamre, L. Nolte and M. Ekman (1996) 'Vi kräver stopp för genförändrad mat. Nu kommer de första lasterna med genmanipulerade sojabönor från USA – Bertil Jonsson och 14 andra organisationsledare kräver förbud', *Dagens Nyheter*, 23 October, p. A2.

11 Brave new sheep – the clone named Dolly

Edna Einsiedel, Agnes Allansdottir, Nick Allum, Martin W. Bauer, Anne Berthomier, Aigli Chatjouli, Suzanne de Cheveigné, Robin Downey, Jan M. Gutteling, Matthias Kohring, Martina Leonarz, Federica Manzoli, Anna Olofsson, Andrzej Przestalski, Timo Rusanen, Franz Seifert, Angeliki Stathopoulou and Wolfgang Wagner

Her name is Dolly. She is a Finn Dorset sheep that happens to be the most famous farm animal in the world. It was in late February 1997 that the world was stunned by the announcement of her birth, and she was proclaimed to be the first animal cloned from an adult cell. Other animals had been cloned before, but the creation of a sheep from a single cell from a six-year-old ewe electrified both the scientific community and the general public.

The piece in the journal *Nature* was given the rather prosaic title, 'Viable offspring derived from fetal and adult mammalian cells', with not a mention of the word *clone* or *cloning* (Wilmut et al., 1997). The implications did not escape the science journalism elite, however, and their coverage raised the alarm for the rest of the world's press corps to follow.

We consider the Dolly story to be the first real global[1] and simultaneous news story on biotechnology.[2] In this chapter, we will compare the varied ways in which this story unfolded across eleven European countries and Canada. We focus on the elaboration of the global trigger, recognising that there will be variations in its assimilation and accommodation to local contexts. Although a unique story on its own, our comparative analysis of the media coverage of this event is also intended to further our more general examination of the continuing public elaborations of biotechnology. The question we pose is, on the face of it, a simple one: how was the story of Dolly, the cloned sheep, elaborated in media texts across these twelve countries?

We use the term 'story' intentionally, recognising that journalistic accounts are indeed narrative constructions, or storytelling (Darnton, 1975; McCombs et al., 1991). The story of Dolly is no different from other media stories in the sense that it is subject to similar, basic journalistic routines and news values, reflecting the operating logic of the media.

Nonetheless, within the panoply of journalistic frames, there develops a story-line that may be both unique and timeless. The cloning of Dolly, we intend to show, has its own distinctive narrative elements but, at the same time, elements of the story appeared in a wide range of countries, apparently irrespective of differing national contexts. How this story unfolded in different national contexts may reveal the existence of a common narrative around the complex of biotechnology.

Our examination of the early elaborations surrounding this event attempts to reconstruct and analyse the narrative thread that evolved, as well as its accompanying themes, images and metaphors. All of these elements are tied within a storytelling frame or the dominant story-line used to organise a story (Gitlin, 1980). These elements, in turn, may be utilised as a way of further understanding the dynamics of social representation. We suggest that these representations may provide a window into our continuing reflections on technology in the context of modernity.

Why Dolly?

Indeed, this question must be raised and addressed. Dolly is doubly significant in our large-scale examination of biotechnology. First, the event represents one of the true breakthroughs in the biotechnological path.[3] As a scientific feat that captured international attention, her birth, public announcement and reception represent a milestone that deserves attention in our attempt to portray the life of this technology. Secondly, and more importantly, Dolly as a social phenomenon deserves further examination. As we intend to show, it is a phenomenon around which larger questions of the place of technology in society coalesce.

Dolly's science

Dolly was the first mammal to be cloned from adult tissue, even though there had been many previous attempts at cloning. In the past, successful attempts had usually involved a donor cell taken directly from an embryo in a process involving 'nuclear transfer'. This donor cell, containing all of its DNA, is then fused with an egg cell from which the DNA has been removed. Once fused, the developing embryo is implanted into a surrogate mother.

Dolly was the first instance of a mammal resulting from a stem cell (in this case, from the mammary tissue of a six-year-old ewe), and it took the Roslin Institute team 277 attempts to clone a sheep successfully. Their technique involved the careful coordination of the states of the donor cell and recipient egg. In the early phase of cell division, the Roslin team artificially induced a state of quiescence by starving the cell of its nutrients. This kept the donor cell in tune with the egg cell. When the two cells fused, development then proceeded normally.

The scientific significance of Dolly was three-fold. First, she brought closer the distinct possibility of cloning humans. Second, the technology responsible for her creation made the alteration of the genetic make-up of animals much simpler. It was in this context that the Roslin team was doing its work, as they attempted to clone sheep for the production of pharmaceuticals. Third, she expanded scientists' understanding of the role of DNA in the development of animals to adulthood. In this connection, the implications for understanding processes associated with ageing were evident.

Dolly in a social context

In many ways, Dolly's story may also be a tale of technology in modernity. The American cultural historian, Marshal Berman, has described being 'modern' as finding ourselves in an environment that promises us adventure, power, joy, growth and transformation of ourselves and the world, 'and, at the same time, that threatens to destroy everything we have and everything we know' (Berman, 1982: 15). Modernity, he maintains, means constant change that is both promising and threatening, tempting yet terrifying, exhilarating as well as exhausting. Living in this world of high modernity is in essence 'riding a juggernaut' (Giddens, 1991: 28).

One of the hallmarks of this world, particularly in late modernity, is the double-edged nature of science and technology, offering beneficent promises for humankind at the same time as they create new definitions of risk and danger at every turn (Giddens, 1994). In this 'risk society' (Beck, 1992), what we previously saw as 'natural' we now subject to control, making us, in turn, worry about what we are doing to nature rather than what nature could do to us (Beck, 1998; Giddens, 1998). In this society, in which risk is an everyday concern and where the flaws of technology have been experienced, a disenchantment with experts ensues and, with it, an erosion of the authority of science.

In this context, we intend to show that the social conversations and questions that might surround a technological event such as Dolly may also represent the way we continually negotiate the place of technology in our world.

The context of social representations

We consider events such as Dolly as phenomena of social representations (see, for example, Moscovici, 1984; Farr, 1987; Wagner, 1996; Bauer and Gaskell, 1999), a theoretical framework that may be helpful to structure our analysis. 'A social representation is the collective elaboration of a social object by the community' (Moscovici, 1984: 25). Such an elaboration becomes a social reality for the community. That is, one cannot distinguish between subject and object; an object exists only in the context of how a community constructs this object through talk and action (Moscovici, 1976: 251). The media offer one forum in which such community conversations take place. These conversations, in turn, may focus on stories of the moment that capture community attention. In this sense, we view media texts (used broadly to include not just words but also images) as an outcome of social processes. That is, they are social in their production, their circulation and their consumption (Wagner et al., 1999).

Two processes are important to the elaboration of social representations. The first is a process of *anchoring* – making the unfamiliar familiar by means of classifying or naming so that new ideas are put into a familiar context. This is not an arbitrary process, nor is it simply an attempt to pigeon-hole for the sake of order or clarity. Anchoring, while facilitating interpretation by classifying and naming, can also communicate a social attitude (Moscovici, 1984: 35).

The second process is called *objectification* – making the abstract explicit. Objectifications unfold in explicit arguments in public or, more implicitly, in pictorial or other materials, conferring clarity on an abstract notion. The process is an attempt to capture the essence of a phenomenon by making it comprehensible, at the same time as it plays on social schemas already in place (Moscovici, 1976; Wagner et al., 1999).

In the context of the present study, we compared newspaper coverage of Dolly over a period of eleven days across twelve countries. In this instance, we were observing the phenomenon of representation in the making in media texts. We considered two modes of representation: the linguistic and the pictorial modes. We further compared emotional associations and metaphors with respect to the anchoring processes and looked at pictures of Dolly and their contexts to elucidate some of the objectification at work in the public discourse of cloning.

We examined newspaper stories from the following countries: Austria, Canada, Finland, France, Germany, Greece, Italy, the Netherlands, Poland, Sweden, Switzerland, and the UK (although some of the material is incomplete for the Netherlands and Poland), covering an eleven-day time-frame from 23 February to 5 March 1997. The sample of newspapers and a more detailed description of the method are described in the methodological appendix to the chapter. A qualitative description of the day-by-day coverage was provided by each country team to allow the charting of patterns in the story's evolution. Themes and metaphors were also identified and described. A quantitative analysis of a set of variables was employed to further describe thematic structures, actors, consequences in terms of risks and benefits, and general portrayals of science. Finally, visual images were analysed separately.

Anchoring: from the esoteric to the common

The basic feature of anchoring is naming the unfamiliar thing or event. By giving it a name, we link the 'thing' to a multitude of references, out of which it takes a shape that is familiar and understandable, identified as enemy or friend, or with ambivalence until further clarification.

The translation of both the scientific technique and the journal's language began almost immediately. The first step in this translation was to name or label the event by means of its most obvious implication, with a term that was already familiar. We had already noted that 'cloning' was not part of the vocabulary of the research paper but, keen to highlight its most important article, *Nature* itself was quick to use as its cover theme 'A flock of clones'. The press release from the Roslin Institute and the newspaper headlines trumpeted cloning from the very beginning, with the term 'clone' or 'cloning' nailing down the event in a network of references that unfolded in the next ten days. Cloning is a word loaded with connotations reaching far into the world of fiction. It is a word that evokes negative expectations, if not fear and loathing.

'Dolly' introduced a second name into the event. The product of the scientific process, an animal, was christened after an American pop singer and movie star, Dolly Parton, perhaps in a moment of whimsy among the Roslin team.[4] At once, the sheep that emerged from a cold laboratory process was humanised.

Once these key anchoring markers were in place – the sheep and the cloning that brought her about – the process of making their meanings concrete was woven into the tapestry of the narrative that subsequently unfolded. We will first trace this narrative development, then identify the attributes of the representational field that emerged.

The evolution of the story

Evolution is used metaphorically. It points to the fact that the social representation of the cloning event takes on a different shape with time and as contexts differ.

The public story of Dolly began with the submission to the journal *Nature* by Ian Wilmut and his colleagues of the account of their experiment. It was written in typical dry scientific prose but, even without referring to cloning, there was no mistaking the import of their account. Scheduled for publication on 27 February, the journal imposed its usual embargo on an important story.[5] Given the fierce competition among news outlets, it was not surprising that the embargo was broken. Two

Italian newspapers and a British newspaper, the *Observer*, relying on its own sources, published news of Wilmut's feat.[6] The *Observer* proclaimed the feat thus: 'Scientists clone adult sheep: triumph for UK raises alarm over human use.' Perhaps missing the import of the story on their hands, the Italian newspaper, *L'Unita*, buried the story in its inside pages and factually declared, 'A cloned lamb is born' (*L'Unita*, 22 February 1997).

Following on the heels of the *Observer* were the world's press. Here, we provide a synopsis of the story as it evolved in written text and visual images over the eleven-day period in our study newspapers. Preliminary analysis of the pictures indicated a remarkable homogeneity across the countries in this study, linked to the practices of the distribution of images through international press agencies. The commentary on the pictorial material is based on an initial determination of the homogeneity of pictures across the twelve countries. A more detailed analysis of the complete set of images from Canada, Greece, Germany and Italy, with references to images from other countries, was then made.

Day 1 (22/23 February)

The Italian newspapers broke the embargo on Saturday, 22 February. *L'Unita* provided a calm and composed descriptive account of the cloned lamb. *Il Giornale* echoed the announcement: 'The first cloned sheep is born.' Given the tone and lack of play, it is likely that the full significance of the story was missed by these Italian newspapers (see Wilkie and Graham, 1998). The UK's *Observer* account appeared on Sunday, the 23rd, and was instantly picked up by the international wire services. Robin McKie, the science editor, wrote a 635-word piece, opening his report with remarks on the triumph of UK science, a breakthrough that could lead to further advances in work on ageing, genetics and medicine. But, with a clear sense of the cultural imaginary behind cloning, he quickly provided references that countered this scientific optimism: Huxley's *Brave New World* and *Boys from Brazil*, a horror movie about producing clones of Adolf Hitler that had been adapted from a book of the same title (Levin, 1977). These references clearly framed the dark side of the technique's possibilities for humans. The main body of the article then elaborated on what had happened at the Roslin Institute, going in some detail into the processes involved and mentioning the more salutary implications: the possibility of new drugs and new insights into the ageing process. The article closed with the image of flocks of medicine-producing sheep and the observation that the vision of creating armies of dictators would attract most attention because the technique, in principle, was suitable for cloning humans.

Day 2 (24 February)

This was really the first day of coverage for the rest of the newspapers in most of Europe and North America. Whereas most newspapers trumpeted the announcement as a major breakthrough, some remained restrained in their coverage, focusing on the announcement of the event. Some countries seemed to miss or downplay the significance of the event. Finland simply reported the announcement, while Sweden buried the story in the inside pages. Perhaps because the achievement was considered a national triumph, and because it had been following the studies more closely, the UK *Guardian* downplayed the fears and highlighted its positive aspects: 'Scientists scorn sci-fi fears over sheep clone: a British breakthrough which has brought new hope to the study of presently incurable genetic diseases.'

Other countries, however, immediately coupled the announcement of the breakthrough with alternating amazement and trepidation. In Canada, Dolly's announcement was heralded as 'a dazzling technological leap and a conundrum for ethicists'. Another called it 'a genetic marvel spawning an ethical nightmare'. The spectre of human cloning was raised right from the start in all the countries under study, as the headlines suggested: 'Will cloned humans follow?' (*Kurier*, Austria); 'Research breakthrough involving female sheep may mean that humans can be duplicated as well' (*Globe and Mail*, Canada); 'Cloning, yes, but not on humans!' (Italy, *Il Giornale*). Coincidentally, in Greek, the word 'shocking' is almost indistinguishable from the word 'cloning' ('klonismos' and 'klwnismos', respectively), and *Apogeumatini* made much of this juxtaposition.

The metaphors that accompanied this announcement and naming process were pronouncements of doom. Although the magnitude of the achievement was recognised, this scientific breakthrough was also equated with the development of the nuclear bomb and continued to draw on the imagery of Huxley's *Brave New World* and the film *The Boys from Brazil*.

Day 3 (25 February)

Without exception, the focus on human cloning in newspapers became even more pronounced. The Finnish as well as the British, Greek, Swedish and Swiss press developed this theme even further: 'In this year, someone will try to clone a human' (Finland, *Ilta Sanomat*); 'First a sheep, then a human?' asked the Swedish paper *Aftonbladet*. Suggestions for who might be cloned varied: in the UK, it could be 'the rich'; in Greece and Switzerland, it would be 'some women'.

On this day, a crescendo of ethical concerns was heard. Many papers interviewed ethicists and philosophers for their views. *Figaro* quoted a psychiatrist and ethologist, Boris Cyrulnik, who said the ethical ramparts man had constructed for moral guidance were collapsing. A Viennese gene expert and philosopher, Johannes Huber, observed in *Kurier*: 'It does not help if we create a new Einstein and remain Stone Age people in ethical terms.'

After a day to catch their collective breath, media storytellers further drew from their bag of metaphors, and this time those related to 'copies' and 'duplication' were widely used. Related to this, another feature emerged, one that seemingly contradicted the moral outrage expressed over the prospect of cloning humans: puns, cartoons and other witticisms began to appear. Cartoons of cloned politicians and sports and entertainment figures were featured. Was this a way of relieving the anxieties of the moment, like gallows humour or jokes around a surgical operating table?

The articles in the early days were accompanied by photographs of Dolly, two of which were widely printed. The first was a picture of Dolly looking straight into the camera (see plate 11.1). This photo became the standard image in the press coverage. The photograph was remarkable for its neutrality, simply picturing a sheep standing on barnyard hay, its body filling the frame with nothing else to catch the eye. A cloned sheep is still

Plate 11.1 Dolly, the cloned sheep, facing the world's media cameras

much like a normal sheep, the photo suggested, but the visual shot also attested to the reality of a clone as an embodiment or the objectification of the abstract idea of cloning.

The other common image in the first days of newspaper coverage was that of Dr Ian Wilmut, standing in a pen with non-distinct sheep in the foreground and talking to Ron James, the managing director of PPL Therapeutics on his right. The connotations primarily concern the business of sophisticated animal husbandry and breeding. All the conventional references to science were notably absent from this picture: both men were without white coats; the scene was set outside, without a glimpse of anything related to the laboratory; and they appeared to be talking business. We might say that the picture anchors the cloner in the world of business and breeding, but not in the world of science.

Day 4 (26 February)

The then US President Bill Clinton's first appearance on the scene was made with his announcement that he had ordered his National Bioethics Advisory Commission to report to him on the issue of cloning. British physicist, and Nobel peace prize winner, Joseph Rotblat was also highlighted when he made a call for an international ethics committee to review the cloning issue. Added impetus to these expressions of concern about the possibilities of human cloning came from a variety of reputable actors at the national and international level. Domestic journalists were finding their own local sources to interview (generally, scientists and government officials) but, at the same time, prominent sources' views began circulating beyond their domestic markets. Officials in the German Bishops' Conference condemned cloning and their views were carried in German and other countries' papers (for example, Austria and Switzerland). Clinton's action was widely reported by all papers.

The science of cloning received further attention. This happened in Canada (*Globe and Mail*), France (*Le Monde*, *Libération*), Greece and Sweden (*Aftonbladet*), with the steps involved in the cloning process being diagrammed or outlined. Positive aspects of cloning in agriculture, particularly with respect to animal breeding, were discussed in Canadian, French, Greek and Swedish papers.

At this stage, close-ups, or the equivalent of facial portraits, of the cloned sheep were widely published. These images are a part of the functional process of identifying protagonists in newspaper stories. The effect is not only to humanise or personalise the cloned sheep but also a way to attribute status to a protagonist or a celebrity. The existence of the sheep had already been visually testified to and it was taken for granted that the reader would recognise the sheep; but at this point Dolly, the object, took

Plate 11.2 Professor Wilmut contemplating a test tube

on new meanings through associations. The media were visibly reflecting upon the irony of attributing celebrity status to a sheep and there were various images where Dolly is being photographed with a microphone held up to her, as though she were being interviewed. The world's media were paying homage to a sheep, another reflection upon the irony of modernity.

At the same moment in the story, cartoons and photomontages began to proliferate. Most of the cartoons contained representations of politicians and cultural celebrities, but also common were armies of fictional clones, or a multiplicity of identical politicians or actors from the world of sport or fashion. We can postulate that one of the primary functions of cartoons is to use irony to reflect on, if not subvert, reality. The same holds for the explicit manipulation of images in photomontages. Both kinds of visual storytelling fall within the realm of narrating possible worlds.

Another important development in the visual storytelling is the picture of Wilmut in profile, holding a test tube at eye level (plate 11.2.). Wilmut, the cloner, has now moved out of the pen and into the laboratory. Although clearly a photograph, it also produced a somewhat nightmarish effect. There were two important signs in this image: the profile of Wilmut and his hand holding the test tube. This image was the first to assert strong associations between the cloner and science, and the test tube as a sign can be read as a metonym, as a part that stands for the whole, evoking associations with life sciences. In particular, the suggestion of 'test-tube babies' was implicit. The emphasis given to his hand

and to his eyes gazing into the liquid in the test tube further strengthened associations with the manipulation of life.

Days 5 and 6 (27–28 February)

Demands for ethical guidelines and laws against human cloning were now being articulated very clearly. Legal regulations should be introduced at an international level, suggested a number of newspapers from Austria, particularly after doubts were expressed regarding the integrity of scientists. Despite Wilmut's denouncement of human cloning, he was portrayed by some as self-interested and motivated by greed. 'After all, this story is not just about scientific knowledge, it is about cash as well' (*Die Presse*). Swiss papers pointed out that human cloning is forbidden at the national level, while female Swedish parliamentarians called for an ethical debate on cloning and embryo experiments. The Vatican also weighed in, calling for immediate laws prohibiting human cloning, an event reported in France, Switzerland, Austria and Italy.

On days 5 and 6, the dominant metaphorical images revolved around humans having crossed boundaries and science spinning out of control. Metaphors that suggested humans were 'playing God' or fooling around with nature evoked fears of humankind overstepping its proper limits. The metaphor of Frankenstein was especially prevalent, and portrayed scientists and the scientific enterprise as having gone wild.

Days 7 and 8 (1 and 2 March)

Over the weekend, a number of articles appeared reflecting on Dolly and the possibilities of human cloning. 'Are there some doors that science should leave closed?' asked the *Calgary Herald* (Canada). In Germany, weekend reflections discussed research limits, regulatory considerations and economic advantages (*FAZ*). The alternative intellectual paper *Taz* devoted two articles to calls for further debate and discussion. Attention was drawn to the media circus around Dolly and the potential implications for human independence and self-determination, as well as potential redefinitions of 'mother', 'family' or 'parenthood'. Greek papers referred to domestic cloning research already under way.

The theme of regulation was also evident in the French coverage, especially after President Jacques Chirac was cited as seeking advice from the country's Ethics Committee, and the UK's 'Nine Wise persons' committee was spotlighted. Greek coverage continued to reflect shock over the cloning feat and speculation about what this might mean for the future.

At this juncture, the announcement was made by the UK Ministry of Agriculture, Fisheries and Food that funding for the Roslin cloning project was to be cut, a reaction to the continuing furore caused by the event and widespread calls for further controls. This was reported in Canadian, Greek, French, Swedish, Swiss and British newspapers.

Perhaps because the weekend provides more time for reflection and extended debate, the metaphors at this time covered the entire gamut of emotions. A few alluded to science marching inexorably onwards and the difficulty of containing the search for knowledge. However, the predominant imagery again involved copies, monsters and magic, the crossing of boundaries into the realms of the unnatural and the Divine, and science unbridled.

The pictorial story became more complex over this weekend, partly reflecting the conventions of the weekend editions of newspapers. The main new development was the appearance of images of science, such as pictures of scientists in their white coats going about their chores in laboratories, or pictures of laboratory equipment. The effect of such images was to move the locus of action back into the world of science. At the same time, most of the press published more or less elaborate diagrams that explicated the technique of cloning. The imagery at the weekend moved the cloning of Dolly out of the pen, away from the political context and celebrity attention, and into the heart of science.

Days 9 and 10 (3 and 4 March)

'Monkey embryos cloned.' The news was carried by virtually every newspaper, and everyone pointed to this step as being inescapably closer to human cloning. From a scientific point of view, the cloned monkeys were in no way pioneering, since the method of embryonic cell cloning had already been applied many times in the past. Even human cloning was attempted back in 1993 with human embryonic cells.[7] Though not all texts discussed the difference between the monkey cloning approach via embryonic cells and Dolly's cloning from an adult somatic cell, virtually all emphasised the diminishing gap between cloning in 'lower' forms and cloning humans. 'The example of Oregon and the manipulation made on primates show that there is basically no biological barrier to undertaking cloning with humans' (Switzerland, *Tages-Anzeiger*). 'Clone-sheep, clone-monkeys, soon, perhaps, clone-humans' (Austria, *Neue Kronen Zeitung*). 'This time, it's about primates; man's turn seems close' (France, *Libération*). 'It is not possible to come closer to humans than this' (Sweden, *Aftonbladet*).

In addition, four countries – Austria, Finland, Greece and Switzerland – printed extracts from an interview Wilmut gave to the German magazine *Der Spiegel*. Wilmut had received numerous calls, primarily from women who wanted to be cloned, and he warned: 'The fear of misuse is justified.

With our technique, you can produce genetic copies of humans. Only clear laws can prevent misuse' (Switzerland, *Le Matin*). On day 10, the same theme continued to dominate coverage: 'Closer to the mass production of humans after the cloning of the sheep and the monkey in the USA', warned Greece's *Eleutherotypia*.

Pope John Paul II expressed concerns regarding 'dangerous experiments' and criticised 'the merchants of life'. Italian and Austrian papers and Canada's French-language newspaper reported this event, as did the German and British press.

Photographs of the Oregon monkeys were printed in most countries. Again, the primary function of the image was to attest to the reality or veracity of the clones by showing the embodiment of an earlier idea, but the most significant aspect of this picture was the difference from the standard Dolly image. The monkeys are huddled together in an otherwise empty corner and the image bears strong emotive connotations through the apparently frightened look on their faces, whereas the standard Dolly image depicts a neutral-looking sheep. The embodiment of the cloning technique was taking on new meanings and associations along the way. The message appeared to be that cloning was really getting too close for comfort, but, at the same time, images were also being used that seemed to suggest that cloning had already moved some way towards acceptability. An interesting indication of this emerging process of naturalisation was the images of natural identical twins that some newspapers employed as illustration, associating clones with what nature was already doing without any manipulation.

Day 11 (5 March)

The fear of misuse continued to be articulated in the newspapers. Would scientists stick to laws? Pros and cons were highlighted and Wilmut's statements about scientists needing boundaries and laws were a focus of all papers except Finland's press. Doubts and cynicism about scientists' integrity were expressed: 'What's technically possible will sooner or later be done' (Switzerland, *Appenzeller-Zeitung*); 'Someday, somewhere, some madman will do cloning experiments with humans, that's for sure' (Austria, *Kurier*).

President Clinton announced the imposition of a ban on federal funds being used for human cloning experiments and called for a voluntary moratorium on human cloning research, a development reported in all of the countries studied.

An analysis of the issues most emphasised in the various countries showed similarities and distinctions (see table 11.1).[8] Human cloning

Table 11.1 *Dolly day by day, 24 February – 5 March 1997*

Issues	Day 2	Day 3	Day 4	Day 5	Day 6	Weekend: Days 7/8		Day 9	Day 10	Day 11
Peak day: production of articles		Can	CH I DE	F	Fin		Gr S UK			A
Human cloning	A Can CH DE F Fin Gr S UK	A Can CH DE F Fin Gr I S UK	A Can CH DE F Fin Gr I S UK	A Can CH DE F Fin Gr I S UK	A Can CH DE F Fin Gr I S UK	A Can CH DE F Fin Gr I S UK	Can CH I S UK	A Can CH DE Fin Gr IS UK	A Can CH DE F Fin Gr I S UK	A Can CH DE Gr I S
Human fallibility and fear of misuse	Can DE S UK	Can CH F Fin Gr I UK	A CH F Fin Gr I UK	A Can CH DE F S UK	A Can CH Fin Gr I UK	A Can CH F Fin IS UK	Can Gr I UK	A Can CH DE Fin Gr 1 UK	A F Fin Gr I	A Can CH DE Gr I
End of sex	Gr	Can F S	Can	Can F Fin	Can F Gr	DE	Can	DE Gr		Gr S
Scientists have no morals	UK	A I UK	A Can DE I UK	A I	A Can DE I UK	A Can I	Can Gr I S UK	F DE I S UK	A Can I S UK	Can CH I S
Dolly started ethical debate		F	Can DE F I	I	CH I	CH I	Gr	Can I	Can F	CH
Scientists are ethically responsible	Can CH UK	CH DE F Gr S	Can CH DE F Gr	F UK	CH I UK	Can CH I UK	I	CH DE F Fin Gr I	A DE	CH Gr
Dolly a media event	CH	Can	I	A Can CH DE F	CHIUK	A Can CH DE I UK	Can Gr		A Can I	A Can Gr
International actors		Can DE	A Can CH DE F Fin Gr I UK	A CH F I S UK	A CH DE F Fin I S UK	Can DE F Gr UK	Can DE Gr UK	Can DE F I S UK	A CH F Gr	A Can CH F Fin I
Religious sources	UK		A DE I	A CH F	A DE F I UK	DE Fin Gr I	Gr I S	Can DE F I	A DE I UK	A I
The monkey story							Can	A Can CH F Fin Gr I S UK	A CH F Gr I UK	A CH Fin
Call for international laws	Fin	A Can UK	A CH DE F Fin Gr I UK	A CH F I S	A CH F I UK	A CH DE F Gr I UK	Gr I UK	F I S UK	A CH F S	A CH Fin I

References to national themes, laws	Can	A Can DE F	Can DE F CH I	Can DE F S	Can CH DE Gr I S UK	A Can F I	Can CH Gr I S	CH DE F I	A CH DE F I UK	A Can CH DE Gr I
Roslin funding cut						Can	Fin S UK	Gr CH UK	F UK	CH
Dolly: business aspects		Can F UK	Can CH DE F UK	F Gr S UK	A F	A Can DE S UK	I UK	Can F I	F Gr I UK	A DE Gr I
Agricultural and medical benefits	A Can CH F S UK	A Can CH DE F Fin Gr I S UK	A Can CH DE F Fin Gr I UK	Can F I S UK	A Can DE F I UK	Can CH Fin I S UK	Can Gr I UK	CH DE I UK	A F Gr I S UK	A CH Gr I
How does cloning work?	A Can CH DE F Gr UK	A Can DE F Gr S UK	CH DE F Fin I	CH F Fin I S	A CH DE Fin Gr I UK	A F I	Can Gr I S	A Can DE F I UK	F I UK	A Fin I
Scientific limitations	Can F	F S	A Can CH DE F	Can F Fin	Can CH F I	Can F I	Can S UK	A Can DE F I UK	Can F	A CH I
Do not reduce human to DNA	CH	CH S	CH	Can F Gr	Can DE I UK	Can CH DE	Can Gr S	I	Can I	A DE Gr I
Scientific meaning of Dolly: totipotent cells		F	CH F	F	CH DE	A CH	I	F	F	A
History of cloning	Can CH Gr UK	A F S	Can DE F	F	A CH DE Gr	Can I	Can Gr S UK	F UK	F Gr I	A CH Gr
Cloning has nothing to do with biotech			A CH I		CH	A			CH	A

Note: A = Austria; Can = Canada; CH = Switzerland; DE = Germany; F = France; Fin = Finland; Gr = Greece; I = Italy; S = Sweden; UK = United Kingdom.

was mentioned in every country in the course of our sample period. Although the usefulness of the cloning technique was described (usually in areas of medicine and animal breeding), this was often accompanied by concerns about potential misuse. These worries were underlined by arguments concerning human fallibility (references to the dark side of human nature or the inability of resisting the temptation to clone). The trustworthiness of scientists (or lack of it) was also tied to this theme.

Most countries discussed the science behind Dolly, although it was less prominent in Finland and Greece. Also generally covered were the scientific limitations around cloning, including the inefficiency of the technique. At the same time, Dolly's immediate positive impact in economic terms was duly noted by most countries.

An examination of the story's evolution suggested the emergence of a distinct story-line: although triggered by the cloning of a sheep, the story was essentially that of human cloning and the horrifying possibilities it presented. In parallel to these early discussions of how the scientific 'genie in the bottle' could be contained, more positive interpretations emerged (for instance, that the bottle contains a beneficial magic potion). Yet, at this stage, these voices were overwhelmed by the voices of those who forecast danger. Although the story of Dolly obviously did not end here, the outlines of the tale are sufficiently discernible. In the next section, we will elaborate on the representations of this story that have emerged.

Framing the narratives

The meanings of Dolly were fixed by major social actors, including the media, in argumentations and iconography. Themes are major threads in the arguments concerning the cloning event, while metaphors are tools that help to amplify these themes.[9] These elements, in turn, tend to be organised within a 'frame'. The concept of 'frame' refers to the patterns of interpretation, presentation, selection and emphasis (or exclusion) employed in organising stories, which are all aspects of developing a general story-line (Gitlin, 1980; Gamson and Modigliani, 1989). 'To frame is to select some aspects of a perceived reality and make them more salient in a communication text, in such a way as to promote a particular problem definition, causal interpretation, moral evaluation, and/or treatment recommendation' (Entman, 1993: 52). These frames arise from media practices, as well as from the claims-making activities of other social actors. These framing practices, in turn, are part of the processes of social representation, illustrating our earlier point about the social nature of news production.

Actors

Who are the relevant social actors who help in framing a technology's emergence in the public sphere and who provide suggested interpretations of its meanings and implications? Our comparative quantitative analysis of sources showed that the scientific voice was dominant in all countries except Canada and Germany, where actors from industry were most frequently cited. Politicians were second in prominence in Switzerland, Germany, Greece, Sweden and the UK, and industry came close behind.

While newspapers highlighted local actors (scientists, ethicists, industry representatives), prominent international actors circulated their messages, which, in turn, were amplified as a result of recirculation through international news agency channels. These actors included the US President, the Pope and well-known Nobel prize-winning scientists, all proclaiming the dire consequences of this event and calling for a moratorium and/or additional controls.

The media arena as a forum for these claims-making activities can, in turn, be viewed as the site of a competitive struggle within and between communities of interest. Scientists, for example, do not necessarily belong to a unified community. The competing claims-making activities even among scientists is suggestive of the process of 'boundary work' that characterises controversial areas of science (Gieryn, 1995). Boundary work refers to claims-making activities occurring around the institution of science – who its practitioners are, what are considered its appropriate methods, its accepted stocks of knowledge, its recognised values. The objective is to construct a social boundary to distinguish science from non-science, or to make claims about acceptable versus unacceptable science, as part of a legitimation process (Gieryn, 1983: 782). These activities occur as different social actors contend for, challenge or validate the cognitive authority of science, including the power, credibility, prestige, other material resources or public acceptance that accompany this authority.

If science is viewed as a cultural map, with each new scientific discovery resulting in new demarcations and reconfigurations, the process of drawing boundaries to define or explain each new 'discovery' constitutes a cultural and political exercise. The view that human cloning would be abhorrent to the general public was marked by one set of claims – made, for example, by Ian Wilmut and his colleagues – which separated the work on Dolly (defined as acceptable) from human cloning (defined as unacceptable and abhorrent). On the other hand, claims by other scientists, from industry or other institutions, that the potential benefits of this technology are too invaluable to humanity to justify constraints, expressed the conviction that science and progress should not – and could not – be contained.

Calls for the regulation of cloning work illustrated further attempts to draw these boundaries. Reminders that certain countries already had rules in place to govern or restrict human cloning were part of this effort. The drawing of moral boundaries similarly evoked attempts to delimit how far science might go or how far it was seen to have gone.

Frames

Two general frames pervade the Dolly narrative: a frame of doom and a frame of progress, with the former initially more dominant and pervasive than the latter.[10] The frame of doom is characterised by thematic concerns that centre around threats to identity, the dangers of crossing boundaries (specifically the step into domains hitherto identified with 'nature' and 'God') and runaway science. The last of these themes refers to the gap between science and the rest of society, and the lack of social control over the exercise of scientific power.

The frame of progress, on the other hand, revolves primarily around specific utilitarian arguments. However, an identity theme also runs through this discourse but with a different take – rather than loss, it speaks of the retention of unique identities. A third strand of this frame is that science is a predictable enterprise comprised of incremental steps toward a laudable goal. Underlying this is the suggestion that it is folly to try to contain scientific activity; science, it is claimed, has its own momentum and its own set of controls. Yet these frames are not always mutually exclusive, and at times they overlap.

The analysis of themes and metaphors helps to elaborate further the way the narrative was framed. Metaphors function to anchor unfamiliar notions, events or things, to include them in ordinary categories and place them in a familiar context. They transfer what we know about some area of life onto the unknown one (Lakoff and Johnson, 1980).

Threats to identity Identity is the very essence of self. We equate identity with uniqueness, with a singular personality, with the core of our humanity. The word 'clone' is the antithesis of identity: it evokes the sub-human, the zombie-like state of the replicant. These ideas were captured in the chorus of dismay that greeted the announcement of Dolly. 'Each life is unique, born of a miracle that reaches beyond laboratory science' (Clinton, quoted in a number of newspapers). Austria's *Kurier* declared: 'Life is arbitrarily copied and the respect for individuality and dignity of man is lost'; France's *La Croix*: 'Cloning goes against the certainty that every human is unique, issuing from the singular meeting of one man and one woman, each with their own particular history'; Canada's *Toronto Star*: 'There should only be one of any of us.'

Identity is also conferred by parentage; and in Dolly's case the notion of lineage was certainly questioned. 'This is Dolly the clone, daughter of none', declared the *Toronto Star* (Canada). Austria's *Die Presse* suggested: 'Dolly's father is no ram but the scientist Ian Wilmut.'

Metaphorical extensions of this theme on identity could be found in allusions to *mechanical reproduction* of two types: references to the photocopying process and references to the assembly-line. For example, 'Life by production line' (UK); 'Cloning, the factory of life', 'Photocopied mammals' (Italy); 'Carbon copies' (Canada); 'Mass human production is not far' (Greece); 'The singular individual is put into question by genetic copy machines' (Austria); 'Mass-produced nature', 'The molecular copying machine rotates to spew out duplicates like the devil' (Germany); and 'We turn animals into factories' (Switzerland).

Crossing boundaries If God is omniscient and represents perfection, being human means accepting imperfection. It also means occupying a particular place in the perceived 'natural order' of things. This, in turn, presumes the presence and acceptance of boundaries. Dolly's cloning and intimations of human cloning blurred these boundaries. Thus, playing God or meddling with creation and nature were common expressions of discomfort: Canada's *Calgary Sun*, 'Morally, [cloning] ruffles all sorts of religious feathers.... Will doing such an end run around God's divine order result in a final and ignominious end to the game?'; Austria's *Salzburger Nachrichten*, 'This technique is an unauthorized intrusion into creation.' 'An absolute transgression', cried France's *Le Figaro*. The German newspapers suggested that dreams of immortality and resurrection could now be realised.

Other concerns were expressed in terms of the disappearance of normative sexual reproduction. 'Sperm is passe; so are men. Only eggs are needed. It's as if the birds and the bees have suddenly been rendered irrelevant' (Canada, *Calgary Herald*)

'Cloning bares a spectacular and symbolic aspect: the disappearance of sexual reproduction' (France, *Le Figaro*)

'Has the time come to move from Homo sapiens to Homo xerox? The only sure thing is that we are mathematically driven to the abolishment of sex as the only means of reproduction' (Greece, *Eleutherotypia)*

'A nation of amazons gets closer!' (Germany, *Taz*)

References to God and religion constituted another metaphorical category reflecting dismay at humans overstepping the proper boundaries of intervention and existence:

'Compared with man, God is just a beginner' (Italy)
'This is meddling with creation. God's sacrosanct make-work project' (Canada)
'Virgin birth' (UK)
'God created man and man created the clone' (Austria)
'Science keeps on taking a bite from the forbidden fruit in paradise' (Italy)

References to lambs, or little lambs, in the English papers were clear allusions to William Blake's 'little lamb, who made thee?' Equally likely, playing on the religious symbol of innocence, a quality now perhaps lost, was emblematic. In Germany, *Der Spiegel* was blunt in this regard when it declared that 'the lamb, epitome of devout nativity, has become the symbol of the lost innocence of science'.

Runaway science A large part of modernity involves living with the benefits – and risks – of science and technology. The authority of science, however, is most often questioned when we are first made aware of some startling or profound discovery, or an event that generates challenges and calls for control. The commentary from various newspapers serves to illustrate this.

'Gene research out of control: cloned sheep and then?' (Austria, *Die Presse*)
'Man is being left behind by the technology he created' (Greece, *Apogeumatini*)
'Scientists and the general public have gone in very different directions. Morality is always one step behind technology' (Canada, *Globe and Mail*)
'Bestialisation of science' (Germany, *FAZ*)
'We are treating nature as a continuous guinea pig without appropriate thought or reflection'(Germany, *Taz*)

Eleutherotypia denounced scientists as 'crazy people now able to make their dreams come true!'

The metaphors depicting this theme evoked images of the scientist or science as *monstrous*, with Frankenstein the dominant image:

'Projects out of Dr Frankenstein's horror cabinet' (Austria)
'Frankenstein's monster' (UK)
'The ghost of Frankenstein' (Finland)
'The preliminary stage of Frankensteinian labour creatures' (Germany)
'Human Frankensteins' (Greece)
'We do not know if we wish to play God or Dr Frankenstein' (Canada)

The press also had recourse to several other popular cultural resources, most notably *The Boys from Brazil*, a best-selling novel-turned-film about the Nazi doctor Joseph Mengele and his attempts to raise clones of Hitler, and Huxley's *Brave New World*. More current references to *Jurassic Park* were also made.

In addition, military metaphors were commonly invoked to depict the unbridled power of science. The development of the nuclear bomb was frequently used as an analogy, with the same allusions to the double-edged sword of ultimate power and destruction:

'Dolly, the winning stage of gene technology . . . with the drastic effect of the atom bomb' (Germany)
'Like having an atom bomb in the house' (Italy)
'The bio-engineering equivalent of the first nuclear weapon' (Canada)

Other military metaphors were also prominent, especially in the context of replicating armies of robot-like foot soldiers.

'Whole armada of identical humans could be produced' (Austria)
'Slave armies' (UK)
'Phalanxes of identical Hitlers' (Canada)
'Army of clones' (Finland)
'Legions of cloned Rambos and dictators' (Switzerland)

The world of myth and magic provided a further wellspring of imagery with which to suggest that science was out of control:

'The human clone stands on the horizon like Chronos who eats his own children' (Austria)
'Genetic voodoo'; 'the genie is irretrievably out of the bottle' (Canada)
'Sorcerers' apprentices of science' (France)

The progress frame It was noteworthy that, in virtually all the coverage in our sample, the progress frame, although present, was secondary to the frame of impending doom. However, this frame was still much in evidence in the shape of four different themes: the utilitarian argument; the defence of identity; the limitations on science and scientists; and the more general argument of the inexorable process of discovery in science.

Claims for the utility of cloning focused on agricultural and health benefits, themselves deriving from the production of pharmaceuticals (which was the Roslin team's initial concern); a better understanding of cell development (with consequences for understanding the ageing process); and the development of more efficient animal husbandry practices. All countries in the sample, at one point or another, mentioned these potential advantages.

The frame of progress was also promoted by way of attempts to diminish the concerns related to cloning. This was particularly evident for the theme of personal identity. In response to the chorus of voices that alleged cloning to be antithetical to the individual's sense of identity, a few sources made the counter-argument that environment plays an equally important role in human development; they evinced their claim by observing that identical twins are still different in many ways.

Arguments about the limits already imposed upon science and scientists were also heard. In eight of the ten countries, these varied from pointing to controls already in place to mentioning ethical norms to which science routinely adheres. The unlikelihood of human cloning was also discussed by reference to the inefficiency of the cloning process ('it took 277 attempts').

Finally, the progress frame was also discernible in the occasional – but noteworthy – attempts to place Dolly in the context of previous scientific advances. For example, German, British, Canadian, French and Swiss papers provided brief historical forays into earlier cloning attempts, with the milestone of Dolly portrayed as simply another step in the inevitable progress of science.

At the level of metaphors, this theme was sometimes couched in terms of the military ('a battle won') or sport ('race'). The metaphors of the 'wheel of progress' or of discovery ('opening doors') were employed and the term 'leap' was also often used.

On the whole, however, progress was characterised in direct terms rather than with the aid of metaphorical imagery. The references to the production of more health care products, the greater potential for studying and understanding genetic diseases, and improvements to animal husbandry were straightforward rather than allusive. This is possibly a sign that technological advancement, equated with progress, is taken to be the norm and no longer needs 'translation' and familiarisation.

Consequences Discussions of the potential consequences of cloning included several basic themes. In order to assess their relative importance we content analysed the different risks and benefits emphasised. From this evaluation, it emerged that health benefits predominated, followed by research gains. Economic benefits were touted in only three of the ten countries analysed (table 11.2). The perception of risk, on the other hand, is more homogeneous, with moral risks underlined in every single country. This ubiquity supports the frame of doom within which the story of Dolly is usually couched.

Portrayals of science The preceding discussion of frames was complemented by an additional quantitative analysis of the perceptions of science contained within the press coverage of the Dolly event. This involved quantifying, by means of semantic differential scales, the overall image of science projected in news stories.[11] As table 11.3 notes, science as an

Table 11.2 *Dolly: type of benefit and risk evaluations*

	Benefits			Risks				
Country	Economic	Health	Research	Health	Inequality	Moral	Other	Environment
UK		58	20			50		16
Germany		35	48			43	28	
Netherlands		34	44			87		
Austria	50	20	20			95		
Sweden		33	20			87		13
Greece		37	40		13	53		
Finland		27	33	18	18	37		
Switzerland	28	46				94		
Poland		62	21			82		
Canada	37	49				71		

Note: Only scores larger than 10 per cent are listed. Data for France and Italy are not available.

Table 11.3 *Dolly: the image of science*

	Image of science					
Country	Creative vs. uncreative	Successful vs. unsuccessful	Moral vs. immoral	Conscious vs. unconscious of responsibility	Constructive vs. destructive	Public welfare vs. egoistic
UK	6.2	6.6	**4.7**	**5.2**	**5.6**	**4.3**
Germany	7.3	8.1	**4.0**	**4.0**	**5.2**	**4.1**
Austria	6.0	6.7	**3.9**	**4.2**	5.7	**4.7**
Sweden	5.6	6.1	**4.2**	**5.1**	**5.4**	**5.1**
Greece	8.0	7.1	**4.5**	**5.1**	5.8	6.0
Finland	7.5	7.0	**4.8**	**5.0**	**4.9**	**5.0**
Switzerland	**5.1**	6.4	**5.5**	5.8	**5.0**	**5.5**
Poland	7.3	7.3	**5.3**	**5.4**	**5.5**	**5.5**
Canada	7.5	7.3	**4.4**	**5.4**	**4.8**	**4.9**

Note: Scores represent rating of image on a semantic differential scale of 1–10.

enterprise was rated by different country teams according to criteria relating to scientific and social orientation. Aspects of the former included notions of creativity and success, while the latter incorporated morality, responsibility, consideration of public welfare, and constructiveness. The picture of science that emerged was of a highly creative and successful enterprise, but at the same time a socially irresponsible and unresponsive institution (table 11.3).[12] This finding, which was confirmed by a factor analysis of the entire sample of stories,[13] showed that, though science was framed in different ways, in this context negative imagery was clearly dominant.

The use of humour We have identified the use of humour as another element in the evolution of the story on Dolly. It was notable that no country was immune to the need to joke. Cartoons and photomontages were frequent vehicles for humour (see plate 11.3), but wordplay was another humour tool. Humour was a common approach to anchoring the Dolly event for readers, and we discuss it separately here if only to point out another unique attribute of the Dolly story.

Textual humour was often conveyed through the use of puns or other forms of wordplay. In Canada's *Globe and Mail*, these puns were illustrative: 'Are "ewe" ready?' or 'Send in the clones' (a play on the popular song 'Send in the clowns'). Cartoon subjects focused on the theme of copying by duplicating politicians or sports or entertainment figures. The Swedish papers suggested, for instance, that Ian Wilmut might make a good house doctor for the Swedish hockey team by replicating the best players.

What was it about this cloning story that so easily – and so frequently – lent itself to joking? Was it because the dark possibilities of human cloning seem so distant that one can have easy recourse to reassuring humour? Or was it a reflection of unease or nervous fascination at what humankind had wrought? Was it the protective cover that joking so often affords, providing a release from the anxieties and paranoias evoked by the dangers cloning might generate, which place in jeopardy our most cherished notions – of who we are, the sacredness and uniqueness of life, the importance of sex? Going to an extreme, were we feeling perhaps 'that this joking in the face of a new possibility for mass-produced life is in fact joking in the face of death, death of the spirit and death of the (male) body' (Miller, 1998: 80)?

Reflecting on the same theme of joking in the context of cloning, Miller (1998: 81) maintains that 'the intense presence of the urge to joke is as sure an indication as there is that we are approaching the dangerous, the sacred, and the magical. Pious and grave talk about human dignity is so often untrustworthy . . . , so unfelt, so by rote, so safe and predictable that some feel it necessary to retreat to the joke to pay serious homage.'

Different motivations may be at play at different times, for different papers or countries, but it is not inconceivable that the varied dimensions of cloning and its implications could evoke a range of humorous reactions – from the laughter of conceit to that of resistance, from the laughter of amusement to laughter in the face of doom.

Plate 11.3 Cover page of *Der Spiegel*, March 1997
Note: The captions read 'Science marching towards the cloned human being' and 'The Fall of Man'.

Iconography

All contemporary news media rely heavily upon visual communication through pictorial representations. These images, along with headlines, are important seduction devices to capture readers' attention and to guide their reading of the text. They are also important tools in fixing meaning and anchoring elements in more global contexts. Finally, images capture, crystallise and render visible the abstract or ambiguous. Out of all potential images of cloning and clones, the standard picture of Dolly was really the only constant in the story told through iconography. This image can be understood as an instance of objectification by which the abstract, ambiguous or threatening idea of cloning was embodied in the image of the sheep. It was present throughout the period, at times subjected to explicit manipulation in photomontages (such as on the cover of *The Economist*) but always recognisable as embodying the core of the issue. The sheep came to embody the image of all clones. By its omnipresence in the process of storytelling, this image absorbed other meanings and associations made in the course of narration, and it maintained and carried with it traces of the political and ethical debate. It can be argued that the image of Dolly became a symbol for cloning and, moreover, a motivated symbol for biotechnology by encapsulating ambivalent public attitudes. The image of an innocent-looking sheep captured the fleeting moment of moral consensus when the world seemed to agree on the boundaries of acceptable interference in life. Its frequent appearance in subsequent stories was a reminder of that fleeting moment. The image of Dolly retained the connotations of an innocent creature and of the necessity for, but at the same time the impossibility of, drawing stable boundaries between manipulation and unacceptable interference in the nature of life (see Kolata, 1998).

From a historical point of view, images invoked in relation to science and scientists displayed considerable constancy. Turney (1998), for example, traces the evolution of biologists' image in the media and shows that shifts in the way they are portrayed correspond to changes in biologists' practices. As they moved from description and classification to experimentation, from a focus on the organism to the more abstract realm of molecular biology, the image evolved from biologists' looking through microscopes to posing with molecular models or a double helix. Turney maintains that life has become less and less recognisable and, over time, has more frequently been reduced to abstractions (1998: 43). In the Dolly story, Wilmut with the test tube may be viewed in this latter context.

Local differences in elaborations

Although we have focused on many commonalities across the twelve countries, there are, of course, notable distinctions as well. Not surprisingly, national pride is clearly evident in British coverage, with frequent emphasis on Dolly being a *British* scientific accomplishment. The usefulness of the cloning procedure was also frequently highlighted in the UK, as were the economic benefits.

Finland was noteworthy for its almost unquestioning representation of the event. Scientists escaped the censure evident in other countries' coverage and calls for control or regulation were muted. This non-critical representation corresponds with more positive Finnish attitudes toward biotechnology in general (see chapter 7 in this volume). At the other extreme, Italy and Greece focused enormous attention on its negative aspects and connotations of aberrancy.

In the case of Germany, the shadow of not-too-distant history may account for the absence of allusions to Hitler (as a symbol of the potential for 'copying' evil) that were so prominent in all the other countries. *Der Spiegel*'s photomontage displayed in plate 11.3 is the exception.

A strong emphasis on regulatory control was conspicuous in the Swiss coverage. This was both a reaffirmation of the ban on cloning already in place but also a reflection of the debates surrounding the impending national referendum on genetic engineering research.

In Canada, Italy and Sweden, some of the rhetoric on cloning was framed in the context of recent domestic debates about reproductive technologies and human embryo experiments. In the case of Italy, the announcement of the birth of a healthy cloned sheep was made in a very particular discursive environment. In the weeks and months before, a heated political debate on the status of the embryo and embryo research had been waged, involving criticisms of the absence of existing legislation on reproductive technologies. It did, however, take a few days for the two stories to fuse. In Canada, a Royal Commission's three-year-long examination of reproductive technologies had already proposed, among other things, a ban on human cloning. In the Canadian press this received renewed attention.

Conclusions

In tracing the evolution of this narrative, we have shown the chief storyline emerging as that of human cloning. Dolly the sheep became the motivated symbol for humans as clones. In this story, the initial consensus was one of moral outrage and condemnation.

Why did Dolly hit the headlines worldwide? It is not the scientific event as such that explains the story's popularity, but the fact that the Dolly issue had profound cultural resonances. That it occurred *across* national imaginations reveals a social construction process of a large-scale mosaic.

One can point to journalistic routines that are evoked in any major media story, and Dolly was no exception. Journalists drew on similar metaphorical images; they highlighted the same key events and actors; they tried to 'balance' their accounts of those who denounced the breakthrough and those who exalted it, though the former were clearly in the ascendant. In terms of media practices, it is interesting that the account of Dolly was treated as a unique event, an enormous surprise, a 'technological leap', despite the fact that the idea behind her creation had been around for more than half a century. This is perhaps because journalism is encoded in short-term memory. Various cloning efforts over several decades, from frogs to mice to cattle, have been detailed in historical accounts of modern biology and its portrayals in the popular press (see, for example, Kolata, 1998; Turney, 1998). With few exceptions, however, this historical context went unutilised.

In viewing the Dolly story as an episode of social representation, the activities of anchoring and objectifying emphasised the media's role in the processes of diffusion and the propagation of representations. The acts of highlighting the scientific advance and classifying it in terms of its human cloning implications helped to anchor the issue, while the uses of imagery, the patterns of thematic emphasis and argumentation, and the elucidation of the scientific processes involved assisted in making concrete or objectifying the event. The questions Moscovici (1984: 23) posed earlier remain highly relevant: how do we explain these patterns and why do we create such representations? In this instance of cross-national comparison, why has there been such a convergence of narratives and representations?

Social representations are said to reveal themselves with greater clarity in instances of crisis or during an upheaval of thought (Moscovici, 1984). Dolly as a technological event provides an important blip on the evolutionary landscape of biotechnology that has instigated – even necessitated – an important restructuring of mental maps. Where the notion of cloning ourselves had been strictly confined to the realm of fiction, Dolly transposed it into the real world.

If the purpose of representations is to make the unfamiliar familiar, what the story of Dolly has done is to acquaint us with a new dimension of our technological prowess. One function of this is to advance a preferred vision of some ideal world, what Moscovici has called 'the hypothesis of desirability' (1984: 23). One version of this desirable world – the world of scientific progress, of conquered ills – is often packaged in 'the rhetoric of hope' (Mulkay, 1993). It is a tale that has been told countless times before and employs devices common to audiences across cultures.

At the same time, this vision has to compete with a version that unequivocally rejects it. This rival frame is replete with intimations of doom and employs an uncompromising 'rhetoric of fear' (Mulkay, 1993). In this version, the evocations of monsters, myth and magic and of 'mad science' have been found across various retellings and across different technologies (Toumey, 1992; Turney, 1998). The use of this 'rhetoric of fear' is functional when science and technology can be represented as violating basic cultural categories and moral codes (Mulkay, 1993: 724). Mulkay argues that the voices of fear represent 'a culturally subordinate discourse of science' (1993: 736). Images of monsters (commonly Frankenstein), of mad scientists and the need for controls, of the distasteful aspects of industrial assembly-line and cookie-cutter uniformity, and of the violation of identity and humanity are all highly serviceable weapons in the struggle to define a technology and its proper limits.

This struggle to define the more enduring representation of a technological advance is at the heart of the story of Dolly. It is a story that illustrates a fleeting moment of moral consensus on this particular scientific endeavour. At the same time, it richly encapsulates the continuing ambivalence we have about the project of modernity and its notion of 'progress'. It encompasses our railings against the increasing chasm between a runaway science and the rest of society, and it involves the projection of our worst fears about the unbridled power of scientists. It incorporates our ambivalence about science and embodies our deepest fears about threats to our identity. Through the allusions to the crossing of boundaries between the 'natural' and the 'artificial', it has brought out in sharp relief our deepening anxieties about the usurpation of divine power. In short, this may be the narrative of our ongoing, reflexive project of ourselves in modernity.

Postscript

About a year after the Dolly story broke, she was reported as giving birth (a natural pregnancy) to a healthy lamb, named Bonnie. This event declared the cloned sheep to be no different from a 'normal' sheep. Not long after, scientists at the University of Hawaii replicated the Roslin effort, this time on mice. The mice were successfully cloned by transferring donor nuclei (not whole cells) into eggs. With this, *Time* magazine declared: 'Dolly, you're history.'

Methodological appendix

Sample

Austria *Neue Kronen Zeitung* (mass readership, opinion leader); *Die Presse* (small readership, quality, conservative Catholic); *Kurier* (mass readership); *Salzburger Nachrichten* (high quality, local readership).

Canada *Globe and Mail* (quality, conservative, national newspaper); *Calgary Herald, Toronto Star, Calgary Sun* (all local city English newspapers); *Le Devoir* (quality French paper).

Finland *Helsingin Sanomat* (independent, centrist); *Savon Sanomat* (independent, centrist); *Demari* (left wing, represents Social Democratic Party); *Suomenmaa* (centrist, Centre Party); *Kotimaa* (religious, Lutheran); *Ilta Sanomat* (independent afternoon paper); *Kauppalehti* (business paper).

France *Le Monde* (centre, liberal); *Libération* (centre-left); *Le Quotidien du Médicin* (specialist press); *Ouest France* (regional); *France Soir* (popular); *Le Figaro* (conservative); *La Croix* (Catholic).

Germany *BILD* (national daily, 5 million circulation); *Frankfurter Allgemeine Zeitung* (*FAZ*, largest national daily, right); *Tageszeitung* (national daily, left wing); *Leipziger Volkszeitung* (regional east German daily paper); *Der Spiegel* (national weekly news magazine); *Taz* (alternative, intellectual).

Greece *Ethnos* (socialist); *Apogeumatini* (popular, right wing); *Ta Nea* (independent, socialist); *Eleutherotypia* (opinion leader); *Kathimerini* (quality Sunday paper).

Italy *Corriere della Sera* (opinion leader, centrist, largest-circulation daily); *La Repubblica* (alternative opinion leader, centre-left); *L'Unita* (centre-left, formerly Communist Party daily); *Il Giornale* (right wing); *La Nazione* (semi-popular, regional, right wing).

The Netherlands *Volkskrant* (quality, left of centre); *NRC/Handelblad* (quality, right of centre); *Trouw* (quality, left of centre); *Algemeen Dagblad* (popular, right of centre); *Telegraaf* (popular, right of centre).

Poland *Gazeta Wyborcza* (quality daily, liberal); *Rzeczpospolita* (quality daily, centre); *Zycie* (quality daily, right); *Trybuna Demari* (left, represents Social Democratic Party); *Slowo* (Catholic daily); *Przeglad Tygodniowy* (weekly, centre); *Zycie Warszawy* (Warsaw daily, national circulation, centre to apolitical); *Gazeta Poznanska* (local Poznan daily, centre to apolitical); *Glos Wielkopolski* (local Poznan district daily, apolitical); *Expres Poznanski* (local Poznan daily, apolitical).

Sweden *Aftonbladet* (tabloid).

Switzerland *Blick* (German popular daily); *Tages-Anzeiger* (German, opinion leader); *Neue Zurcher Zeitung* (elite, international reputation, German); *Basler Zeitung* (German daily, regional); *Le Matin* (French daily); *Le Nouveau Quotidien* (French daily, urban readership); *Journal de Genève* (French daily, quality); *Appenzeller Zeitung* (German daily, rural).

The UK *Guardian* (left, quality), *Independent* (centrist, quality), *The Times* (conservative, quality).

Procedure

A census of all stories about Dolly in the selected newspapers was made by the country teams. The selection of newspapers for analysis was left to each country team, with an eye to having a diverse set for analysis.

Both qualitative and quantitative analyses were carried out. The qualitative analysis was conducted through close reading of the textual material by each country team. A day-by-day synopsis was then prepared in English for each paper. The second coding stage involved developing a list of themes from the text. From this data set, a summary matrix was created showing which themes were covered by which countries, providing a picture of similarities and differences across countries.

A third stage involved listing all the metaphors employed in each story and categorising this data set. The fourth stage incorporated analysis of the images (photographs, diagrams, cartoons) on the subject.

Quantitative analysis was also carried out. A coding sheet was drawn up with a variable set that included actors, risks and benefits, and an overall image of science as determined from the tone of the article. This data set was analysed with SPSS.

NOTES

1. To put the term 'global' in context, there is considerable evidence of coverage of Dolly in many parts of the world. Our study, however, focuses on a small sample of Western industrial countries.
2. It should be noted that the term 'biotechnology' is not exclusive to the process of recombinant DNA, more commonly known as genetic engineering. Its broader definition includes processes such as protoplast fusion and cell and tissue culture. It is in this respect that the notion of cloning can be considered to be a biotechnological activity. Nor is the term 'cloning' entirely precise, covering a number of different scientific processes. For example, it has been used to refer to molecular cloning (the duplication in a host bacterium of DNA strings containing genes); cellular cloning (where copies of cells are made indefinitely to create 'cell lines'); embryo twinning (where a sexually developed entity, an embryo, is split into identical halves); and nuclear somatic transfer, the process by which Dolly was created (see Pence, 1998).

3. We make this judgement not just in terms of the media attention accorded to the event, but also on the basis of the judgement from within the scientific community itself. The journal *Science* deemed the cloning story one of the major 'breakthroughs' at the end of 1997 (*Science*, 19 December 1997). It described a breakthrough as 'a rare discovery that profoundly changes the practice or interpretation of science or its implications for society' (p. 2029).
4. Dolly, the sheep, was created from a mammary cell. This udder lineage inspired a moment of wry frivolity, with the well-endowed Dolly Parton providing the naming inspiration, according to Wilmut (Kolata, 1998: 3).
5. An 'embargo' (literally, an official suspension of an activity) is a press relations tool to hold a story announcement or keep its contents from public dissemination until a specified date. Important journals such as *Science* or *Nature* will send out press releases on an important upcoming article and will sometimes put an embargo on the story until the day of publication, ensuring maximum publicity for the journal.
6. The UK *Observer*'s science reporter was able to break the embargo because he used a different source – an upcoming TV documentary on the Roslin research (see Wilkie and Graham, 1998; also Kolata, 1998). As for the Italian newspapers, the national news agency, Anza, may have provided the initial story (see Wilkie and Graham, 1998), but it is not clear where the agency's information originated. In any case, the sensationalised coverage in Italy did not begin until Monday, the 24th, when everyone else raised the human cloning theme in alarm.
7. In 1993, American scientist Jerry Hall was reported to have cloned seventeen human embryos into forty-eight. This was an attempt to increase the embryo supply in fertility clinics. This was not a nuclear somatic transfer procedure, however, but a case of embryo twinning (see Pence, 1998).
8. We recognise that there are obvious differences between newspapers within any country (for example, the differences between quality and popular newspapers). However, our analysis glossed over these differences, emphasising instead between-country distinctions or similarities. More in-depth analyses of media coverage of this event have been carried out in Italy, Canada and France (see Berthomier, 1999; Di Palma, 1998; Downey, 1999; Manzoli, 1998; and Rizzo, 1998).
9. We rely on the work of Lakoff and Johnson (1980), who maintain that our conceptual system is largely metaphorical, that is, our ways of thinking and experiencing and doing are structured by metaphor. When metaphors are used as linguistic expressions, they succeed largely because these are usually present in the recipients' conceptual systems.
10. In identifying the general frames of doom and progress that have surrounded representations of the Dolly story, our findings correspond with similar framing approaches that emerged around the Human Genome Project. According to Durant and colleagues (see Durant et al., 1996), a discourse of progress or hope and a discourse of concern were embedded in the British media coverage, but were similarly echoed in focus group discussions with the general public.
11. Semantic differential scales are rating scales using polar adjectives to tap evaluative dimensions of an object, particularly quality and intensity dimensions

(Osgoode et al., 1971). These measures are used to analyse 'meaning' by tapping representational processes in language. In this instance, 'science' as an enterprise might be rated according to how it is portrayed in a news story as 'socially responsible' at one end of the scale and 'socially irresponsible' at the other end.

12. This observation is restricted to science mentioned in respect to Dolly and covers only the first ten days of the newspaper coverage.
13. Two components explain about 60 per cent of the variation: 'ingenious science' and 'morality of science' (these terms are derived from the factors with the highest loading).

REFERENCES

Bauer, M.W. and G. Gaskell (1999) 'Towards a Paradigm for Research on Social Representations', *Journal for the Theory of Social Behaviour* 29: 163–86.

Beck, U. (1992) *Risk Society: towards a New Modernity*, London: Sage.

(1998) 'Politics of Risk Society', in J. Franklin (ed.), *The Politics of Risk Society*, Cambridge: Polity Press, pp. 9–22.

Berman, M. (1982) *All That Is Solid Melts into Air: the Experience of Modernity*, New York: Simon & Schuster.

Berthomier, A. (1999) 'Discours médiatiques sur la biotechnologie en France, 1973–1996', unpublished PhD thesis, Ecole Normale Supérieure de Fontenay/Saint-Cloud, France.

Darnton, R. (1975) 'Writing News and Telling Stories', *Daedalus* 104: 175–94.

Di Palma, V. (1998) 'Tra Sacro e Profano: biotecnologie e clonazione nell stampa cattolica q in quella laica. Tesi di Laurea in Scienze della Communicazione', University of Siena.

Downey, R. (1999) 'The Social Construction of Technology in Late Modernity: a Case Study of Cloning in the Canadian Media', unpublished MA thesis, University of Calgary.

Durant, J., A. Hansen and M. Bauer (1996) 'Public Understanding of the New Genetics', in T. Marteau and M. Richards (eds.), *The Troubled Helix: Social and Psychological Implications of the New Human Genetics*, Cambridge: Cambridge University Press.

Entman, R. (1993) 'Framing: towards Clarification of a Fractured Paradigm', *Journal of Communication* 43 (4): 51–8.

Farr, R. (1987) 'Social Representations – a French Tradition of Research', *Journal for the Theory of Social Behaviour* 17: 343–70.

Gamson, W. and A. Modigliani (1989) 'Media Discourse and Public Opinion on Nuclear Power: a Constructionist Approach', *American Journal of Sociology* 95: 1–7.

Giddens, A. (1991) *Modernity and Self-Identity: Self and Society in the Late Modern Age*, Stanford, CA: Stanford University Press.

(1994) 'Living in a Post-Traditional Society', in U. Beck, A. Giddens and S. Lash (eds.), *Reflexive Modernization: Politics, Tradition and Aesthetics in the Modern Social Order*, Stanford, CA: Stanford University Press.

(1998) 'Risk Society: the Context of British Politics', in I.J. Franklin (ed.), *The Politics of Risk Society*, Cambridge: Polity Press, pp. 23–34.

Gieryn, T. (1983) 'Boundary work and the demarcation of science from non-science: strains and interests in professional ideologies of scientists', *American Sociological Review* 48: 781–95.

(1995) 'Boundaries of Science', in S. Jasanoff, G. Markle, J. Petersen and T. Pinch (eds.), *Handbook of Science and Technology Studies*, Thousand Oaks, CA: Sage Publications, pp. 393–443.

Gitlin, T. (1980) *The Whole World Is Watching: Mass Media in the Making and Unmaking of the New Left*, Berkeley: University of California Press.

Kolata, G. (1998) *Clone: the Road to Dolly and the Path Ahead*, New York: William Morrow.

Lakoff, G. and M. Johnson (1980) *Metaphors We Live by*, Chicago: University of Chicago Press.

Levin, I. (1977) *The Boys from Brazil*, New York: Random House.

McCombs, M., E. Einsiedel and D. Weaver (1991) *Public Opinion: Issues and the News*, New Jersey: Lawrence Erlbaum.

Manzoli, F. (1998) *Divulgazione o finzione? La clonazione rappresentata sui quotidiana. Tesi di Laurea in Scienze della Comunicazione*, Siena: University of Siena.

Miller, I.W. (1998) 'Sheep, Joking and the Uncanny', in M. Nussbaum and C. Sunstein (eds.), *Clone and Clones: Facts and Fantasies about Human Cloning*, New York: W.W. Norton, pp. 78–87.

Moscovici, S. (1976) *La Psychanalyse – son image et son public*, 2nd edn, Paris: PUF.

(1984) 'The Phenomenon of Social Representations', in R. Farr and S. Moscovici (eds.), *Social Representations*, Cambridge: Cambridge University Press, pp. 3–70.

Mulkay, M. (1993) 'Rhetorics of Hope and Fear in the Great Embryo Debate', *Social Studies of Science* 23: 721–42.

Osgoode, C., G. Suci and P. Tannenbaum (1971) *The Measurement of Meaning*, Urbana, Il: University of Urbana Press.

Pence, G. (1998) *Who's Afraid of Human Cloning?* Lanham, MD: Rowman & Littlefield.

Rizzo, B. (1998) 'La construzione sociale della notizia scientifica: Il caso di Dolly, pecora clonata. Tesi di Laurea in Scienze della Comunicazione', University of Siena.

Toumey, C.P. (1992) 'The Moral Character of Mad Scientists: a Cultural Critique of Science', *Science, Technology and Human Values* 17: 411–37.

Turney, J. (1998) *Frankenstein's Footsteps: Science, Genetics, and Popular Culture*, New Haven, CT: Yale University Press.

Wagner, W. (1996) 'Queries on Social Representations and Construction', *Journal for the Theory of Social Behaviour* 26: 95–120.

Wagner, W., G. Duveen, R. Farr, S. Jovchelovitch, F. Lorenzi-Cioldi, I. Marková and D. Rose (1999) 'Theory and Method of Social Representations', *Asian Journal of Social Psychology* 2: 95–125.

Wilkie, T. and E. Graham (1998) 'Power without Responsibility: Media Portrayals of Dolly and Science', *Cambridge Quarterly of Healthcare Ethics* 7 (2): 150–9.

Wilmut, I., A.E. Schneike, J. McWhir, A.J. Kind and K. Campbell (1997) 'Viable Offspring Derived from Fetal and Adult Mammalian Cells', *Nature* 385: 810–13.

Part IV

The transatlantic puzzle

12 Worlds apart? Public opinion in Europe and the USA

George Gaskell, Paul Thompson and Nick Allum

Introduction

Through the 1990s research and applications of modern biotechnology were making impressive strides in the United States. Viewed from the perspective of the biotechnology lobby in Europe, the United States was enjoying an enviable and apparently effortless assimilation of the fruits of the technology of the twenty-first century. By the middle of the decade, the idea of 'Life Sciences Conglomerate' integrating agri-chemicals and genetically modified (GM) seeds, foods and pharmaceuticals became a reality. Monsanto emerged as the Microsoft of gene technology and appeared to be on the threshold of global domination. New GM strains of soya, maize and cotton, along with a number of other GM crops, were planted across millions of hectares of the United States and were taking larger and larger market shares. After the failure of the 'Flav'r Sav'r Tomato®' (1995–6) tomato, it seemed as if the potential of modern biotechnology to revolutionise food production was to be realised.

On the other side of the Atlantic, to the chagrin of the biotechnology lobby, Europe was in turmoil as biotechnology became increasingly controversial and political. In 1997 the cloning of Dolly the sheep (see Einsiedel et al., chapter 11 in this volume) became an international, and largely polemical, news event. In response to public hostility over the importing of GM crops, Monsanto launched a European public relations campaign in 1998. The company's own research indicated that the campaign failed to persuade the sceptical public.

Following widespread controversy, six European member states, in contravention of EU regulations, introduced a moratorium on the commercial planting of GM crops. Prompted by commercial rather than scientific considerations, supermarkets in a number of countries announced boycotts of GM foods. In 1999, prompted by the Pusztai research,[1] which claimed to demonstrate that rats fed on a diet of GM foods suffered ill-health, the media turned against modern biotechnology.

As a result a fault-line opened between Europe and the United States in a World Trade Organisation (WTO) dispute, bringing threats of a trade war. The US companies and the Department of Agriculture asserted that GM products were safe, judged on the criterion of scientific risk assessment. In Europe, confronted by a sceptical public operating along the lines of an intuitive version of the precautionary principle, many national governments dithered.

The United States, however, could not insulate itself from the impact of the European revolt against GM crops and foods. Confronted by a consumer opposition in Europe that spread to other parts of the world, US farmers worried about markets for their produce and the shareholders of the life sciences companies took flight. As stock market valuations fell, the integrated life science concept was in tatters and there followed the break-up of agri-chemicals and GM seeds from the still successful pharmaceutical divisions.

With the benefit of hindsight, a shift in the European public's response to biotechnology can be identified. Before the 'watershed years' of 1996 and 1997 it was an esoteric issue, the concern of specialists and of relatively little interest to the majority of the public. From 1997, biotechnology entered a second 'political' phase of development. As products came into the market, the traditional criteria for technological assessment of risk and safety were challenged on both scientific and ethical grounds. Bioethics, environmental impacts and public participation had featured in the European debate since 1975, but now these issues achieved greater prominence.

In this chapter we contrast public perceptions of biotechnology in the United States and Europe and investigate whether the controversy of the late 1990s might have been anticipated. Can we find at least partial accounts for the dramatic differences between the reception of biotechnology in Europe and the United States in the mid-1990s and for the subsequent change in fortunes of the US life science companies at the close of the decade? Were there harbingers in the pre-political phase of the problems to come?

Our analysis starts with a comparison of public perceptions of five applications of modern biotechnology in Europe and the United States and then explores the roots of these differences in the context of media coverage, scientific knowledge and trust in the regulatory processes.

Public perceptions of biotechnology in Europe and the United States

Our primary data source is the 1996 Eurobarometer survey (46.1) on biotechnology (Durant, Bauer and Gaskell, 1998). Many of the questions

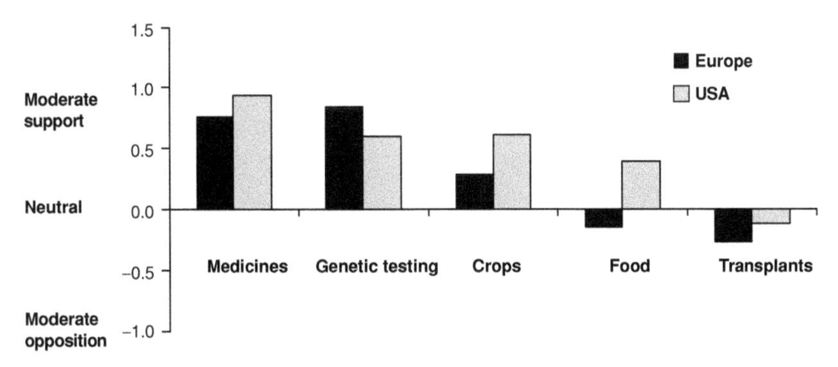

Figure 12.1 Mean support for five applications of biotechnology: Europe and USA

Note: Support is measured on a scale from −2 to +2. The USA and Europe differ significantly for each application (f-values from one-way ANOVAs for each application were all significant with $p < .05$).

from the Eurobarometer survey were also used in a US survey in late 1997.[2] These surveys provide a historical snapshot of public perceptions in 1996–7. Of course, with the rapid advance of food biotechnologies and other developments in the life sciences we would not expect to find the same opinions and attitudes today. But the use of similar questions in the surveys makes it possible to look at comparative structural differences in the pattern of public perceptions that may hold clues to understanding the contrast between the USA and Europe in more recent years.

Respondents were asked whether they thought each of five biotechnologies – genetic testing, GM medicines, GM crops, GM food and GM animals for use in human transplantation ('xenotransplantation') – was useful, risky, morally acceptable and to be encouraged. Figure 12.1 shows the mean levels of support (encouragement) on a scale from +2 to −2 for all the applications.

People in Europe and the USA showed varied levels of support across the different applications. GM medicines and genetic testing received the highest levels of support, GM crops and GM foods received intermediate levels of support, and xenotransplantation received least support. And there was not always strong support for biotechnology in the USA: for example, the average US respondent was opposed to xenotransplantation. Furthermore, they were not always more supportive than Europeans; for example, Europeans were more supportive of genetic testing. However, people in the USA were significantly more supportive of GM crops and GM foods than were people in Europe.

When the surveys were conducted, biotechnology was a relatively unfamiliar topic. On the questions about the five applications, 19 per cent

Table 12.1 *Three common logics in relation to attitudes to five applications of biotechnology*

	Attitude			
Logic	Useful	Risky	Morally acceptable	Encouraged
1. Support	YES	NO	YES	YES
2. Risk-tolerant	YES	YES	YES	YES
3. Opposition	NO	YES	NO	NO

of people in the USA and 27 per cent of Europeans did not give a complete set of responses. With this level of unfamiliarity we can assume that some people responded to the questions with poorly informed and unintegrated 'non-attitudes' (see Midden et al. in chapter 7 of this volume). Such responses would be likely to be volatile if, for example, the issue became more controversial. In the absence of a filter question allowing us to exclude people with 'no opinion', the following analysis uses only those who gave a full set of responses, on the assumption that they were more likely to have better formed opinions. Judgements of use, risk, moral acceptability and encouragement were each collapsed into a dichotomy (useful/not useful, etc.) in order to model patterns of response (henceforth 'logics') over the four dimensions of attitude. This produces sixteen possible combinatorial 'logics', but empirically only three were widely used (see table 12.1). Logics 1 and 2 are similar in being supportive, but they display different perceptions of risk. For the 'supporter', risk is not an issue. The 'risk-tolerant supporter' sees, but then discounts, the risk. Opponents take a position exactly opposite to that of supporters.

Table 12.2 shows the distribution of these three prevalent logics for each application. For GM medicines and genetic testing, supporters constituted the single largest category. Levels of risk-tolerant support were also relatively high, and levels of opposition were relatively low. Higher opposition to genetic testing in the USA ($p < .05$) than in Europe may indicate a sensitivity about genetic privacy in the context of work, credit or insurance. In contrast, for xenotransplantation, supporters and risk-tolerant supporters totalled only 36 per cent in Europe and 42 per cent in the USA, with about 33 per cent in opposition.

Turning to GM crops and GM foods, we see a considerable contrast between Europe and the USA. Both GM crops and GM foods were better supported in the USA than in Europe (for both contrasts, $p < .05$). The contrast is greatest in the case of GM foods, to which 30 per cent of Europeans with a logic were opposed, compared with only 13 per cent

Table 12.2 *Distribution of logics for five applications of biotechnology: Europe and USA*

	Europe		USA		
Application	% of respondents with complete set of responses[a]	% of total sample[b]	% of respondents with complete set of responses[c]	% of total sample[d]	T-value[e]
Medicines					
Support	41	30	54	44	4.76
Risk-tolerant	37	27	29	23	1.52
Opposition	8	6	5	4	
Genetic testing					
Support	50	37	51	41	−6.08
Risk-tolerant	33	24	21	17	−9.38
Opposition	7	5	14	11	
Crops					
Support	35	26	51	41	8.17
Risk-tolerant	26	19	22	18	3.07
Opposition	18	13	10	8	
Food					
Support	22	16	37	30	11.89
Risk-tolerant	21	15	24	19	8.13
Opposition	30	22	13	11	
Xenotransplants					
Support	16	12	23	19	2.86
Risk-tolerant	20	15	19	15	−1.47
Opposition	33	24	35	28	

Notes: Loglinear modelling on each application, with 'opposition' as the reference category, shows that the probability of being a 'supporter' or 'risk-tolerant supporter' differs significantly ($p < .05$) for the USA and Europe, with the exception of 'xenotransplants' and 'medicines' where there is no significant difference in the probability of risk-tolerant support.

[a] $N = 12{,}178$.

[b] $N = 16{,}500$.

[c] $N = 863$.

[d] $N = 1{,}067$.

[e] *T*-values of > 1.96 indicate significance at $< .05$.

of Americans. By the same token, whereas over 60 per cent of people in the USA were supporters or risk-tolerant supporters of GM foods, the comparable figure for Europe is just over 40 per cent.

That, as early as 1996, such a large percentage of Europeans were concerned about the usefulness, risks and moral acceptability of GM foods is of particular note. What is clear is that there was a groundswell of people opposed to GM foods, for whom assurances about the absence of scientific risks would have been unlikely to alleviate their concerns. For them the tortuous discussions and long delay in introducing a labelling scheme for GM foods may well have created further anxieties. People may have thought 'If *they* say it is safe, why should *they* hesitate to label it? *They* must be hiding something' – *they* being a mixture of the industry and anonymous regulators.

A fourth possible logic – 'moral opposition' (in terms of table 12.1, answers = yes, no, no, no) – counts for no more than 3 per cent on any applications. That so few people adopted the converse of the logic of risk-tolerant support implies that respondents with concerns about gene technology tended to think principally in terms of moral acceptability rather than risk – a significant difference from the way in which experts normally judge the acceptability of new technologies. But we must not assume that the opposition was purely of a moral kind without considerations of benefits and costs. In the public mind the representation of potential danger is a complex of perceived benefits and risks. The first generation of GM foods offered producer rather than consumer benefits, yet any health risks were to be borne by the consumer. In these circumstances it seems not unreasonable to take the position, 'Why should I take any risk with this GM food, if there is no apparent benefit for me?'

In summary, with 30 per cent of Europeans opposed to GM foods there is clear evidence that the foundations of the consumer revolt were laid before 1997. This raises the question about the origins of the contrasting views of GM foods in the United States and Europe. In some respects the perplexing issue is why Europe was relatively more negative – perplexing because, when the surveys were conducted, there was relatively little penetration of GM food products in Europe (save for the introduction of GM tomato puree in Britain in 1996, voluntarily labelled as genetically engineered and greeted with little interest).

It seems as if negative opinions were based on individual cognitions and preferences rather than being the product of debates in the public domain. In this sense the opposition may have been implicit, articulated only in the context of a survey question, rather than reflecting the explicit contents of prior conversations and discussion.

Exploring the bases of the transatlantic differences

We consider three factors that may help to account for the differences in public attitudes between Europe and America.

Media coverage

First, consider the influence of the press. One popular view suggests that the content (either positive or negative) of press coverage shapes public perceptions in the corresponding direction. Another hypothesis suggests that in technological controversies it is the sheer quantity of press coverage that is decisive: the greater the coverage, the more negative the public perceptions (Leahy and Mazur, 1980).

In order to compare US and European press coverage, we analysed a longitudinal sample of articles drawn from elite national newspapers in twelve European countries and the USA (see Gutteling et al., chapter 3 in this volume). We assume not that these newspapers are widely read, but rather that they inform politicians and other journalists and, over time, reflect the tone of the national debate. In order to compare Europe and the USA (and because there is no trans-European press), figure 12.2 shows the average of the twelve European national newspapers compared with the *Washington Post*. Since we are exploring post hoc explanations, strict comparability of measures is not essential.

Between 1984 and 1991 there is a broadly similar trajectory in Europe and the USA. Thereafter, however, the European trajectory rises more

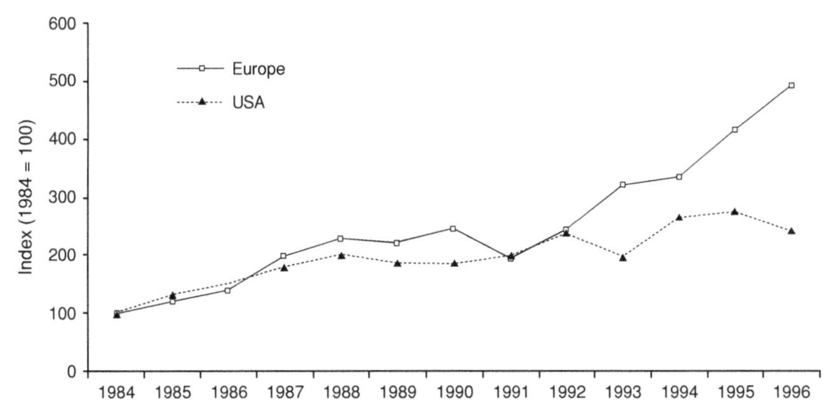

Figure 12.2 Number of articles about biotechnology in opinion-leading press, 1984–1996 (index: 1984 = 100)
Note: $N = 12$ newspapers for Europe; $N = 1$ newspaper for USA.

steeply than that in the USA. The comparison is consistent with the hypothesis concerning the importance of the quantity of media coverage. The relatively greater increase in coverage in Europe goes together with greater public concern.

Based on coding categories designed to facilitate systematic comparison of media coverage (Durant et al., 1998), table 12.3 shows the content of coverage in Europe and the USA. From 1984 to 1990, there are relatively few differences between the European and the US press. 'Progress' and 'economic prospect' are the dominant frames in both cases, and the important themes are 'health', 'basic research' and 'economics'. From 1991 to 1996, differences between Europe and the USA are evident. The *Washington Post* moves from 'progress' to 'economic prospect', while in Europe 'progress' remains dominant. The emerging frames are 'public accountability' and 'nature/nurture' in the USA, but 'ethics' in Europe. In the USA, we see fewer 'benefit' stories and more 'risk and benefit' stories. There is no evidence of increasing 'risk' stories in Europe.

These results do not confirm the view that public perceptions reflect the content of press coverage. On the contrary, whereas the trend in European press coverage was more positive than that in the USA, by 1996 public opinion in Europe was more negative. Instead, our evidence supports the hypothesis that increased amounts of press coverage of technological controversies are associated with negative public perceptions. What may have happened is that, on reaching a critical threshold, the press coverage acted as a catalyst, transforming private concerns into more visible public attitudes.

Scientific knowledge

Another factor is the role of knowledge in public perceptions. A common belief is that scientific literacy generates support for science and technology. Two types of knowledge of biology and genetics were tested by nine items. Six true/false items tested general knowledge:

1. More than half of human genes are identical to those of chimpanzees.
2. The cloning of living things produces exactly identical offspring.
3. Yeast for brewing beer consists of living organisms.
4. It is possible to find out in the first few months of pregnancy whether a child will have Down's syndrome.
5. Viruses can be contaminated by bacteria.
6. It is possible to transfer animal genes into plants.

Three true/false items tested images of food biotechnology:

7. Ordinary tomatoes do not contain genes while genetically modified tomatoes do.

Table 12.3 *Content of press coverage in the USA and Europe (%)*

	Frames[a]	USA	Europe	Themes[b]	USA	Europe	Risk/benefit	USA	Europe
1984–1990[c]	Progress	50	49	Health	29	24	Benefit	39	43
	Economic prospect	17	18	Basic research	14	12	Risk and benefit	34	30
	Nature/nurture	10	1	Economics	13	10	Risk	6	12
	Ethical	8	12	Regulation	11	9	Neither	21	15
	Public accountability	8	13	Safety and risk	11	7			
1991–1996[d]	Progress	30	50	Health	37	30	Benefit	27	38
	Economic prospect	29	15	Economics	12	9	Risk and benefit	44	24
	Nature/nurture	15	4	Regulations	12	10	Risk	9	10
	Public accountability	15	10	Safety and risk	11	7	Neither	20	30
	Ethical	6	16	Basic research	8	10			

Notes: The figures show the average of twelve European national newspapers compared with the *Washington Post*.

[a] Frames are the perspectives in which biotechnology is discussed.

[b] Themes are specific topics within the area of biotechnology.

[c] Europe: $N = 1{,}769$; USA: $N = 117$.

[e] Europe: $N = 2{,}861$; USA: $N = 89$.

8. By eating a genetically modified fruit a person's genes could become modified.
9. Genetically modified animals are always bigger than ordinary ones.

For the textbook items, an incorrect answer is presumed to reflect a lack of scientific knowledge. For the image items, an incorrect answer is presumed to reflect both a lack of scientific knowledge and the willingness to entertain an image of threatening possibilities of food adulteration, infection and monstrosities.

The textbook and image items formed two scales. For the textbook items, the scale records the number of correct responses (0–6). For the image items, the scale records the number of threatening images (0–3).

On textbook knowledge the mean score for Europe is 2.9, which is not significantly different from the US score of 3.0.

Thus, textbook knowledge does not explain the more positive attitudes of people in the USA. By contrast, the mean score for threatening images of food biotechnology in the USA is 0.24, significantly lower than the European mean score of 0.88 ($T = -36.24$, $p < .0005$). The lowest score for threatening images in any European country is more than twice as great as the US score. If more Europeans think that GM foods are the only foods containing genes, that eating GM foods may result in genetic infection and that GM animals are always bigger, it is hardly surprising that they approach modern food biotechnology with greater suspicion.

There are several possible explanations of the greater prevalence of menacing images of agricultural and food biotechnology in Europe. First, food has strong cultural connotations; it is a part of national identity. It is possible that European culinary traditions are more resistant to technological change. Secondly, the recent series of food safety scares in Europe, most notably BSE, may have sensitised large sections of the European public to the potential dangers inherent in industrial farming practices.

Beyond these agriculture and food-related issues, it is possible that other factors shape contrasting perceptions of biotechnology in Europe and North America. The new genetics touches upon deep-rooted beliefs about the boundaries between the natural and the unnatural and about the differences between 'good genetics' and 'bad genetics'. In this context, GM crops and GM foods may be hard to classify. Are these 'natural' or 'unnatural', 'good genetics' (like medicines) or 'bad genetics' (like eugenics)? As a result of different histories and different assumptions about the boundaries of the natural, it may be that Europeans are less inclined than Americans to embrace food biotechnology.

Trust in regulatory systems

The third putative explanation for the differing levels of support for biotechnology in the United States and in Europe concerns public trust in the processes of regulation. In an increasingly complex world, trust functions as a substitute for knowledge (Luhmann, 1979). Essentially people need to act on the assumption that systems of regulation and control supporting everyday life will not fail. With trust in the regulatory process, people can behave, more or less, as if the future were certain, and with confidence that unforeseen problems will be sorted out. Under conditions of distrust, however, the future is uncertain, perceived risks and resulting anxieties may be accentuated, and there may be doubts about the ability of the authorities to counter emergent problems. For this reason, the extent to which the public has trust and confidence in the regulatory processes may be a further factor contributing to public opinion on the new developments in biotechnology.

To put the issue of trust into context it is necessary to understand the rather different histories of biotechnology regulation in Europe and the United States. The European history is outlined in detail by Torgersen et al. in chapter 2 of this volume. In Europe there has been a protracted public debate and great difficulty achieving a viable multi-level consensus. Biotechnology has been treated as a novel process requiring novel regulatory provisions, leading to a complex series of national and European initiatives embracing a wide range of both known and unknown risks, including risks to the environment. The development of regulatory arrangements has been further complicated by the competing agendas of different directorates of the European Commission (Cantley, 1995). Although it is fair to conclude that biotechnology was not as controversial in the United States as in Europe during the 1990s, for most of the period since 1975, policy activity in the USA has been lively and hotly contested.

Excursion: US biotechnology policy, 1975–2000

Since the history of US policy-making on biotechnology is rather different from that of Europe, this section provides a brief overview. Following a summary of the general trends and events of the period, three subsections trace policy decisions undertaken by the US government.[3]

Few Americans outside the leading graduate research programmes in biology would have known much about recombinant DNA (rDNA) in 1975. Although Watson and Crick's discovery of DNA was, by this time, part of every well-educated American's knowledge set, the technological possibilities and the attendant risks and benefits were wholly unknown. This changed with the Asilomar Conference, called to debate whether transfer of genes from one organism to another – something that was already known in 1975 as genetic engineering – was an inherently dangerous activity. The conference was widely covered by US news media and, within a few years, popular books on genetic engineering began to appear.

One of these was *Who Should Play God?* by Howard and Rifkin, published in 1977. The book did not create a sensation. It was one of many that attempted to raise awareness about ethical questions concerning the prospects for human eugenics in the coming world of biotechnology. It is primarily noteworthy as the first entry for Rifkin, who went on to publish a bestseller entitled *Entropy* in 1980. *Entropy* was a highly accessible introduction to a broad range of environmental issues, and it was published with a postscript by Georgescu-Roegen, the distinguished theorist who is given credit for the first effective integration of economics and ecology. *Entropy* gave Rifkin prestige and an audience, and it laid the basis for the formation of an environmentally oriented non-governmental organisation (NGO) entitled the Foundation on Economic Trends (FET).

It was from this platform that Rifkin launched a series of attacks on biotechnology, beginning with *Algeny* in 1983. Though not as widely read as *Entropy*, the new book characterised genetic engineering as the natural extension of a reductionist philosophical programme launched by Bacon and Descartes some three hundred years earlier. The book linked environmentalists' interest in holistic thinking with postmodern critics of science, and opened the door for an interest in matters of the spirit. The link to spirituality was made more explicitly in Rifkin's 1985 book, *Declaration of a Heretic*, in which the Darwinian elements of ecology were replaced with a Biblical account of creation and a broad denunciation of modern biology. The specific content of these three books may be less important than the fact that they allowed Rifkin to assemble a loose coalition of environmentalists, postmodern intellectuals and fundamentalist Christians – strange bedfellows indeed – in a focused attack on the nascent products of biotechnology.

Rifkin's public activism during this period consisted in college speaking engagements, popular magazine and radio interviews, and coalition-building with his constituency groups. During these activities, the theme was frequently to highlight the ethical concerns associated with human eugenics that he had first raised in 1977, and to suggest that American science was on a slippery slope that would lead inevitably to this result. Rifkin told his audiences that they would be tempted by extremely attractive possibilities – drugs and therapies that would provide hope for people afflicted with horrible disease. Nevertheless, these temptations would tend toward ecological collapse and corruption of our respect for the integrity of life.

Behind the scenes, the Foundation on Economic Trends was launching a series of activities that would bedevil the scientists and companies that were hoping to develop new products of biotechnology. A coalition of religious leaders remarkable for its breadth was induced to sign a statement protesting against animal patenting. Meanwhile, FET filed an application for a patent on human beings (with the full expectation that the patent would be denied), partly for publicity, partly to make a point and partly to establish some legal precedents in patent law. The most successful action was a lawsuit (discussed below) that delayed agricultural experiments on ice-nucleating (so-called 'ice minus') bacteria for several years, and tested the US government's entire regulatory apparatus for agricultural biotechnology.

Rifkin found himself with a number of allies when it came to the environmental and agricultural implications of biotechnology. In 1985, Doyle published *Altered Harvest*, a book that alerted the attentive educated public to the importance of genetic diversity in maintaining an ecologically robust agriculture. In 1986, noted veterinarian and animal activist Fox published *Agricide*. In 1987, the Biotechnology Working Group, a consortium of environmental and agricultural activists, published 'Biotechnology's Bitter Harvest' (Goldburg et al., 1990). In 1993, Vandana Shiva, the well-known advocate for women of the developing world, published *Monocultures of the Mind*. Although none of these publications captured the attention of more than a fraction of Americans, all of them were sharply critical of agricultural biotechnology. Cumulatively, they created a climate of suspicion about the likely social and ecological impact of biotechnology in agriculture.

Thus, although concerns about human eugenics were never too far distant, the decade of the 1980s was a time when a number of Americans were feeling sceptical about biotechnology. It was, however, also a time when there were no products of agricultural biotechnology on the market. As such, it was difficult for this sceptical minority to raise much concern among ordinary Americans. In the meantime, a number of organisations undertook efforts to deflect the brunt of the criticism being levelled by sceptics. The first large-scale effort was conducted by the Keystone Foundation, a US NGO oriented toward finding non-violent consensual solutions to contentious issues. Keystone funded and conducted a series of workshops around the USA between 1995 and 1998 involving leadership and representatives from environmental organisations, as well as scientists and representatives from the US biotechnology industry. These workshops allowed issues and concerns to be aired, and became the forum in which many environmental leaders learned about biotechnology.

Following the completion of the Keystone project, a consortium of US and Canadian universities and not-for-profit research centres formed the

National Agricultural Biotechnology Council (NABC), which held its first meeting in Ames, Iowa, in 1989. This meeting also brought critics together with university and industry scientists, as well as government regulators. NABC has held similar meetings every year since. Ralph Nader, US Green Party presidential candidate and frequent critic of biotechnology, was the keynote speaker in 2000. NABC meetings have been structured with ample small-group discussion time, and a process that generates a report on areas of agreement, areas where research or education efforts are needed, and some recommendations, which may be directed to US government officials or to NABC members themselves. In addition to holding annual meetings, NABC publishes a newsletter and an annual report and holds a congressional briefing on consensus recommendations each year in Washington D.C.

The outcome of the Keystone and NABC projects, beyond a parade of books critical of biotechnology, has been that US policy on agricultural biotechnology has been formed amidst a rich and often contentious atmosphere of debate. Other activities, such as series of workshops on teaching ethical issues in biotechnology conducted by Iowa State University, augmented the Keystone and NABC efforts. Issues in human genetics and medicine have been in the background of those debates, but there have arguably been fewer policy decisions in the human/biomedical area in any case. By the mid-1990s, US government agencies had made formal policy decisions for regulating biotechnology, which are described below. Although many of the individuals who were most active in criticising biotechnology between 1985 and 1995 were deeply dissatisfied with the direction of US policy, the decade of workshops and hearings had apparently exhausted the energy and interest of the broader community of politically active US citizens.

Undoubtedly, many environmental leaders who participated in Keystone, NABC and other activities decided that, whatever environmental issues might be associated with biotechnology, other problems were more pressing and more worthy of their limited time and effort. Similarly, few NGOs representing consumer interests placed biotechnology on their agenda for public action. By 1995, when products began to appear in large numbers, Rifkin's odd coalition of environmentalists, advocates of social justice, animal activists, postmodern intellectuals and fundamentalists had apparently had enough of working together. When the US Food and Drug Administration (FDA) announced its intent to implement a policy that would discourage labelling of genetically engineered foods in 1996, few of the groups even bothered to comment. Rifkin himself had gone on to write books about the beef industry and the end of work. All apparently had other fish to fry.

By 2000, this situation had changed in a manner that surprised many scientists, government regulators and the biotechnology industry itself. Rifkin was back with a new book, *The Biotech Century*, in 1998. In 1997 and 1998, a US Department of Agriculture proposal to allow genetically engineered foods to be labelled as 'organic' was soundly criticised – the mirror image of the FDA's experience in 1996. In 1999, controversy in Europe began to be reported in US newspapers, and this spawned stories critical of biotechnology in the widely read Sunday supplements of the *Washington Post* and the *New York Times.* Protestors at the 1999 WTO meetings in Seattle got biotechnology on the nightly news, and leading news outlets gave sensational coverage in 2000 to a report that a leading brand of taco shells contained samples of genetically engineered maize that were not approved for human diets.

Although US policy on biotechnology seemed stable and complete in 1995, it is not at all clear that it will remain so. No key policy changes have been made during the period of controversy, but the apparatus for policy change is beginning to move. Hearings are being held and politicians are expressing outrage. However, it is important to have a general appreciation of the US policy process in order to understand the significance of these events. Some elements of current US policy on biotechnology were developed through normal procedures – the assimilation of the new technology into existing policies – and this is likely to continue in the future. Other elements were developed as a response to controversy of all sorts – the processes of accommodation to challenges – and the revived controversy of the waning years of the 1990s may lead to a new round of policy change as a form of political reaction.

Here, 'normal procedures' include activities of review and regulation that occur as a matter of course under US law. Following Kuhn's distinction between normal and revolutionary science, we can classify policy developments that occurred in response to the novelty of the new biotechnology as 'revolutionary'. However, normal procedures are not wholly distinct from revolutionary ones. Indeed, the most contentious issue of the late 1980s and early 1990s concerned whether or not existing regulatory procedures were adequate for biotechnology. During that period, regulatory agencies adapted policies and procedures that had not been previously applied to research and products of gene transfer. Whether this adaptation was 'normal' or 'revolutionary' is largely a matter of interpretation. The value of the normal/revolutionary dichotomy consists primarily in providing a succinct way to characterise two strands of policy development in the USA, and in suggesting an analysis of how and why policies continue to be contested. The following sections build on this framework, and discuss the development of both biomedical and agricultural policy.

Normal regulatory policy in the United States

On the normal side, the United States has fairly well-established regulatory procedures for research and development of activities having the potential for impact on human, animal or environmental health. Pure research activities are self-regulated by both non-profit research organisations and commercial firms, though these self-regulatory activities are themselves overseen by the primary research funding agencies of the US government. Drugs and food technologies are regulated by the US Food and Drug Administration, while chemicals or organisms with a potential for environmental impact are regulated by divisions of the Environmental Protection Agency (EPA) and by the Animal and Plant Health Inspection Service (APHIS). These federal agencies operate under a suite of laws with varying degrees of specificity regarding both the methods used to ascertain risks and the criteria for determining acceptability of risk.

In some cases, such as drugs or pesticides, firms must register a product with the regulating agency; in other cases, such as the development of new foods or food additives, submission of a product for regulatory review is optional. Thus, drugs developed using recombinant techniques must receive FDA approval before being marketed, whereas foods, such as the Flav'r Sav'r Tomato®, are reviewed only on request. In either case, however, the review process requires that organisations requesting approval must generate data demonstrating the safety and efficacy of the product. These data are reviewed by FDA officials, who frequently raise additional questions and request additional studies. Drugs must offer the chance for substantial benefit in exchange for risks in order to be approved, and must be marketed in a manner that informs end users (prescribing physicians, pharmacists and their patients) about the risk of side-effects. Generally speaking, foods must meet a *de minimus* standard of risk, though risk standards for foods have been evolving in the past twenty-five years.

Policies for product approval apply to any developer of a new technology, whether that be an individual, a for-profit firm or a non-profit research organisation (including universities or the US government itself). Traditionally, public programmes such as the US Department of Agriculture and US agricultural universities have obtained approvals for new crop varieties and have made these products available to the public at a nominal cost. However, it would be very unusual for a publicly funded organisation to undertake the extensive testing needed to demonstrate safety for drugs, pesticides or novel foods. It is only because most new crop varieties have been exempt from regulatory review on food safety grounds that public research institutions have been able to develop and register new seeds.

Whenever US agencies take action to approve a product or to alter their procedures for product approval, notice of the action is published in the US Federal Register, and parties that are potentially affected by the action are given an opportunity to comment before the action becomes final. Agencies are required to respond to public comments, also in the Federal Register, providing a public record of the basis on which decisions are made. Clearly, the US public is oblivious to most of what is published in the Federal Register. However, non-governmental organisations representing special interests do monitor the Federal Register for actions that affect their members. These organisations often express dissatisfaction with the US government's responsiveness to their concerns, and this has particularly been true for decisions regarding products of rDNA technology in the food and agricultural arena. Nevertheless, the normal US procedure for setting policy with respect to science and technology is both transparent (one may discern the basis on which decisions are made) and responsive to the interests of affected parties, when compared with that of many other industrialised countries. US agencies have made normal regulatory decisions on hundreds of drug and agricultural products since 1980.

It is important to recognise that Americans also have recourse to the courts to pursue issues of interest to them. Two venues merit special note. First, it is possible to bring a civil action against a regulatory agency in an attempt to overturn or reverse an agency's action on a particular issue (though the US government limits the opportunities to bring such action). Agencies permit and even encourage such actions, especially in cases where agency administrators believe that there may be flaws in existing legislation or procedures for conducting regulatory review. Secondly, Americans have extraordinarily broad access to legal venues for product liability actions, and settlements can be quite large. As such, firms operating in the United States have significant incentive to limit their exposure to product liability lawsuits. This leads them to seek regulatory approval whenever there could be any question about the safety of a product and to be assiduous in submitting the data needed to demonstrate the safety of a product.

Revolutionary biomedical policy and rDNA in the USA: 1975–2000

In normal circumstances, the US press would not take an interest in an issue unless agency decision-making (including its checks and balances) had clearly failed. As noted, many products (especially pharmaceutical and diagnostic devices) have moved through the normal process without exciting comment during the past twenty-five years. Thus, the fact that an issue receives coverage in national media outlets is evidence that something out of the ordinary is going on. It is fair to say that many biotechnology policy decisions had become normalised in the USA by 1997, and this may have led to a general perception that biotechnology was uncontroversial and accepted by the US public. However, biotechnology made the news regularly between 1975 and 1995, and the resurgence of reporting on biotechnology issues in 1999 and 2000 suggests that the revolutionary period of US policy may not be entirely over.

Controversy over policy on biotechnology research began with the Asilomar Conference in 1975, a meeting of scientists working at the frontiers of science of recombinant DNA. The immediate policy result of this meeting was that scientists voluntarily agreed to a moratorium on genetic transformation of organisms until procedures that would limit the risks of such research could be established. A set of guidelines for conducting such research was developed under the auspices of the US National Institutes of Health (NIH), a government agency that both conducts and funds research in biomedical science and public health. NIH also established the Recombinant DNA Advisory Committee (RAC) to review proposals for research using rDNA and to advise administrators on agency-wide policy for future research (Krimsky, 1982).

The Asilomar Conference sparked several years of speculative debate about the risks and likely applications of gene transfer. Much of this debate centred on the question of whether such research is inherently dangerous or immoral (Goodfield, 1977). The city of Cambridge, Massachusetts, home of Harvard University and the Massachusetts Institute of Technology, attempted to ban rDNA research within its boundaries, initiating what some have called the first serious attempt to regulate science in the United States (Krimsky, 1982). The Cambridge controversy also led to the 1983 creation of the Committee for Responsible Genetics, the first US NGO focusing on biotechnology policy. However, by the early 1980s most US biotechnology research was required to meet only the least onerous of NIH safety guidelines. By the end of the 1980s, review for the safety of rDNA had been thoroughly normalised, being conducted primarily by local institutional review boards at the organisations (universities, government laboratories and private firms) employing principal investigators.

As rDNA research led to the development of valuable drugs and diagnostic procedures, media coverage of the science shifted to the financial pages and toward stories that portrayed the science in non-controversial terms. Through the 1980s the NIH RAC debated the scientific and ethical issues associated with new protocols for gene therapy (Anderson, 1987). However, these debates neither engaged a large segment of the American public nor created an environment in which biotechnology policy evolved in response to controversy or public pressure. It would be more accurate to say that, by 1990, the NIH process for soliciting advice through the RAC and for making judgements about the risks and ethics of genetic research had evolved into an entirely normal process.

One key (and hence *somewhat* revolutionary) development in the normalisation of policy at NIH was the creation of the Ethical, Legal and Social Issues (ELSI) program of the Human Genome Initiative in 1992. NIH initiated this program to support both social science and philosophical research on issues in human genetics that would become problematic or controversial as more information about the location and function of genes became available. The original proposal was to dedicate 5 per cent of human genome funding to ELSI projects, though the figure dropped to considerably less than half that amount. Even at this level, however, the ELSI program alone exceeded the total amount of federal grant money that had been available for research on ethical topics before its inception. Over its history, ELSI has funded dozens of research projects and conferences on topics such as genetic discrimination, genetic screening, privacy, implications for insurability and access to health care, and the genetic basis of behavioural traits. The result is a fairly large scholarly literature on these topics and frequent coverage of these issues in the US press. A compendium of this scholarly work, *The Encyclopedia of Legal and Ethical Issues in Biotechnology*, was published in the summer of 2000 (Murray and Mehlmann, 2000).

A final spurt of revolutionary policy-making for biomedical biotechnology was sparked by the announcement that a sheep had been cloned at the Roslin Institute in February 1997. Within days of the announcement, President Clinton commissioned the National Bioethics Advisory Commission (NBAC) to study the ethical issues raised by the possibility of human cloning and to issue a report within a ninety-day time horizon. The NBAC is not a standing committee of the US Executive Branch, but was in fact an ad hoc committee of experts in reproductive medicine, embryology and medical bioethics that had been convened to review the federal government's policy on research using human stem cells. However, the NBAC launched into an intensive study on likely uses of human cloning and its risks and on ethical issues raised both by secular bioethicists and by representatives of the principal religious denominations. In addition to hearings conducted by the NBAC, the US House and Senate both conducted hearings on cloning, in which they heard from many of the same people who had prepared testimony for NBAC.

Some of the religiously oriented respondents expressed strong opposition to cloning based on their views on human embryos – views that had already surfaced in the debate over stem cell research. However, the NBAC ultimately decided not to revisit these issues, and the result was a consensus on a somewhat narrow set of issues concerning the safety of human cloning research. Specifically, the NBAC noted that many questions remain about the safety and efficacy of the procedure that produced Dolly, and recommended a moratorium on any attempt to use adult cell nuclear transfer for the purpose of producing a human child (National Bioethics Advisory Commission, 1997). As of 2000, the US Congress had not acted on the NBAC recommendations.

There is no US law that prohibits or regulates mammalian cloning of any kind. The NIH would rule on any proposal to use federal funding in such a research project, hence there is a 'normal' mechanism for considering the ethical implications of human cloning when public funds are involved. It is clear that NIH would apply many of the same considerations recommended in the NBAC report. However, NIH policies do not apply to research conducted solely with public funds, hence it is possible that human cloning could be undertaken in the USA without violating any laws.

Revolutionary agricultural biotechnology policy: 1975–2000

Immediately after the initial controversies associated with Asilomar, the next round of public controversy over biotechnology was associated with agriculture. A University of Wisconsin researcher submitted a proposal to conduct experiments using bacteria that had been transformed in a manner that could affect the freezing point for crops such as strawberries or potatoes. In 1994, Lindow, the scientist who proposed this research on so-called 'ice-minus' bacteria, duly sent his research protocol to the NIH RAC, which approved the experiment. However, as noted already, activist Rifkin and the Foundation on Economic Trends successfully sued NIH and the University of Wisconsin to block the experiment, arguing that the NIH – a biomedical research agency – lacked the scientific expertise to assess environmental risks. Rifkin's lawsuit sparked public protests and destruction of field plots where ice-minus research was being conducted. The lawsuit and public controversy eventually led to the formation of the Agricultural Biotechnology Research Advisory Committee (ABRAC) in 1986 within the US Department of Agriculture (USDA). ABRAC was intended to perform an advisory function similar to that of the NIH RAC, but focusing on applications of food and agricultural biotechnology (Thompson, 1995).

As controversy over ice-minus was subsiding, controversy was beginning over the next product of agricultural biotechnology, recombinant bovine somatotropin (rBST), which can be used to stimulate milk production in dairy cows. As an animal drug, rBST clearly came under the approval process for veterinary drugs administered by the FDA. However, a 1995 paper on the socio-economic impact of rBST had predicted that it would exacerbate economic trends that were affecting both the farm size and geographic location of dairy production in the United States (Kalter, 1985). The FDA took an unusually long time to conduct its review of rBST. The agency needed to be satisfied that the drug was efficacious in stimulating milk production, that it posed no risk to human health and that it posed acceptable risks to animal health.

Only the first of these decisions was straightforward. With respect to human health, the FDA ultimately developed one of the first versions of the policy that has come to be known as 'substantial equivalence'. Based on a comparison of the structure of the rBST protein with that of natural BST, and on the fact that the proteins are indistinguishable in any of the studies that FDA scientists were able to devise, the agency decided that evidence for the safety of natural BST (present in all milk) could be extended to rBST as well (Juskevich and Guyer, 1990). That this should be regarded as a revolutionary rather than a normal policy decision is supported by the fact that FDA officials went well beyond standard agency practices to publish the basis for this evaluation (Krimsky and Wrubel, 1996: 173). Debate over the animal health implications of rBST became protracted. High-yielding dairy cows are at greater risk of mastitis, whether or not they have been treated with rBST, but rBST can clearly increase milk yields, moving animals into this higher-risk group. In the end, the FDA decided that, since careful herd management could control the risk of mastitis, there was not sufficient reason to withhold approval of rBST, and rBST was approved for use in US dairy herds in November 1993.

However, the US Congress intervened in the process, declaring a ninety-day moratorium on the use of rBST and requiring the White House to complete a study on the impact of rBST. Although rBST did come on the market after the ninety days had expired, the fact that Congress took this extraordinary measure testifies to the level of public outcry and controversy that surrounded rBST during the long, nine-year history of FDA review. Although debate over rBST may never have risen to the level of public awareness associated with so-called 'Frankenfoods' in the UK during the late 1990s, rBST was extremely controversial in the USA for well over a decade. Debate over rBST split the US farm community, with some farm NGOs coming out strongly against, and others equally strongly in favour. Animal protection NGOs joined with farm groups – an unprecedented and unrepeated event in US NGO politics – to oppose 'bovine growth hormone' (or BGH), the term that opponents of rBST preferred to use (see Krimsky and Wrubel, 1996). The debate surfaced in political cartoons and in the comedy routines of American television performers. Retailers such as Ben and Jerry's ice cream continue to print anti-BGH statements on their product labels as of this writing.

The US debate over rBST has both formal and informal policy implications. The formal implications are largely confined to the key years of 1993 to 1996, while the informal implications cover a fifteen-year period (continuing to this day) in which US NGOs, the biotechnology industry and scientific organisations such as universities, disciplinary groups and government agencies jockeyed for control. In terms of formal policy decisions taken by the US government, the significance is three-fold. First, the fact that the FDA decision was ultimately allowed to stand indicates a decision in favour of 'normalising' regulatory decision-making for biotechnology by the Clinton administration. The ABRAC was closed in 1996 and biotechnology policy was given over wholly to the normal processes of review under the auspices of the FDA, the EPA and APHIS. Secondly, the White House explicitly rejected the legitimacy of regulating a biotechnology on the basis of its socio-economic impacts. None of the agencies listed above is allowed to consider socio-economic impact when making a regulatory decision. Thirdly, the FDA and other US government agencies adopted policies that placed heavy legal and economic burdens of proof on anyone who attempted to impose labelling requirements on food products in which modern biotechnologies had been used (Thompson, 1995).

Public trust in the regulatory processes

With this background we can now explore trust in the regulatory systems in Europe and the United States as evidenced in the surveys conducted in 1996 and 1997.

European respondents were asked to select, from a list of national and international institutions, the one best placed to regulate biotechnology. The results show most confidence in international organisations such as the United Nations or the World Health Organisation (34.5 per cent), followed by scientific committees (21.6 per cent) and national public bodies (12 per cent). Secondly, Europeans were asked, 'Which of the following sources of information do you have most confidence in to tell you the truth about genetically modified crops grown in fields?' Here the vote of confidence went to environmental, consumer and farming organisations (23 per cent, 16 per cent and 16 per cent, respectively), whereas national public bodies (4 per cent) and industry (1 per cent) commanded little support. These results appear to confirm the trend, observed by others (Samuelson, 1995), of an increasing lack of confidence

Table 12.4 *Trust in the US Department of Agriculture and the federal Food and Drugs Administration (%)*

	Lot of trust	Some trust	No trust	Don't know
FDA	22.9	61.1	15.4	0.6
USDA	21.8	68.3	9.3	0.6

Note: $N = 1{,}067$.

in national political institutions. They also suggest that biotechnology is seen by many Europeans as having transnational consequences that national bodies are powerless to influence.

US respondents were asked, 'If the US Department of Agriculture and the Food and Drugs Administration [separate questions] made a public statement about the safety of biotechnology, would you have a lot, some or no trust in the statement about biotechnology?' The USDA carried the support of 90 per cent of respondents, the FDA 84 per cent (see table 12.4).

These findings clearly indicate that trust in the regulatory system is considerably greater in the USA than in Europe. However, the American public's confidence in its regulatory organisations should probably be interpreted to mean that people are satisfied with the performance of this system *in toto.* The normal process for policy formation in the United States involves much more than a simple process of scientific judgement. The fact that organisations such as the American Diabetes Association or the Union of Concerned Scientists can and do monitor the Federal Register should be viewed as part of the total system. In addition, Americans' access to the courts reinforces the accountability of agency performance, as well as providing economic incentives for industry compliance. Hence the US system of checks and balances in the regulatory process inspires confidence that applications of biotechnology will be appropriately monitored. It is perhaps Europe's failure to establish a similarly coherent and stable system that has created the conditions of distrust.

Conclusion

In conclusion, no single explanation accounts for the greater resistance to food biotechnology in Europe. Media intensity, knowledge of biotechnology and trust in the regulatory process are implicated and interrelated. Different histories of media coverage and regulation go together with different patterns of public perceptions; and these in turn reflect deeper cultural sensitivities, not only towards food and novel food technologies but also towards agriculture and the environment. It is also apparent that the European public's concerns about GM foods were brewing some years before the consumer revolt of 1999. What is perhaps of note is that many of the European concerns about agricultural biotechnologies were also expressed, sometimes much earlier, in the United States. But differences between the European and American political systems led to different outcomes. The two-party system in the USA filtered out the sceptical voices once the regulatory regime had been established. By contrast, the unsettled pluralism of multi-party state systems and the evolving integration of Europe offered more effective channels to those challenging the scientific and industrial actor network.

The events of the watershed years of 1996 and 1997 point to an uncertain and contentious future for biotechnology on both sides of the Atlantic. For Europe, Torgersen et al. (chapter 2 in this volume) see the prospects of more conflicts arising from national diversity and European integration. In the USA, an increasing level of controversy over biotechnology may be a harbinger of difficulties to come. Events such as the USDA's difficulties with organic labelling (1999), a rise in critical media reportage spilling over from Europe, the protests at the WTO meetings in Seattle (1999), expressions of concern by politicians and the calling of public hearings suggest that Europe and the USA may not be worlds apart. How European and American policy and regulation respond to the likely challenges will be watched with interest around the globe.

NOTES

1. See Pusztai's website: http://www.freenet pages.co.uk/hp/A.Pusztai.
2. The US survey of public perceptions of biotechnology was conducted under NSF award 9732170, Principal Investigator: Jon D. Miller.
3. Note that the review omits one extremely important policy domain, that of intellectual property. There is little doubt that the US Supreme Court decision in *Diamond vs. Chakrabarty* in 1980 and the US Animal Patent Act of 1986 have had a crucial influence over the direction of US biotechnology research and development. Furthermore, US religious groups have taken a particular interest in these issues (Lesser, 1989).

REFERENCES

Anderson, W. French (1987) 'Human Gene Therapy: Scientific and Ethical Issues', in R. Chadwick (ed.), *Ethics, Reproduction and Genetic Control*, London and New York: Routledge, pp. 147–63.

Cantley, M. (1995) 'The Regulation of Modern Biotechnology: a Historical and European Perspective', in D. Brauer (ed.), *Biotechnology*, vol. 12 (New York: VCH), pp. 505–81.

Doyle, Jack (1985) *Altered Harvest: Agriculture, Genetics, and the Fate of the World's Food Supply*, New York: Viking Press.

Durant, J., M.W. Bauer and G. Gaskell (eds.) (1998) *Biotechnology in the Public Sphere: a European Sourcebook*, London: Science Museum Publications.

Fox, Michael W. (1986) *Agricide: the Hidden Crisis That Affects Us All*, New York: Shocken Books.

Goldburg, R., J. Rissler, H. Shand and C. Hassebrook (1990) 'Biotechnology's Bitter Harvest. Herbicide-Tolerant Crops and the Threat to Sustainable Agriculture', a report of the Biotechnology Working Group, New York: Environmental Defense, March, at http://www.environmentaldefense.org/pubs/Reports/biobitter.html, accessed November 2001.

Goodfield, June (1977) *Playing God: Genetic Engineering and the Manipulation of Life*, New York: Random House.

Howard, Ted and Jeremy Rifkin (1977) *Who Should Play God? The Artificial Creation of Life and What It Means for the Future of the Human Race*, New York: Dell Books.

Juskevich, J.C. and C.G. Guyer (1990) 'Bovine Growth Hormone: Human Food Safety Evaluation', *Science* 264: 875–84.

Kalter, Robert (1985) 'The New Biotech Agriculture: Unforeseen Economic Consequences', *Issues in Science and Technology* 13: 125–33.

Krimsky, Sheldon (1982) *Genetic Alchemy: the Social History of the Recombinant DNA Controversy*, Cambridge, MA: MIT Press.

Krimsky, Sheldon and Roger Wrubel (1996) *Agricultural Biotechnology and the Environment: Science, Policy and Social Issues.* Urbana, IL: University of Illinois Press.

Leahy, P. and A. Mazur (1980) 'The Rise and Fall of Public Opposition in Specific Social Movements', *Social Studies of Science* 10: 259–84.

Lesser, William H. (1989) *Animal Patents: the Legal, Economic and Social Issues.* New York: Stockton Press.

Luhmann, N. (1979) *Trust and Power*, Chichester: Wiley.

Murray, T.M. and M.J. Mehlmann (eds.) (2000) *Encyclopedia of Legal and Ethical Issues in Biotechnology*, New York: Wiley.

National Bioethics Advisory Commission (1997) *Cloning Human Beings*, 2 vols. Rockville, MD: NBAC.

Rifkin, Jeremy (1985) *Declaration of a Heretic*, Boston: Routledge & Kegan Paul.

(1998) *The Biotech Century: Harnessing the Gene and Remaking the World*, New York: Jeremy P. Tarcher/Putnam.

Rifkin, Jeremy, with Ted Howard (1980) *Entropy: a New World View*, New York: Viking Press.

Rifkin, Jeremy, with Nicanor Perlas (1983) *Algeny: a New Word; a New World*, New York: Viking Press.

Samuelson, R. (1995) *The Good Life and Its Discontents: the American Dream in the Age of Entitlement*, New York: Times Books.

Shiva, Vandana (1993) *Monocultures of the Mind: Perspectives on Biodiversity and Biotechnology*, London: Zed Books.

Thompson, Paul B. (1995) *The Spirit of the Soil: Agriculture and Environmental Ethics*, London and New York: Routledge.

Part V

Towards a social theory of new technology

13 The biotechnology movement

Martin W. Bauer and George Gaskell

Preliminary considerations

In this chapter we outline a framework for a social scientific analysis of the changing relations between science, technology and the public. Our concern is with modern biotechnology, which emerged in the early 1970s with the development of recombinant DNA (rDNA) techniques. We regard biotechnology as the third strategic technology of the period since the Second World War, following nuclear power and information technology. The framework outlined in this chapter reflects the research presented in this book on biotechnology in the European public spheres. Because it is not yet possible to talk of an integrated European public sphere, we pragmatically assume a multitude of public spheres, more or less related to the nation-states.

Our framework is built on various concepts: the biotechnology movement, the scientific-industrial complex, primary and secondary objectification, activity and the symbolic environment, actor networks, challenge and response, arenas of the public sphere, social milieus, public conversations and perceptions, media coverage and the regulatory process, and assimilation and accommodation. This terminology is not an end in itself, but an attempt to add clarity to the development of new technologies under the conditions of late modernity.

In the first section we refine a model formulated in 1994 in the so-called 'Hydra paper' (Bauer et al., 1994). This served as the rationale for our research activities with our colleagues in the multinational project, the outcomes of which are presented in this volume. The model proposes a triangular model of the public sphere of biotechnology comprising public perceptions, media coverage and regulation. Secondly, we integrate insights from science and technology studies. The analysis of the social shaping of technology gives us a better understanding of the 'system' that, because of our foregrounding of the public sphere, remains in the background of our research activities. However, we find in the research on the social shaping of technology only limited ideas and empirical research regarding the operations and functions of the public sphere. In short, in this chapter we develop a language for grasping the relationship between a technology trajectory and the public sphere within the context of post-industrial societies. We aim to build a framework theory that is capable of integrating middle-range hypotheses on public perceptions, media coverage and the regulatory process and the technological trajectory.

We start with a distinction between 'biotechnology' as a development of the routine activities of production and consumption – the primary objectification – and 'biotechnology' as it is represented or commonly understood in various societal domains and milieus – the secondary objectification. Primary objectification involves putting ideas and visions into products and services, while secondary objectification involves the appropriation of these products and services into everyday life in the form of ideas, symbols and behaviour. It is inappropriate to conceive of objectification as a simple causal process from primary to secondary forms. In our framework, primary and secondary objectification constitute mutual challenges, setting mutual opportunities and constraints. Representations and actions are co-determined.

The producers, observers and consumers of biotechnology, the 'insiders' of the movement, represent the technology in a variety of different ways. Others, for a variety of reasons, are either unaware of it, or ignore it or are indifferent to it; they are the 'outsiders'. But, because biotechnology is a strategic technology and, as such, increasingly challenges different aspects of everyday life, the scope for remaining an outsider is diminishing. The move from outsider to insider is accompanied by the adoption of a position towards the technology: support or rejection, or possibly a conditional approach of weighing up the options for the development of the technology – the 'third way'. We consider outsiders and insiders less as individuals and more as classes of individuals either as actors, as in primary objectification, or as arenas and milieus, as in secondary objectification, for whom the new technology takes on a particular symbolism.

Three ideas are central to the research presented in this volume. First, the development and exploitation of genetic engineering techniques is the focus for a growing 'biotechnology movement' at the core of which is a scientific-industrial complex. This is dominated by a tendency that inherits some of the significance of the older military-industrial complex, 'big science' (Price, 1965) or the techno-structure (Galbraith, 1968). However, in the case of biotechnology, the controls are not in the hands of a secretive state operating with a Cold War mentality, but are dispersed across international business and small to medium-sized companies with R&D functions operating and competing in the global market.

In its development of products and services, the scientific-industrial complex of biotechnology creates a variety of challenges to society that lead to responses. The idea of 'challenge and response' is attributable to the historian Arnold Toynbee (1978: 298), who employed these terms to understand the clashes between and the growth of civilisations, or, in other words, the encounter between large-scale projects. A challenge arises when a change to the status quo affects or threatens to affect the project or rationale of a social group. The response to a challenge is unpredictable. If the challenge is too small, it may go unnoticed; or it may be too great, leading to a crisis and deleterious effects. A response to a middle-range challenge will be creative, and exert a momentum for future development. The challenges of biotechnology may be seen by a particular interest group as a threat, leading to responses that in turn constitute counter-challenges towards the scientific-industrial complex. On the other hand, a challenge may be seen as an opportunity, leading to enhanced pro-tendencies. In our framework, the biotechnology movement comprises actors from not only the scientific-industrial complex but also the groups for whom the challenge of biotechnology leads to a response, whether of a 'pro' or 'anti' nature.

Beyond this biotechnology movement is a heterogeneous public, the 'outsiders' to whom the various 'insiders' in the biotechnology movement appeal for support. Over time this heterogeneous public is increasingly drawn into the movement, not least because the technology expands into more and more areas of life. People may be asked to adopt new products and services as promising investments or consumer goods, or they may be confronted by claims that the technology is a threat to the environment and human dignity. In other words, the 'outsiders' among the public are 'invited' to join the bandwagon, and have to decide whether to join the movement and on which side of the fence to sit.

Secondly, we assume that national public spheres are involved in constituting the counter-challenge of biotechnology by taking a 'position' vis-à-vis biotechnology. Without an understanding of the esoteric scientific principles, the public must rely on other forms of knowledge. This is the process of representation, whereby the primary activities of biotechnology are re-presented in terms of common sense, taking the symbolic form of ideas, stories of key events, historical analogies, metaphors and iconic images. We expect that these secondary objectifications of biotechnology are an eclectic mixture of modern and pre-modern mythological elements (Blumenberg, 1990).

In our research we distinguish three key arenas of the public sphere: public perceptions and everyday conversations, mass media coverage and regulatory processes. In the public sphere, we conceive of public opinion on biotechnology as comprising perceptions and conversations and also mass media coverage. We distinguish these two elements of public opinion from the regulatory processes. An objective of our research is to monitor and to disentangle the complex relations between the elements of public opinion and between these elements and regulation. There are, of course, other arenas of the public sphere that represent biotechnology in their own terms: the capital markets and stock exchanges, for example. But with our focus on the public sphere we are mainly concerned with the symbolic capital, and as such temporarily ignore the areas of business and commerce (e.g. Ernst and Young, 2001).

The representations of biotechnology in the public sphere are not epiphenomenal, but constitute the symbolic environment in which biotechnology operates (Boulding, 1956). This environment forms the 'developmental space', sometimes enabling and at other times constraining the course of biotechnology. To illustrate the point, during the 1980s many people working in the scientific-industrial complex thought that the regulatory process was the primary hurdle or constraint to their activities, only to discover in the 1990s that a second and problematic hurdle of public opinion could no longer be ignored and had to be taken into account.

Thirdly, as biotechnology, the primary objectification, develops and changes in form over time, and in so doing creates new challenges, so there is a parallel and dynamic process attached to the secondary objectifications. In other words, there is an evolution of symbolic representations in the three arenas. This may be related to a variety of contingencies: the changing cycles of public attention; mutual influences among the arenas, e.g. convergence, disjunction and 'causal linkages' at different phases of development; or the emergence of new frames of discourse, such as the emergence of bio-ethics as an issue in the 1990s.

The final issue is in the nature of a question concerning the relations between science, technology and the public sphere. What are the forms of and conditions for collective learning, in the biotechnology movement and in the public sphere? The various actors in the biotechnology movement, whether the pro or anti tendencies, try to influence and react to one another's positions and to what is happening outside the movement in the arenas of the public sphere. Having identified the 'second hurdle', the scientific-industrial complex must decide whether, and by what means, it should assimilate and/or accommodate its strategic project to other actors in the biotechnology movement and to the public sphere. In moving from ignoring to representing biotechnology, the public sphere assimilates and/or accommodates biotechnology in a particular way, thus conditioning the likely development and reception of future technologies.

Technology: politics by other means

Technology, defined as a complex of know-how, people, artifacts and services, is a relatively neglected area within the social sciences. This is rather strange given that technology is the prime example of a socially constructed reality, understood in a literal sense. Any technological innovation is a novel piece of reality. It has been constructed out of the ideas of designers and engineers, and carries with it both intended and unintended consequences for society. In this constructivist perspective a technology must be analysed like social norms and institutions. Any technically constructed device, simple or complex, embodies a 'behavioural script' that tells people what to do and what not to do (Joerges, 1988). This social fact produces a likely pattern of behaviour that is represented by things or services that become affordances for actions (Gibson, 1977; Norman, 1988). People are socialised into using things and services through the acquisition of tacit knowledge or skills or 'know-how' and through the acquisition of explicit symbolic knowledge or 'know-that'.

However, socialisation is never perfect. People may not use certain things, or they may use them in ways that were not intended or even imagined. In a Durkheimian sense, social norms and technology are functionally equivalent 'social facts' constraining behaviour into recurrent patterns (Linde, 1982), making some actions more likely while discouraging others. Whereas norms are enforced by social sanctions, the use of things is channelled by affordances, the action adequate for the thing or the opportunity costs from not using the thing. For example, the rise of the personal computer has increased keyboard writing and has probably decreased the intensity and quality of handwriting. Thus technology structures behaviour, both individually and collectively, bringing both opportunities and threats to human pursuits. This is the basic ambivalence of new technology, and it is why it constitutes a permanent challenge to social life. This is not to imply that technology is independent from social norms. Technology creates the context for reasserting social norms or the creation of new norms. Furthermore, existing social norms influence the process of technology through both informal processes (voice and exit decisions) and formal processes (regulation).

A number of models have been proposed to explain the trajectory of technology.

Technology as rational design sees the process as the outcome of rational choices between alternative designs on the criterion of maximising desired outcomes. The decision-taker may also consider the unintended consequences of his or her earlier decisions, which may carry implications for decisions to ameliorate such problems. This elite model privileges a professional class of experts, who by dint of their higher form of 'rationality' can make better decisions between design options. This elite theory considers interference either from capitalistic owners or from public politics as a limit to efficiency and effectiveness. This is variously described as the technocratic or engineering model (Winner, 1977: 135ff.).

By contrast, *technology as politics* sees the trajectory of a technology as a Machiavellian process of strategic survival based upon the managerial logic of local optimisation of returns (Simon, 1969). The main task for the promoter of a technology is to design things flexibly in order to sign up more and more people and to maintain their support; in other words, to incorporate more and more people. This is achieved by transforming the original designs and by making appropriate concessions to attract resistant users. Thus 'user-oriented' design (Norman, 1998) sets out to make friends, but at the same time operates to keep enemies or sceptics at bay. This may be the work of an actor network that conspiratorially expands its realm into a monopoly of power, or of a multitude of actor networks muddling through in an opportunistic fashion, making the best out of opportunities as they arise. This perspective considers technology as the continuation of politics by other means (Latour, 1988).

In recent years social scientific studies of technology have undergone a paradigm shift. The old paradigm focused on assessing the technology 'after the fact' by monitoring intended and unintended consequences (see Merton, 1936). The technology is a fait accompli; the research questions concern its impacts after the event. Feedback, at most, leads to fine-tuning of the packaging of the product or the service, or the design of the context as considered in diffusion and acceptance research (Rogers, 1996). The new paradigm, by contrast, investigates the processes contributing to the development of the content of a particular technology as it moves in a sequence of stabilised forms (Hughes, 1989; Mokyr, 1992; Pinch, 1996). In the new paradigm the monitoring of impacts is moved to ever-earlier stages of the design process. Resistant users are canvassed at the conceptual or at the proto-typing stage, thus lifting the 'afforded user' from the designer's imagination into empirical reality. This approach considers the shaping of a technology as a continuous feedback loop from relevant but resistant groups onto technical designs, and raises the questions of technological options and alternative trajectories. The historical, sociological or psychological reconstruction seeks to explain why certain trajectories prevailed while others failed, and why sub-optimal solutions succeed in a world of rational aspirations. Why, for example, do all English keyboards follow a QWERTY layout? Or why do bicycles have pneumatic tyres? (See Pinch and Bijker, 1987). Technology is a political learning process, in which actors aim at securing support either by pretend play or by accommodating resistance into the redesign of their project (Bauer, 1995).

In these perspectives, the concern is with a technology and its relevant actors. As yet, however, they provide a limited conceptualisation of the public sphere as a constitutive context of a technology movement.

Specific changes in the context: post-industrial society

The biotechnology movement exists, like previous technologies, in a historical environment that enables and constrains its future. We conceive of this state of affairs as a system–environment relationship (Luhmann, 1995). The colonising system is the biotechnology movement; the environment comprises relatively autonomous spheres of activities such as the economy, the sciences, the law, the polity, religion, education and the arts, each one operating according to a particular logic (Luhmann, 1995). Each sphere of activity represents biotechnology in its particular way. For example, in the sciences, genetic engineering is the latest frontier of progress for the twenty-first century. In law, it is an issue of risk regulation or of intellectual property rights; in business, a high-risk stock, a market opportunity or a means to rationalise production. In the polity it appears as concerned public opinion; in the religious domain as a potential threat to the moral fibre of society and to the sanctity of God's creation; and in the arts world it may be a new medium of expression.

Luhmann presents a long-term view of secular changes in societies away from hierarchical structures to functional differentiation. Science, which during the course of modernisation displaced religion from the apex of the hierarchy, is no longer taken for granted as the dominant form of knowledge about the world. Today it is one among others. Nested within this perspective of social change, a number of sociological imaginations postulate more recent transitions of societal structures that bear on the development of technology in contemporary times. These characterise axial changes in the public sphere within which the actors of the biotechnology movement must operate.

Bell (1973) and Stehr (1994), for example, allude to the increasing importance of knowledge as a productive sector of the economy. And with this comes the privatisation of knowledge as its production moves from public universities and state research institutions to the private sector of corporate R&D (Cohen and Noll, 1994). This development carries consequences for the operations of the public sphere – for example, in public access to knowledge and the availability of independent expertise.

Inglehart (1990) identifies a secular shift in value orientations, particularly among the affluent young generation in advanced countries, away from a commitment to materialistic progress towards a post-materialist focus on a more humane society. This value shift cultivates concerns for sustainable development. For Beck (1992), the axial change is the uncontrollable and globalised risks attributable to human activities; for Giddens (1991), it is the uncertainties of late capitalism without traditions. The success of past technologies undermines the very foundations on which they are based. The unintended consequences of modern technologies go beyond the controls of the traditional institutions, becoming increasingly global and unmanageable. Whereas, in the past, technology was a means to control local dangers and to extend human capacities, technology increasingly has perverse effects with global impacts (Tenner, 1996). Witnessing these trends, modernity turns reflexive, which undermines its foundational belief in 'progress'.

Durant et al. (2000) speculate on the emergence of different cultures of public understanding of science. As scientific knowledge increases among the public with progress in education and economic development, so is there a specialisation even in popular scientific knowledge, a decline in interest in science and technology, and an increasing ambivalence towards science and technology. The potential negative consequences of scientific and technological developments loom larger, giving rise to support for the 'precautionary principle'.

These trends – the private knowledge sector, value shifts, global risk and diversity of popular scientific cultures – bring a change in the dominant representation of technology. Technology is no longer presumed to be equated to *progress* (Touraine, 1995), and is instead part of the problem that it purports to solve.

A consequence of these nested axial changes is that the relationship between specific technology movements and their representations in societal arenas is also changing. From a system theoretical point of view, we expect a plurality of representations of biotechnology in the public sphere arising from autonomous domains of society other than science. From forms of sociological imagination we expect increased challenges to the autonomy of the scientific-industrial complex and calls for national and international regulation (the primary hurdle), and increasing legitimation of resistance by interested groups and the wider public (the second hurdle). These counter-challenges are likely to increase the significance of public representations of biotechnology to the actors in the biotechnology movement. The overall outcome will be the expansion and professionalising of proactive and reactive public relations activities of all actors of the biotechnology movement as they attempt to secure the support of various social milieus for their particular vision of biotechnology – whether utopian, dystopian or some 'third way'.

The activities of the biotechnology movement: the primary objectification

A technology at any moment in history is a complex network of actors, materials and images that objectify ideas for a particular future of society. The anthropological significance of technology is the extension of control over the environment and the shaping of natural processes into some human design (Ropohl, 1979). Every technology represents a particular vision of society through the past and future activities that are embodied in devices and services. It is hardly surprising that there is a genre of literature that explicates these imaginations in the 'atomic state', the 'space society', the 'information society' or the 'bio- or genetic-society'. Ideas are institutionalised into 'object scripts' for dealing with things or services, which people either adopt spontaneously as affordances, or into which people are socialised symbolically as aspirations. Some people will resist this process because their ideas are at variance with the affordances or the aspirations of the new technology. As resistance to a technological project arises, so is the question raised: what does resistance contribute to the process of development? The detached analysis has to see beyond the polemic that diagnoses resistance as technophobia or Luddism, and ignores the deficiencies in design and implementation (Bauer, 1995).

Like its predecessors nuclear power (Weart, 1988; Radkau, 1995) and computerisation (Kling and Iacona, 1988), the biotechnology movement is not free of controversy. All sides in the controversy attempt to cultivate representations in the public sphere in order to rally support. Like any technological movement there is a growing network of interested and concerned insiders. It excludes those who are indifferent to the technology, either because they have been kept ignorant or by their own deliberate choice, but these outsiders are potential members of the movement and constitute the pool of mobilisation.

In the context of the biotechnology movement, three distinctions regarding groups of actors may be made. First, there are those concerned with the creation and exploitation of the new knowledge: the scientific-industrial complex (figure 13.1). This includes scientists, engineers, bankers, managers, marketing executives, stockholders, communicators, patent lawyers and many other actors. All are more or less unified by an economic interest in biotechnology, albeit in competition over market segments for particular products and services.

Secondly, there are those who see opportunities for the project of their group, without being directly involved in the production of biotechnology: for example, patient groups who have high expectations that genetic engineering will eliminate diseases or alleviate the suffering of their members.

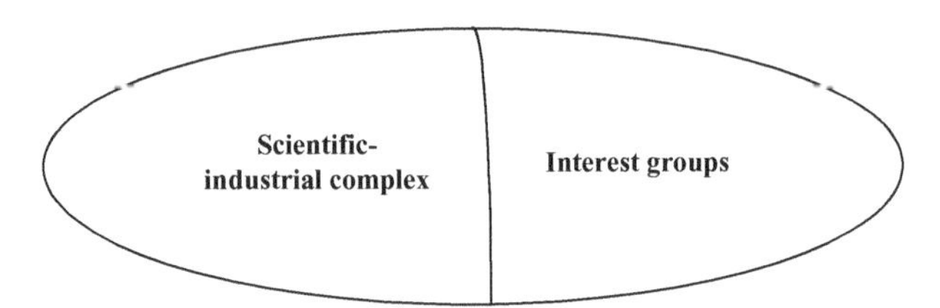

Figure 13.1 The biotechnology movement

Other examples are those who would benefit from genetic testing and a 'genetic information society', such as the police and forensic services in catching and convicting criminals, and insurance companies in their risk assessment strategies.

Thirdly, there are those who voice reservations and cultivate concerns about these new developments. For example, the environmental, consumer and other specialist interest groups mobilise religious believers, entrepreneurs, retailers, scientists, lawyers, protesters, activists, patients, consumers and others to express concerns about genetic drift in the environment, food adulteration, civil liberties, animal rights and threats to privacy or human dignity.

The scientific-industrial complex is at the centre of the biotechnology movement. By 'scientific-industrial' we refer to the ever-closer integration of knowledge production and corporate exploitation of molecular genetics since the 1980s. This is evidenced by various international trends (Haber, 1996; Ernst and Young, 2001): the number of small companies emerging since the mid-1980s with an overwhelmingly R&D orientation; the number of joint ventures between university departments and private companies; and the trends in international mergers of pharmaceutical, agri-chemical and agri-seed companies around a vision of the 'life sciences'. By 'complex' we stress its principal heterarchical nature: there is no necessary national or international unified line of command. However, at times the conspiracy theorist may be close to the truth in assuming an integrated global strategy. The complex comprises ongoing and multifaceted negotiations among various actors over the control of genetic processes in the pursuit of economic imperatives.

The biotechnology movement may be schematised as a field of forces in continuous shifts in the equilibration of demands and concessions, resistance and accommodation (Lewin, 1952; Latour, 1988). 'Progress' is depicted as the meandering movement along a non-linear curve that links alterations to the technology to formerly resistant participants. Each point on this trajectory represents new actors that signed on to a particular development of the original design. Additional users are 'bought' by changes to the original design. The strategy is called 'user-oriented design' (Norman, 1998). For example, biotechnology is developing an increasing range of products and services. In the course of this development certain products are discontinued, such as the 'terminator gene' for agricultural seeds, while others are altered, such as the removal of antibiotic marker gene.[1] The appropriate mix of flexibility and rigidity for the technology is an uncertainty of the process. If a design is inflexible it will fail to sign up new participants, but if it is too fluid few will be in a position to recognise its identity as a technology and be able to integrate it into daily routines. The objectification of an idea 'closes' a technology into a temporary match between materials and actors. Closure represents an equilibrium of the forces within the social-technical system at any moment in time. Progress and development arise through a series of 'closures'. In responding to perturbations, the design is unfrozen, changed and then refrozen.

The process can be considered a Machiavellian one: actors are working both within current regulations and public opinion, and on the limits in order to obtain competitive advantages on multiple frontiers. 'Working on limits' may involve moves to change existing regulations and public opinions, to prevent regulation in line with or contrary to public opinion, or to set selected limits where there are currently none to allay public concerns.

Challenge and counter-challenge

Where a development in biotechnology affords an alteration to the modus operandi of a particular group in society, it is seen as a challenge (figure 13.2). A challenge may be positive in the sense of an emerging new opportunity to be exploited, or negative in the sense that it threatens the status quo or the world-view of the group. Such challenges lead to collective action either to embrace the opportunity or to resist the perceived threat. These responses, which, as they unfold, constitute counter-challenges, are in turn 'internally represented' or 'incorporated' within the biotechnology movement through the activities of interest groups.

All these groups of actors are competing by forming coalitions with the objective of imposing their aspirations for the future of biotechnology

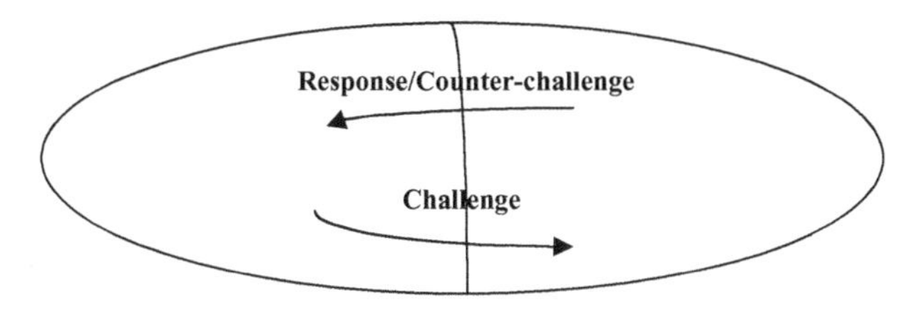

Figure 13.2 Challenge and response within the movement

onto the biotechnology movement. In order to achieve this, actors will attempt to mobilise and incorporate the wider world. In the context of our research, the wider world is the public sphere of regulation, media coverage and public perceptions.

Of the many actors engaged in the biotechnology movement, some have been specifically founded for this purpose. For example, there is the proliferation of small biotechnology companies, often joint ventures with university departments; there are the groups that coordinate and support such efforts ('Bingos', or business NGOs); and there are interest groups that scrutinise the developments in and around biotechnology, such as the group that coordinated the opposition against gene technology in the two Swiss referenda of 1992 and 1998 ('Pingos', or public NGOs).

Beyond these interest-specific groupings are existing actors/groups that were formed for other purposes but discover that biotechnology offers new opportunities. Thus they become issue entrepreneurs using biotechnology as a way of sustaining and building their general project. For example, environmental organisations, consumer groups and human rights groups adopt the emerging biotechnology as an issue within their remit.

The public arenas and social milieus that make up the public sphere of biotechnology represent the new developments in a variety of different ways, that is, as different challenges. It is crucial to recognise that these representations are not epiphenomenal and of merely passing interest. They constitute the symbolic environment of biotechnology, and symbolism is the core of any culture; and culture both constrains and enables activities.

Traditionally, social milieus have channelled their visions for society through party political activities. Traditional segmentation separates the population into distinct milieus of *Weltanschauung* with particular orientations towards the past, present and future of society and with a project for society as a whole. These structures have been the basis for party political affiliation such as conservatism, socialism or liberalism. Nowadays, social milieus constitute the new reserve army for pressure groups emerging within the biotechnology movement. It is unclear when issues raised by specialist actors will become aligned to the party political cleavages. Post-industrial society is in transition, with a mix of hierarchical milieus and modern functional differentiation. Modern segmentation increasingly reflects the spheres of productive activity such as business, law, education, religion, politics or the arts. Although interest groups have emerged from the functional differentiation of society, they appeal to traditional milieus for the mobilisation of support (figure 13.3).

Analysing the structure and function of these varied symbolic activities is the analytic challenge. In our analysis, we focus on the environment

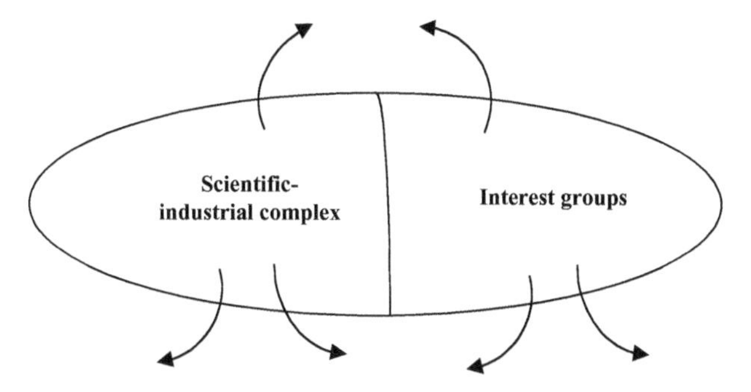

Figure 13.3 The technology movement reaching out to the public

of biotechnology defined by the public spheres of various European countries.

The symbolic environment of the biotechnology movement: the secondary objectification

The birth of modern technology is accompanied by wild imaginings and as yet unwarranted claims as to its significance. The symbolic elaboration of a new technology into utopian or dystopian *Leitbilder*, what we call secondary objectification, is a crucial element of its reality and its future potential (Dierkes, Hoffmann and Marz, 1996). Such imaginings are an integral part of the biotechnology movement. Our research focuses on secondary objectifications of biotechnology as found in public arenas and in social milieus, and the effects that these may exert on the trajectory of biotechnology.

The actors of the biotechnology movement are not only talking among themselves; they are also mobilising sections of the public in support of their views. In this way, a fait accompli, such as the exporting to Europe of genetically modified soya beans by the US company Monsanto in the autumn of 1996 (see Lassen et al., chapter 10 in this volume), offered much scope for reactive mobilisation in society.

Activists of the biotechnology movement are more or less professionally involved in their activities, whereas the public's concerns are much broader. The public at any moment in time needs to be convinced of the relevance of the issue. Advocates and opponents have one thing in common: they push the significance of biotechnology as an issue in the public domain. In consequence they combine forces to draw others into the biotechnology movement.

This leads to tangible representations that can be researched in the three arenas of the public sphere. Representations are likely to be structured as frames that organise the issues into imagery and a range of positions vis-à-vis the technology (Gamson, 1988). A frame delineates what to agree or to disagree about: biotechnology may be a matter of morals, risk, democracy, utility, globalisation, public accountability, etc. Particular sponsors promote particular frames; they match the routines of media production, and resonate more or less with public perceptions and with the regulatory process. The contest is two-fold: the choice of which frame and the position within the chosen frame.

These secondary objectifications familiarise the novelty of the fait accompli in public and co-determine the future of the 'fact'. They anchor the new technology in resonating images and objectify the uncertainties in arguments about the technology. They anchor the challenge of novelty by offering historical analogies and familiar images of the future. In the public sphere, representations are cultivated in and carried by particular social milieus.

Representation involves different modes and mediums. Modes refer to habits, cognition, informal conversations or formal communication such as in mass media or public speaking. Mediums refer to language, images, sounds or movements that embody and familiarise the new technology and establish a relation for or against it (Bauer and Gaskell, 1999).

The public sphere of modern societies is structured and dynamic. On the one hand it comprises audiences, actors and historically established forums and arenas, the typical occasions for public exchanges (Neidhardt, 1993), while on the other hand its quality of functioning may improve or decline (Habermas, 1989). We focus on three arenas of the public sphere, thus considering different modes and mediums of representation: biotechnology represented in the formalised mass media coverage, in everyday conversations and perceptions, and in regulatory routines and regimes (figure 13.4).

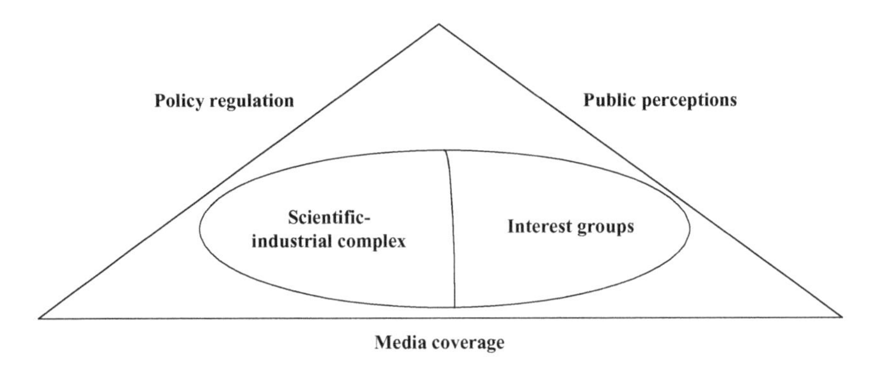

Figure 13.4 The three arenas of the public sphere

Within the public sphere these arenas serve different functions: media coverage and public perception constitute the public opinion of biotechnology, while the regulatory process constitutes the governance of biotechnology. From the point of view of the biotechnology movement, these are the primary and the secondary hurdles, and cannot be ignored.

Advocates and opponents of biotechnology attempt to make inroads in the three arenas in order to make their representation of biotechnology become the 'real future' of routine activity. The outcome for biotechnology is uncertain, as was the case with previous technologies, and is likely to be different from that which is imagined by any of the actors.

We presume that the representations arising from the challenge posed by biotechnology are beyond the control of the scientific-industrial complex or other activist groupings. This is because representations link spheres of action and as such are located in between biotechnology and other societal domains. The law, education, the media, politics, the political public, religion and the arts, although subject to the influences of the scientific-industrial complex, enjoy relative autonomy in their symbolic activities.

The triangular model of the three arenas is a heuristic device that achieves two objectives simultaneously. First, it captures the complexity of the public sphere without privileging, in the sense of uni-directional causal relations, any of the three arenas. Secondly, it allows us to simplify the complexity within the context of research. Each side of the triangle may be taken separately as a perspective on the public sphere. Thus we can ask how public perceptions relate to the policy process and to media coverage; or how media coverage relates to policy process and public perceptions; or, finally, how the policy process is related to public perceptions and to media coverage. Each of these perspectives on the public sphere calls attention to a variety of middle-range models from the social sciences, the formation of public opinion, theories of media production, and models of policy-making. The 'ideal framework' keeps the debate on a certain level of complexity and invites a synthesis of perspectives. However, at present, this may be beyond the scope of current social science theory. The triangular model also works as a shield against simplistic reduction of the 'public sphere' to surveys on public perceptions, or analysis of media coverage, or the analysis of documents on policy-making. Each arena must be understood in the context of the other two and in the context of the activities of the biotechnology movement.

Table 13.1 provides a characterisation of the modus operandi, achievements and functions of the three arenas of the public sphere. The mass media work to a short production cycle. One or two weeks is a long time

Table 13.1 *The operating principles of the three arenas of the public sphere*

	Public arenas		
	Public conversations and perception	Mass media coverage	Regulation
Modus operandi	Medium cycles Some consistency Memory	'Media logic' Short cycles News values Imperative of novelty Little consistency Hardly any memory Framing	Long cycles Bias against novelty Consistency Long memory
Achievements	Opinion Attitude Stereotype Schemata Awareness Skills	Dissemination Propagation Propaganda Advertising Education and training Agenda-setting	Regulatory regimes
Functions	Being able to act on it Communicating with other people	Information Entertainment Linking domains of societal action	Allocation of responsibility Assimilation and accommodating public opinion Enabling technology

in the media world. News values and routines guide the selection of science and technology stories from a large pool of possible stories (Hansen, 1994). Consistency in reportage over time is not a key feature, which indicates a limited memory span; novelty is imperative. The main functions of the media are information, entertainment and the linking of concerns across different societal domains, for example business with politics or science with business (Kohring, 1998).

From the point of view of our research, the mass media act as a 'window' on to biotechnology for the majority of the public. We do not assume a direct influence on public perceptions of media coverage. This relationship is one of mutual constraint. Mass media coverage is a source as well as a constraint for everyday conversation and the regulatory process. Although opinions of the audience or regulations set limits to the coverage in the mass media, we may hypothesise about possible linkages between media styles and public perceptions. For example, diffusion of information forms opinions; propagation, by virtue of its qualified and value-based judgement, fosters attitudes; and propaganda, characterising a black-and-white world in terms of friends and enemies, is likely to cultivate stereotypes (Moscovici, 1976). While advertising seduces awareness and stimulates desires, training by observation establishes new skills and routines where educational functions are retained.

Media coverage for or against biotechnology does not generally map directly onto public perceptions or into policy-making. Most of the time we expect the mass media to disseminate information and/or to entertain ('info-tainment'). Beyond this, we may expect the media to synchronise attention to biotechnology across spheres of life in terms of agenda-setting (McCoombs and Shaw, 1972) and cultivation (Morgan and Shanahan, 1996). Figure 13.5 shows the increasing coverage of biotechnology across Europe aggregated from elite newspapers in twelve countries between 1973 and 1996, and this can be regarded as an index of reduced public indifference and, under certain conditions, of public controversy.

Public conversations and perception may focus on an issue such as biotechnology. But, to the dismay of many in the biotechnology movement, the public at large does not share their level of concern with the issue. The issues of biotechnology are often non-salient, because other concerns are more pressing. Conversations and perceptual topics have a certain degree of consistency and work with some memory of what went on before, and concerns come and go in a medium-term cycle. Awareness, skills, values and stereotypes tend to be rather stable and consistent, whereas opinions and attitudes may change more easily. Attitudes bring general values to bear on concrete issues, and provide categories and evaluative judgements regarding biotechnology. We expect awareness, opinions, attitudes, stereotypes or skills to enable people to communicate and, to different degrees, to influence everyday actions.

The effect of public perception on policy-making and media coverage is uncertain. Although some individuals may have direct access to the regulatory process, most opinions count only in mass. Aggregate opinion may legitimate political action; it may support or resist biotechnology, or remain ambivalent.

Public perception is achieving greater significance for technological trajectories. This is because it is the key legitimisation for both pro- and anti-activism within the biotechnology movement. Without claims of public support, interest groups would wield little influence. However, the relative autonomy and unpredictability of public perceptions is a source of great concern among activists. Hence, the attempt to take control of public opinion. Think of the effort and subsequent failure

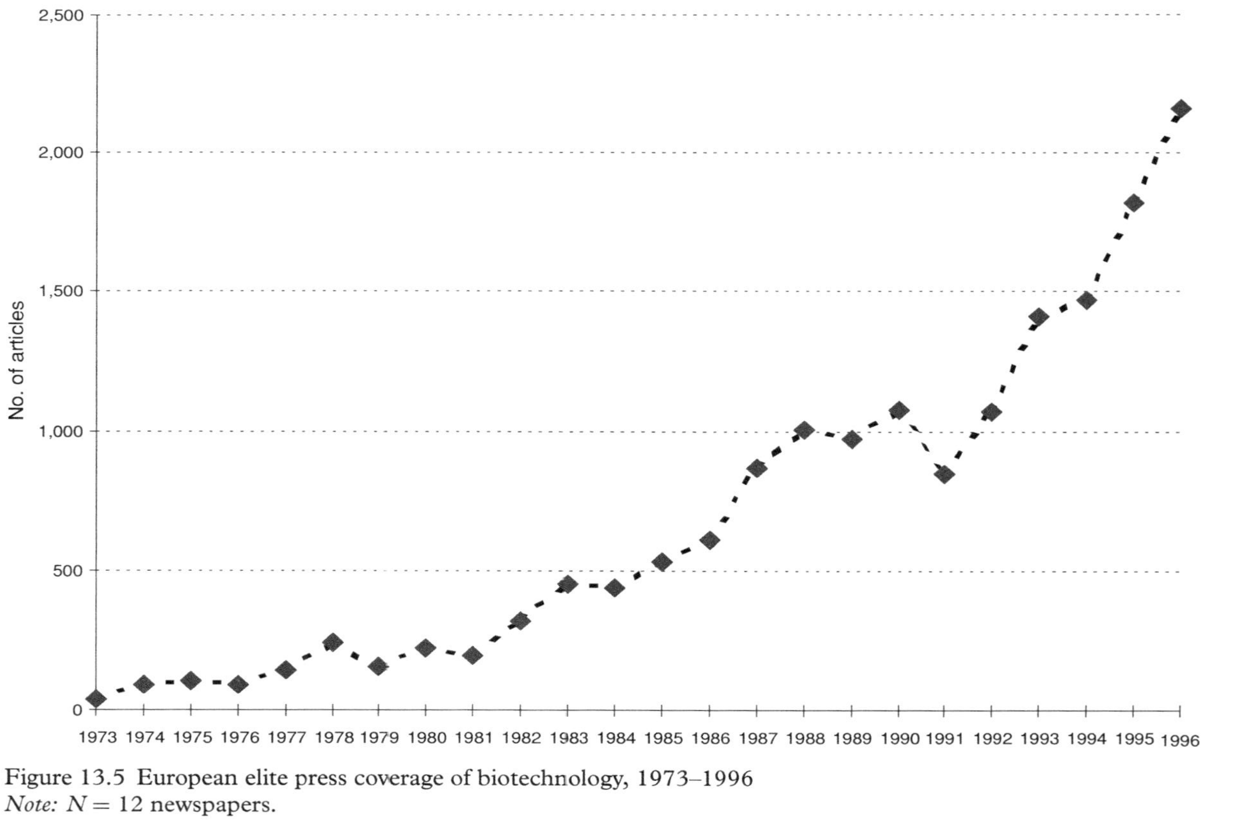

Figure 13.5 European elite press coverage of biotechnology, 1973–1996
Note: $N = 12$ newspapers.

of the 1998 'charm' campaign following the introduction of Round-up Ready soya into the European market (see Lassen et al., chapter 10 in this volume).

The regulation of technology in policy- and law-making is slow, with a long memory of past activities and a strong concern for consistency between levels and across domains of regulation. There is an inherent bias against novel regulation; the working assumption is that existing procedures cover the new things. The new is assimilated into existing regimes and frameworks. There is an inherent resistance to accommodation, in the sense of changing the regulatory processes or regimes in response to challenges from various actors (Hood and Rothstein, 1998). Within this broad operating logic, different regulatory regimes may be identifiable. For example, Hood et al. (1999) identify three ideal-types of policy regimes for risk management. 'Responsive government' reflects public and media opinion, 'client politics' reflects vested organised interests, and 'minimum feasible response' is oriented towards the correction of failures in the market or tort-law processes, in effect avoidance of blame. A regime is a temporarily stable modus operandi. As the biotechnology movement expands, the policy regime may move or switch back and forth between these types of risk management.

Regulation is the primary constraint from the point of view of the biotechnology movement. The representation of biotechnology in the regulatory process reflects the degree of autonomy or incorporation of the policy process by the biotechnology movement, on the one hand, and by public opinion, in terms of media coverage or public perceptions, on the other hand. The main functions of the regulatory process are to enable a new technology to serve economic imperatives, to avoid blame for future failures and to alleviate concerns expressed in the media and in public perceptions. Each of the above regimes seems to highlight one of these functions of the regulatory process.

Our final theoretical concern is how to account for the multiple representations of biotechnology that appear in different societal arenas in many societies. For this we need to go beyond the idea of social groups as the locus of activism to the concept of the social milieu. Social milieus are vector combinations in the triangular framework in which a resonance links media coverage, public perceptions, regulatory regimes and particular actors. The concept of social milieus acknowledges the heterogeneity of the three key arenas, and links divisions of the three arenas to particular representations. A milieu links particular public perceptions and media resources, and projects these into the policy arena. In so doing, a milieu supports a particular regulatory regime advocated by an actor in the biotechnology movement. In their resonance between positions

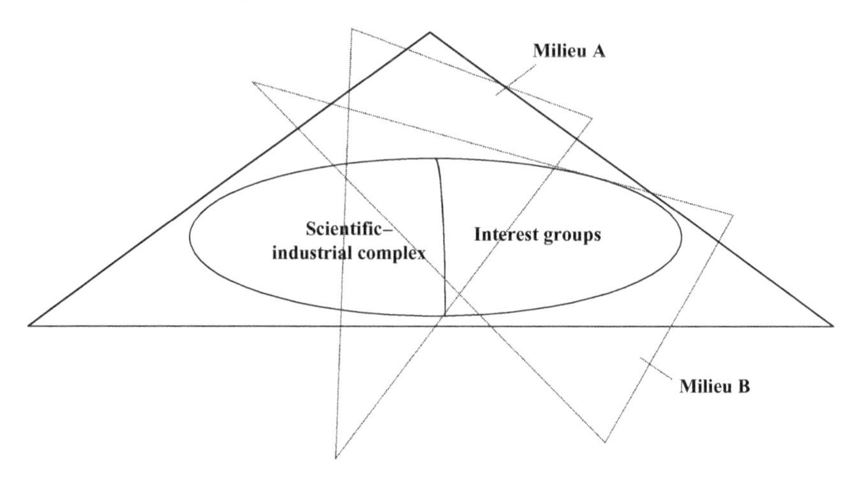

Figure 13.6 The movement mobilising milieus in the three arenas

in the three key arenas, social milieus can be said to carry particular representations of biotechnology in the public sphere (figure 13.6).

Researching the public arenas

With our description of the biotechnology movement, we expect to find multiple representations of biotechnology in the public sphere – that is, a variety of representations in public perceptions, media coverage and regulation. From the point of view of the biotechnology movement, milieus constitute established segments of the public, which are often associated with particular media outlets. In the jargon of public relations, they are target groups for whom effective means of access need to be identified.

Linkages, in terms of content, across the three arenas would be expected to be associated with a particular milieu. Let us illustrate this idea. Consider environmentalism as a relatively new but established milieu in many countries. It is characterised in its relationship with biotechnology by particular media frames and outlets, by particular attitudes and stereotypes and by support for 'responsive government' on the basis of the precautionary principle. Or consider the commercial milieu: this is characterised by other media frames and outlets (growth potential, the business press), different attitudes and stereotypes (investment opportunities) and a preference for minimal response risk regimes. These milieus legitimise actors, for example environmental or commercial, who cultivate their particular representations and seek to incorporate them into the growing biotechnology movement.

A milieu may be observed from the perspective of three viewpoints. Looking in from the regulatory arena, the milieu is the voters; from the media it is readers, listeners or viewers; and from public perceptions it is like-minded people, the community of common interests and lively conversations.

This discussion opens up the following implications and principles for research. Secondary objectifications are structures of common sense cultivated in communication and spanning the three arenas of the public sphere. With the concept of 'social representation', research focuses on the comparative analysis of milieus and their common sense rather than on the representations-for-action developed by actors in the biotechnology movement. The latter would be a different focus of research and require a different methodology (Bauer and Gaskell, 1999).

Empirically, in the research reported in this volume, informal communications are observed in everyday conversations during focus group sessions and in surveys of public perceptions. The formal communications of the mass media are assessed by the analysis of press coverage. And finally the regulatory routines are analysed by recording regulatory activities and by scrutinising official reports and documents (see Gaskell et al., 1998, for methodological details).

Representations of biotechnology vary across the three arenas in two basic dimensions: intensity and content. Each arena makes reference to biotechnology with a certain degree of intensity. The public is more or less aware of these new developments and pays attention to them; the media carry a certain number of biotechnology stories; and the regulatory sphere produces more or fewer codes of practice or statutory regulations. In all arenas, we have to assume a limited capacity of attention, where different issues compete for limited resources. The finding of simultaneous attention and activity across the three arenas may constitute an indicator of a 'crisis'. Secondly, we find different contents, that is references, frames and themes, on the topic of biotechnology. Similarities and differences of representations in time and space, across arenas and across different milieus, are empirical matters.

We attempt to relate intensity or content in one arena to intensity or content in another on the basis of middle-range hypotheses. For example, the 'intensity of coverage hypothesis' relates increasing media coverage (intensity) to the rise in negative public perceptions (particular content) of a new technology. The 'agenda-setting hypothesis' of media effects relates changes in contents of media to changes in public perception. The 'responsive government hypothesis' suggests that public concern (content of public perception and media coverage) will antecede policy-making (content).

Observing learning by assimilation or accommodation

Learning is a useful metaphor for the processes and consequences of socio-technical change. Change may be associated with progress, accumulation and improvement, or with problems, decay and collapse. Learning is based on a dual process of assimilation and accommodation. In assimilation or 'pretend play', changes are absorbed into current routines. Continuity is achieved with minimum disruption as the new is anchored in the old and familiar. By contrast, accommodation is the often painful process of readjustment, involving structural change. The new resists classification into the old categories and resists management within the old routines. When assimilation reaches its limit, the system is forced to change its categories and routines in order to accommodate the new.

This model of learning was developed to explain the cognitive development of children coming to terms with the world of things and adults (Flavell, 1963; Piaget, 1972). In the case of the biotechnology movement or the public sphere, we are working with a system and its environment without a single fixed referent; a co-evolution, with no fixed world of things with which to come to terms. Rather, things as well as people are on the move. In this process of co-evolution the participants are not equally prone to learning. All participants can choose to ignore, assimilate or accommodate the other.

On the side of the biotechnology movement, efforts to assimilate can be observed in evolving actor networks, their increased specialisation and diversification, their lobbying on regulation, public relations activities, and the use of established forms of activism. Accommodation can be observed in new forms of activism, for example the confluence of activism and public relations (such as 'no action without a TV camera or an audience of journalists'), or the setting up of NGOs on a business basis (the so-called 'bingos'); in the abandonment of product and service lines such as the 'terminator gene'; or in the alteration of product and service lines to attenuate public concerns, such as eliminating antibiotic markers.

On the side of the public process, we may observe a growing sensitivity to public opinion in the regulatory process. In some countries such as the USA, assimilation prevails in the regulatory process, as modern biotechnology is considered to be nothing new; whereas in Europe a gamut of new laws and regulation accommodate the development within the different legal frameworks (see Torgersen et al., chapter 2 in this volume). Public accommodation may also be observed in experiments with novel arenas of public debate such as 'consensus conferences' and 'deliberative democracy', or in the introduction of new regulatory regimes and frames such as 'ethics', the precautionary principle or wider notions of 'risk' such as social risk or phantom risks.

Public perceptions are informal, much more diffuse and less organised. However, in some countries we have seen a steady decline in the belief in the traditional links between science/technology and progress. With this has come greater interest in the control of science and technology, and new concerns about labelling and the right to choose, and about issues of a moral and ethical nature. There is a growth in the expectation – whose seeds were sown in the area of conflicts over civil nuclear power – of increased public participation. This public unrest has been called 'the second hurdle'. Much of the activity of actors in the biotechnology movement is directed towards managing the second hurdle, either to lower or to raise it.

The media are probably more likely to assimilate new technology into their existing modus operandi. The scientific or political journalists are covering just another technology within the established news routines of novelty and human interest. It may be that, as in the case of the environment some years ago, a particular journalistic specialisation may appear: the biotechnology correspondent. This would constitute accommodation.

Conclusion

In this chapter we have characterised a model of the reception of a new technology in the public sphere that has informed our research on biotechnology in the European context during the period 1973–96. It has served as a background from which to reflect on what has been achieved, and, as it has been extended and developed, it has pointed to possible new research questions. The model sets out a broad framework within which to understand the many interrelated processes that shape the public sphere. The public sphere at the end of the twentieth century constituted the symbolic environment for current technological developments and for those to come. This environment is not epiphenomal to an autonomous technological process, but constitutes a counter-challenge that will influence the future trajectory of biotechnology. The chapters in this book explore some, but by no means all, of the elements of this environment. Furthermore, we have restricted our gaze to the three arenas of the public sphere – public perceptions, media coverage and regulations – and have not investigated how the biotechnology movement also changes in this process. Although we are aware that both the scientific-industrial complex and other actor groups are in constant change, the dynamics of the biotechnology movement are beyond the scope of this research. In the course of this project we have been systematically exploring the 'learning conditions' up to 1996, rather than the 'learning outcomes' for the biotechnology movement. To document the outcomes we may have to wait some years to assess them with the necessary circumspection. The scientific-industrial complex may be well advised to learn from and to accommodate to public concern and calls for new forms of regulation. Otherwise the public may simply assimilate modern biotechnology to nuclear power, which once had a seemingly unlimited future but whose trajectory radically changed in the last quarter of the twentieth century.

NOTE

1 Monsanto's declaration of intent not to commercialise terminator gene technology was reported by Associated Press, 7 October 1999.

REFERENCES

Bauer, M. (1995) 'Towards a Functional Analysis of Resistance', in M. Bauer (ed.), *Resistance to New Technology – Nuclear Power, Information Technology, Biotechnology*, Cambridge: Cambridge University Press, pp. 393–418.

Bauer, M.W. and G. Gaskell (1999) 'Towards a Paradigm for Research on Social Representations', *Journal for the Theory of Social Behaviour* 29: 163–86.

Bauer, M., G. Gaskell and J. Durant (1994) 'Biotechnology and the European Public. An International Study of Policy, Media Coverage and Public Perceptions, 1980–1996', unpublished manuscript, June, Hydra, Greece.

Beck, U. (1992) *The Risk Society*, London: Sage; German original, 1985.

Bell, D. (1973) *The Coming of Post-Industrial Society*, New York: Basic Books.

Blumenberg, H. (1990) *Work on Myth*, Cambridge, MA: MIT Press.

Boulding, K. (1956) *The Image*, Michigan: University of Michigan Press.

Cohen, L.R. and R.G. Noll (1994) 'Privatising Public Research', *Scientific American* 271(3): 72–7.

Dierkes, M., U. Hoffmann and L. Marz (1996) *Visions of Technology: Social and Institutional Factors Shaping the Development of New Technologies*, Frankfurt am Main: Campus.

Durant, J., M. Bauer, G. Gaskell, C. Midden, M. Liakopoulos and L. Scholten (2000) 'Two Cultures of Public Understanding of Science and technology in Europe', in M. Dierkes and C. von Grote (eds.), *Between Understanding and Trust. The Public, Science and Technology*, Amsterdam: Harwood Academic Publishers, pp. 131–56.

Ernst & Young (2001) *Integration. 8th Annual European Life Sciences Report, 2001*, Cambridge: E&Y International.

Flavell, J.H. (1963) *The Developmental Psychology of Jean Piaget*, Toronto: van Nostrand Corporation.

Galbraith, K. J. (1968) *The New Industrial State*, New York: American Library.

Gamson, W. (1988) 'The Constructivist Approach to Mass Media and Public Opinion', [The 1987 distinguished lecture], *Symbolic Interaction* 11: 161–74.

Gaskell, G., M.W. Bauer and J. Durant (1998) 'The Representation of Biotechnology: Policy, Media, Public Perception', in J. Durant, M.W. Bauer and G. Gaskell (eds.), *Biotechnology in the Public Sphere*, London: Science Museum, pp. 3–14.

Gibson, J.J. (1977) 'The Theory of Affordances', in R.E. Shaw and J. Bransford (eds.), *Perceiving, Acting, Knowing*, Hillsdale, NJ: Erlbaum, pp. 67–82.

Giddens, A. (1991) *Modernity and Self-Identity*, Cambridge: Polity Press.

Haber, E. (1996) 'Industry and the University', *Nature Biotechnology* 14: 441–2.

Habermas, J. (1989) *The Transformation of the Public Sphere*, Cambridge: Polity Press; German original, 1961.

Hansen, A. (1994) 'Journalistic Practices and Science Reporting in the British Mass Media: Media Output and Source Input', *Public Understanding of Science* 3: 111–34.

Hood, C. and H. Rothstein (1998) 'Institutions and Risk Management: Problem Solvers or Blame-Shifters', in Opening Plenary address 'Risk Analysis: Opening the Process', Paris, 11–14 October, pp. 13–22.

Hood, C., H. Rothstein, M. Spackman, J. Rees and R. Baldwin (1999) 'Explaining Risk Regulation Regimes: Exploring the "Minimal Feasible Response" Hypothesis', *Health, Risk and Society* 1: 151–166.

Hughes, T.P. (1989) *American Genesis – a Century of Technological Enthusiasm*, New York: Penguin.

Inglehart, R. (1990) *Culture Shift in Advanced Societies*, Princeton, NJ: Princeton University Press.

Joerges, B. (1988) 'Technology in Everyday Life: Conceptual Queries', *Journal for the Theory of Social Behaviour* 18: 219–30.

Kling, R. and I. Iacono (1988) 'The Mobilisation of Support of Computerisation: the Role of the Computerisation Movement', *Social Problems* 35: 226–43.

Kohring, M. (1998) 'Der Zeitung die Gesetze der Wissenschaft vorschreiben? Wissenschaftsjournalism und Journalismus-Wissenschaft', *Rundfunk und Fernsehen* 46(2/3): 175–92.

Latour, B. (1988) 'The Prince for Machines as well as for Machinations', in B. Elliot (ed.), *Technology and Social Process*, Edinburgh: Edinburgh University Press, pp. 21–43.

Lewin, K. (1952) *Field Theory in the Social Sciences*, London: Tavistock.

Linde, H. (1982) 'Soziale Implikationen technischer Geräte, ihrer Entstehung und Verwendung', in R. Jokisch (ed.), *Techniksoziologie*, Frankfurt: Suhrkamp, pp. 1–31.

Luhmann, N. (1995) *Social Systems*, Stanford, CA: Stanford University Press.

McCombs, M.E. and D.L. Shaw (1972) 'The Agenda-Setting Function of the Mass Media', *Public Opinion Quarterly* 36: 176–87.

Merton, R.M. (1936) 'The Unintended Consequences of Purposive Social Action', *American Sociological Review* 1: 894–904.

Mokyr, J. (1992) 'Technological Inertia in Economic History', *Journal of Economic History* 52(2): 325–38.

Morgan, M. and J. Shanahan (1996) 'Two Decades of Cultivation Research: an Appraisal and Meta-Analysis', *Communication Yearbook* 20: 1–45.

Moscovici, S. (1976) *La psychanalyse – son image et son public*, Paris: Presses Universitaires de France.

Neidhardt, F. (1993) 'The Public as Communication System', *Public Understanding of Science* 2: 339–50.
Norman, D.A. (1988) *The Psychology of Everyday Things*, New York: Basic Books.
(1998) *The Invisible Computer*, Cambridge MA: MIT Press.
Piaget, J. (1972) *The Principle of Genetic Epistemology*, London: Routledge & Kegan Paul.
Pinch, T. (1996) 'The Social Construction of Technology: a Review', in R. Fox (ed.), *Technical Change. Methods and Themes in the History of Technology*, Amsterdam: Harwood Academic Publishers, pp. 17–32.
Pinch, T. and W. E. Bijker (1987) 'The Social Construction of Facts and Artifacts: or How the Sociology of Science and the Sociology of Technology Might Benefit Each Other', in W.E. Bijker, T.P. Hughes and T. Pinch (eds.), *The Social Construction of Technological Systems*, Cambridge MA: MIT Press, pp. 9–50.
Price, D.K. (1965) *The Scientific Estate*, Cambridge MA: Harvard University Press.
Radkau, J. (1995) 'Learning from Chernobyl for the Fight against Genetics?' in M. Bauer (ed.), *Resistance to New Technology – Nuclear Power, Information Technology, Biotechnology*, Cambridge: Cambridge University Press, pp. 335–55.
Rogers, E.M. (1996) *Diffusion of Innovation*, 4th edn, New York: Free Press.
Ropohl, G. (1979) *Eine Systemtheorie der Technik. Zur Grundlegung der allgemeinen Technologie*, Munich: Hanser.
Simon, H.A. (1969) *The Sciences of the Artficial*, Cambridge MA: MIT Press.
Stehr, N. (1994) *Knowledge Societies*, London: Sage.
Tenner, E. (1996) *Why Things Bite back. Predicting the Problems of Progress*, London: Fourth Estate.
Touraine, A. (1995) 'The Crisis of Progress', in M. Bauer (ed.), *Resistance to New Technology – Nuclear Power, Information Technology, Biotechnology*, Cambridge: Cambridge University Press, pp. 45–56.
Toynbee, A. (1978) *A Selection from His Work* [parts 6 and 7: Challenge and Response], Oxford: Oxford University Press [originally published between 1934 and 1954].
Weart, S.R. (1988) *Nuclear Fear. A History of Images*, Cambridge, MA: Harvard University Press.
Winner, L. (1977) *Autonomous Technology. Technology-out-of-control as a Theme in Political thought*, Cambridge, MA: MIT Press.

Index

For EU product safety concerns, contact us at Calle de José Abascal, 56–1°, 28003 Madrid, Spain or eugpsr@cambridge.org.

www.ingramcontent.com/pod-product-compliance
Ingram Content Group UK Ltd.
Pitfield, Milton Keynes, MK11 3LW, UK
UKHW042316180425
457623UK00005B/16

* 9 7 8 0 5 2 1 7 7 3 1 7 1 *